T0254894

# Modeling Change and Uncertainty

**Textbooks in Mathematics**
Series editors:
Al Boggess, Kenneth H. Rosen

**The Elements of Advanced Mathematics, Fifth Edition**
*Steven G. Krantz*

**Differential Equations**
Theory, Technique, and Practice, Third Edition
*Steven G. Krantz*

**Real Analysis and Foundations, Fifth Edition**
*Steven G. Krantz*

**Geometry and Its Applications, Third Edition**
*Walter J. Meyer*

**Transition to Advanced Mathematics**
*Danilo R. Diedrichs and Stephen Lovett*

**Modeling Change and Uncertainty**
Machine Learning and Other Techniques
*William P. Fox and Robert E. Burks*

**Abstract Algebra**
A First Course, Second Edition
*Stephen Lovett*

**Multiplicative Differential Calculus**
*Svetlin Georgiev, Khaled Zennir*

**Applied Differential Equations**
The Primary Course
*Vladimir A. Dobrushkin*

**Introduction to Computational Mathematics: An Outline**
*William C. Bauldry*

https://www.routledge.com/Textbooks-in-Mathematics/book-series/CANDHTEXBOOMTH

# Modeling Change and Uncertainty

## Machine Learning and Other Techniques

**William P. Fox**
College of William and Mary, USA

**Robert E. Burks**
Naval Postgraduate School, USA

CRC Press is an imprint of the
Taylor & Francis Group, an **informa** business
A CHAPMAN & HALL BOOK

First edition published 2022
by CRC Press
6000 Broken Sound Parkway NW, Suite 300, Boca Raton, FL 33487-2742

and by CRC Press
4 Park Square, Milton Park, Abingdon, Oxon, OX14 4RN



CRC Press is an imprint of Taylor & Francis Group, LLC



*Library of Congress Cataloguing-in-Publication Data*
Names: Fox, William P., 1949- author. | Burks, Robert, author.
Title: Modeling change and uncertainty : machine learning and other techniques / authored by William P. Fox, College of William and Mary, USA, Robert E. Burks, Naval Postgraduate School, USA.
Description: First edition. | London ; Boca Raton : C&H/CRC Press, 2022. | Series: Textbooks in mathematics | Includes bibliographical references and index.
Identifiers: LCCN 2021061944 (print) | LCCN 2021061945 (ebook) | ISBN 9781032062372 (hbk) | ISBN 9781032288437 (pbk) | ISBN 9781003298762 (ebk)
Subjects: LCSH: Mathematical models. | Mathematical analysis.
Classification: LCC QA401 .F696 2022 (print) | LCC QA401 (ebook) | DDC 511/.8--dc23/eng20220415
LC record available at https://lccn.loc.gov/2021061944
LC ebook record available at https://lccn.loc.gov/2021061945

ISBN: 978-1-032-06237-2 (hbk)
ISBN: 978-1-032-28843-7 (pbk)
ISBN: 978-1-003-29876-2 (ebk)

DOI: 10.1201/9781003298762

Typeset in Palatino
by MPS Limited, Dehradun

# Contents

# Preface

Mathematical modeling is a craft that requires practice if you want to maintain and enhance your mathematical skills and abilities. Bottomline, the more practice that you are willing to put in, the better you will become in executing the art of mathematical modeling. We will cover many topics in this book that are typical of the mathematical modeling courses that we have taught over the past 30 years and still teach today. Where appropriate, we will introduce a problem to help motivate the learning of a particular mathematical modeling topic. The problem provides the issue or the "what" of what we need to solve using an appropriate modeling technique. We then apply mathematical modeling principles to that problem and present the steps in obtaining an appropriate mathematical model to solve the problem. We do utilize technology to build, compute, or implement the model and then analyze the results of the model. However, the focus, throughout this book, is the mathematical model and not the technology. Technology is a tool, but we will present the technology method we use to solve the mathematical model or perform sensitivity analysis. Over the past 15 years, the authors have taught a three-course mathematical modeling sequence to students where Excel was chosen as the software technology of choice because of its availability, ease of use, familiarity to students, and student access in their future jobs and careers. In this book, we extend beyond just Excel © and will introduce valuable techniques using both MAPLE© and R©.

## Audience

We caution that this book is not intended to serve as a textbook for an introduction to mathematical modeling course. It is designed to be used for individuals who already have a foundational or introductory level of mathematical modeling knowledge and have been introduced to technology at some level. *Modeling Change and Uncertainty* would be of interest to mathematics departments that offer mathematical modeling courses focused on decision making or discrete mathematical modeling and by a student looking for an opportunity to practice the craft of mathematical modeling.

*Modeling Change and Uncertainty: Machine Learning and Other Techniques* is designed to target the following groups of students:

- Undergraduate: quantitative methods courses in business, operations research, industrial engineering, management sciences, industrial engineering, or applied mathematics.

- Graduate: discrete mathematical modeling courses covering topics from business, operations research, industrial engineering, management sciences, industrial engineering, or applied mathematics.

## Objectives

The objective of *Modeling Change and Uncertainty: Machine Learning and Other Techniques* is to illustrate advanced applied mathematical modeling techniques that are accessible to students from many disciplines. The goal is that this book is used as part of the craft of mathematical modeling and support fostering a desire for lifelong mathematical learning, habits of mind, and competent and confident problem-solvers for the 21st century. Chapter 1: Perfect Partners helps set the tone for the incorporation of mathematical modeling and technology. This chapter provides a process for thinking about the problem and illustrates many scenarios and examples. We establish the solution process and will provide solutions to these examples in later chapters.

## Technology

We have selected Excel, Maple, and R to support, where applicable to the content, the mathematical modeling process because of their wide accessibility to most students. We believe it is fundamental that individuals have a basic foundational understanding of at least one of these three technologies to get the most out of this book and to build their mathematical modeling skills. We provide the appropriate code for the relevant technologies that we think are applicable to solve a problem in each chapter. In this way, the chapter can be covered in general discussion and the technology chosen from those provided to illustrate the models. However, we do not provide a discussion covering the basic setup of the technology.

As a reminder, technology, such as Excel, Maple, and R, is just a means to an end. The process of model building, and the interpretation or analysis of the results are the two essential elements of mathematical modeling, and technology is merely a tool to perform the necessary calculations. Throughout the book, we will present features of the selected technology that anyone can use in their modeling efforts. Many of the figures in this book are generated with one of the three selected technologies. It is important to note that one limitation of Excel is its limited graphics capabilities, especially in 3Ds. Although we attempt to use all features available in Excel, occasionally we felt the need to create Excel templates, macros, and programs to solve the problem. This material is available upon request. Maple and R are both great compliments to Excel, and both have great graphics capabilities and many integrated commands to assist with mathematical modeling. Maple templates and example solutions are also available, upon request, from the authors.

## Organization

This book contains information that could easily be covered in an advanced semester course focused on mathematical modeling or a semester-long survey course of the various topics in the book. The book is designed to provide instructors the flexibility to pick and select material to support their course. Chapter 2 through Chapter 11 provide material and solution techniques to address the problems introduced in Chapter 1.

- Chapter 2 provides a foundation in using both linear and nonlinear models of discrete dynamical systems. The chapter covers important concepts such as equilibrium and stability interrelated models. In addition, the chapter provides technology examples to determine and understand the long-term behavior of these systems.
- Chapter 3 and Chapter 4 are designed to introduce statistics and probability modeling to gain understanding and insights from data. If the world were perfectly predictable and contained no variability, there would be no need to understand statistics. Chapter 3 introduces critical statistical concepts to handle both univariate and multivariate data.
- Chapter 4 builds upon the concepts introduced in Chapter 3 and extends them to establish a foundation in probability modeling. This includes a look at both discrete and continuous probability models. Probability is a measure of the likelihood of a random phenomenon or chance behavior occurring. This chapter focuses on using probability models to describes the long-term proportion with which a certain **outcome** will occur in situations with short-term uncertainty.
- Chapter 5 discusses the concepts of ordinary differential equations (ODEs) and how to use them to develop a more robust solution to problems. The chapter provides an understanding of analytical solution techniques for separable and linear ODEs. The chapter also addresses methods for homogeneous and non-homogeneous system and the use of technology to solve these type problems.
- Chapter 6 covers forecasting with mathematical programming (linear and machine learning) to solve problems in support of decision making. We start with defining the mathematical programming methods and illustrate various techniques to include regression (linear and nonlinear) exponential smoothing models and auto-regressive integrated moving average (ARIMA). Technology is used to solve the formulated problems. This chapter also sets the foundation for future chapters covering linear programing and queuing models.

- Chapter 7 builds upon concepts developed in Chapter 2 and introduces the reality of uncertainty and randomness that is all around us and that we must take into account during the modeling process. This chapter focuses on explicitly representing this uncertainty in our mathematical models. We will introduce some of the most important and commonly used stochastic models to include Markov chains, transition matrices, and Bayes' Theorem.
- Chapter 8 discusses the use of linear programing to solve common economic-related problems that frequently occur in modern industry. In fact, areas using linear programming are as diverse as defense, health, transportation, manufacturing, advertising, and telecommunications. The reason for this is that in most situations, the classic economic problem exists – you want to maximize output, but you are competing for limited resources. The chapter provides multiple techniques to handle these problems and the importance of conducting sensitivity analysis to better understand the impact of the developed solutions.
- Chapter 9 covers the power and limitations of simulations in developing solutions to problems. The chapter focuses on queuing models, which are common in many instances in modern industry. We will develop an understanding of the concept of algorithms while building both deterministic and stochastic simulations to solve problems.
- Chapter 10 introduces the mathematical methods and formulas that are used in many businesses and financial organizations. The chapter extends the DDS concepts introduced in Chapter 2. We will begin our discussions with a quick review of some sequences from DDS as it gives essential understanding to the basic principles that we will develop for these formulas. We also present some advanced modeling in finance using previous algorithms from past chapters.
- Chapter 11 addresses how we can determine the reliability or expected failure time of common systems. The chapter demonstrates how we can use the probability concepts introduced in Chapter 4 to determine the reliability of equipment. We will develop models for system components that are in series or parallel to determine the reliability of the system. The chapter will also review how to determine the reliability of active and redundant systems.
- Chapter 12 extends some of the machine learning concepts we introduced in Chapter 6 and expands them to tackle more difficult models. Machine learning can range from almost any use of a computer (machine) performing calculations and analysis measures to the application of artificial intelligence (AI) or neural networks in

> the computer's calculations and analysis methods. The defining characteristic of machine learning is the focus on using algorithmic methods to improve descriptive, predictive, and prescriptive performance in real-world contexts. In this chapter, we will discuss some common techniques, such as the gradient method, genetic algorithm, and simulated annealing to handle our problems.

The length of this book prevents us from addressing every potential nuance in modeling real-world problems. We attempt to provide a set of models and potential appropriate techniques to obtain useful results for a common problem the reader may encounter as a guide. We do assume a basic or fundamental background in mathematical modeling and therefore only spend a little time establishing the procedure before we return to providing examples and solution techniques.

While the focus of this book is on utilization in a mathematical modelling course, it does have application for decision-makers in any discipline. The book provides an overview to the decision-maker of the wide range of applications of quantitative approaches to aid in the decision-making process. As we constantly remind our students, mathematical modeling is not designed to tell you what to do, but it does provide insights and is designed to support critical thinking and the decision-making process. We view the mathematical modeling process as a framework for decision-makers. This framework consists of four key components: the formulation process, the solution process, interpretation of the solution in the context of the actual problem, and sensitivity analysis. The users of mathematical modeling should question the procedures and techniques and assumptions used in the analysis during every step in the process. At a minimum, you should always consider, "Did I use an appropriate technique to obtain a solution?" and "Why were other techniques discounted during the process?" Another question could be, "Did I oversimplify the modeling process so that any solution I develop will not really apply in this situation?"

We thank all the mathematical modeling students that we have had over the last 30 years as well as all the colleagues who have taught mathematical modeling with us during this adventure. We particularly single out the following who helped in our three-course mathematical modeling sequence at the Naval Postgraduate School over the years: Bard Mansger, Mike Jaye, Steve Horton, Patrick Driscoll, and Greg Mislick. We are especially appreciative of the mentorship of Frank R. Giordano over the past 30 plus years.

**William P. Fox**
*College of William and Mary*
**Robert E. Burks**
*Naval Postgraduate School*

# Authors

**Dr. William P. Fox** is currently a Visiting Professor of Computational Operations Research at the College of William and Mary. He is an Emeritus Professor in the Department of Defense Analysis at the Naval Postgraduate School and teaches a three-course sequence in mathematical modeling for decision making. He received his Ph.D. in Industrial Engineering from Clemson University. He has taught at the U.S. Military Academy for 12 years until retiring and at Francis Marion University, where he was the Chair of Mathematics for 8 years. He has many publications and scholarly activities, including 20 plus books and 150 journal articles.

**Colonel (R) Robert E. Burks, Jr., Ph.D.** is an Associate Professor in the Defense Analysis Department of the Naval Postgraduate School (NPS) and the Director of the NPS' Wargaming Center. He holds a Ph.D. in Operations Research from the Air Force Institute of Technology. He is a retired logistics Army Colonel with more than 30 years of military experience in leadership, advanced analytics, decision modeling, and logistics operations who served as an Army Operations Research Analyst at the Naval Postgraduate School, TRADOC Analysis Center, U.S. Military Academy, and the U.S. Army Recruiting Command.

Other book by William P. Fox and Robert E. Burks: *Advanced Mathematical Modeling with Technology*, 2021, CRC Press.

Other books by William P. Fox from CRC Press:

*Mathematical Modeling in the Age of the Pandemic*, 2021, CRC Press.

*Advanced Problem Solving Using Maple: Applied Mathematics, Operations Research, Business Analytics, and Decision Analysis* (w/William Bauldry), 2020, CRC Press.

*Mathematical Modeling with Excel* (w/Brian Albright), 2020, CRC Press.

*Nonlinear Optimization: Models and Applications*, 2020, CRC Press.

*Advanced Problem Solving with Maple: A First Course* (w/William Bauldry), 2019, CRC Press.

*Mathematical Modeling for Business Analytics*, 2018, CRC Press.

# 1

# *Perfect Partners: Combining Models of Change and Uncertainty with Technology*

**OBJECTIVES**

1. Understand the mathematical modeling process.
2. Understand the process of decision modeling.
3. Understand that models have both strengths and limitations.

Decision making under uncertainty is incredibly important in many areas of life, but it is particularly important in business, industry, and government (BIG), especially in today's big data environment. BIG decision making is essential to success at all levels, and we do not encourage "shooting from the hip". However, we do recommend good analysis for the decision maker to examine and question in order to find the best alternative course of action. So, why mathematical modeling?

A **mathematical model** may be defined as a description of a real-world system using mathematical concepts to facilitate the explanation of change in the system or to study the effects of different elements of the system and to understand changes in the patterns of behavior.

Mathematical models are used not only in the natural sciences (e.g. physics, biology, earth science, meteorology) and engineering disciplines (e.g. computer science, artificial intelligence), but also in the social sciences (e.g. business, economics, psychology, sociology, political science); physicists, engineers, statisticians, operations research analysts, and economists use mathematical models most extensively. A model may help to explain a system and to study the effects of different components, and to make predictions about behavior.

Mathematical models can take many forms, including but not limited to dynamical systems, statistical models, differential equations, or game theoretic models. These and other types of models can overlap, with a given model involving a variety of abstract structures. In general, mathematical models may include logical models, as far as logic is taken as a part of

DOI: 10.1201/9781003298762-1

mathematics. In many cases, the quality of a scientific field depends on how well the mathematical models developed on the theoretical side agree with results of repeatable experiments. Lack of agreement between theoretical mathematical models and experimental measurements often leads to important advances as better theories are developed.

## 1.1 Overview of the Process of Mathematical Modeling

Consider for a moment a basic real-world search situation that we have used for years to introduce students to mathematical modeling. Two observation posts 6.50 miles apart pick up a brief radio signal. The sensing devices were oriented at 115° and 120° respectively when a signal was detected. The devices are accurate to within 3° (that is ±3° of their respective angle of orientation). The signal reading is believed, by intelligence, to come from a region of active terrorist exchange, and it is believed that there may be a boat waiting to pick up a group of terrorists. The sun is setting, the weather is calm, and there are no ocean currents. A small helicopter leaves a pad from Post 1 and is able to fly accurately along the 115° angle direction. This helicopter has only as a searchlight for detection. At 200 feet, it can just illuminate a circular region with a radius of 30 feet. The helicopter can fly 250 miles in support of this mission due to its fuel capacity. Some basic prelaunch decisions would include: Where do you search for the boat? How many search helicopters should you use to have a "good" chance of finding the target?

Photochemical smog permeates the Los Angeles basin most days of the year. While this problem is not unique to the Los Angeles area, conditions in the basin are well suited to this phenomenon. The surrounding mountains and frequent inversion layers create the stagnant air that gives rise to these conditions. Can we build a model to examine this? Can we use such a model to study or analyze the poor air quality in Los Angeles due to traffic pollution or measuring vehicle emissions?

In the sport of bridge jumping, a willing participant attaches one end of the bungee cord to themself, attaches the other end to a bridge railing, and then drops off a bridge. Is this a safe sport? Can we describe the typical motion?

You are a new city manager in California. You are worried about earthquake survivability of your city's water tower. You need to analyze the effects of an earthquake on your water tower and see if any design improvements are necessary. You want to prevent catastrophic failure.

If you have flown lately, you may have noticed that most airplanes are full. As a matter of fact, many times an announcement is made that the plane is overbooked, and the airlines are looking for volunteers to take a later flight. Why do airlines overbook? Should they overbook? What impact does this have on the passengers? What impact does it have on the airlines?

These events all share one common element – we can model them using mathematics to support making decisions. This textbook will help you understand what a mathematical modeler might do for you as a confident problem-solver using the techniques of mathematical modeling. As a decision-maker, understanding the possibilities and asking the key questions will enable better decision to be made.

## 1.2 The Modeling Process

In this chapter, we turn our attention to the process of modeling and examine many different scenarios in which mathematical modeling can play a role.

Mathematical modeling requires as much art as it does science. Thus, modeling is more of an art than a science. Modelers must be creative and willing to be more artistic or original in their approach to the problem. They must be inquisitive and question their assumptions, variables, and hypothesized relationships. Modelers must also think outside the box in order to analyze the models and their results and to ensure their model and results pass the common sense test. Science is very important, and understanding science enables one to be more creative in viewing and modeling a problem. Creativity is extremely advantageous in problem solving with mathematical modeling.

To gain insight we should consider one framework that will enable the modeler to address the largest number of problems. The key is that there is something changing for which we want to know the effects. We call this the system under analysis. The real-world system can be very complicated or very simplistic. This requires a process that allows for both types of real-world systems to be modeled within the same process.

Consider striking a golf ball with a golf club from a tee. Our first inclination is to use the equations about distance and velocity that we used in high school mathematics class. These equations are very simplistic and ignore many factors that could impact the fall of the ball, such as wind speed, air resistance, mass of the ball, and other factors. As we add more factors, we can improve the precision of the model. Adding these additional factors makes the model more realistic and more complicated to produce. Understanding this model might be a first start in building a model for such situations or similar situation such as a bungee jumper or bridge swinger. These systems are similar for part of the model: the free fall portion has similar characteristics.

Figure 1.1 provides a closed loop process for modeling. Given a real-world situation like the one above, we collect data in order to formulate a

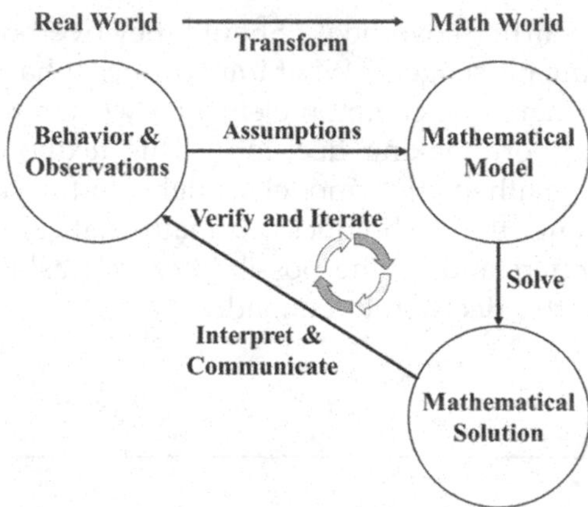


**FIGURE 1.1**
Modeling real-world systems with mathematics.

mathematical model. This mathematical model can be one we derive or select from a collection of already built mathematical models depending on the required level of sophistication. We then analyze the model that we used and reach mathematical conclusions about it. Next, we interpret the model and either make predictions about what has occurred or offer explanation as to why something has occurred. Finally, we test our conclusion about the real-world system with new data. We may refine or improve the model to improve its ability to predict or explain the phenomena. We might even reformulate a new mathematical model.

### 1.2.1 Mathematical Modeling

We will build some mathematical models describing change in the real world. We will solve these models and analyze how good our resulting mathematical explanations and predictions are. The solution techniques that we employ in subsequent chapters take advantage of certain characteristics that the various models enjoy. Consequently, after building the models, we will classify the models based on their mathematical structure.

When we observe change, we are often interested in understanding why change occurs the way it does, perhaps to analyze the effects of different conditions, or perhaps to predict what will happen in the future. Often, a mathematical model can help us understand a behavior better, while allowing us to experiment mathematically with different conditions. For our purposes, we will consider a mathematical model to be a mathematical construct designed to study a particular real-world system or behavior. The model allows us to use mathematical operations to reach mathematical conclusions about the model, as illustrated in Figure 1.1.

### 1.2.2 Models and Real-World Systems

A system is an assemblage of objects joined by some regular interaction or interdependence.

Examples include sending a module to Mars, handling the United States' debt, a fish population living in a lake, a TV-satellite orbiting the earth, delivering mail, or locations of service facilities. The person modeling is interested in understanding not only how a system works but also what interactions cause change and how sensitive the system is to changes in these inputs. Perhaps the person modeling is also interested in predicting or explaining what changes will occur in the system as well as when these changes might occur.

Figure 1.2 suggests how we can obtain real-world conclusions from a mathematical model. First, observations identify the factors that seem to be involved in the behavior of interest. Often, we cannot consider, or even identify, all the relevant factors, so we make simplifying assumptions excluding some of them. Next, we conjecture tentative relationships among the identified factors we have retained, thereby creating a rough "model" of the behavior. We then apply mathematical reasoning that leads to conclusions about the model. These conclusions apply only to the model and may or may not apply to the actual real-world system in question. Simplifications were made in constructing the model, and the observations upon which the model is based invariably contain errors and limitations. Thus, we must carefully account for these anomalies and test the conclusions of the model against real-world observations. If the model is reasonably valid, we can then draw inferences about the real-world behavior from the conclusions drawn from the model. In summary, we have the following procedure for investigating real-world behavior:

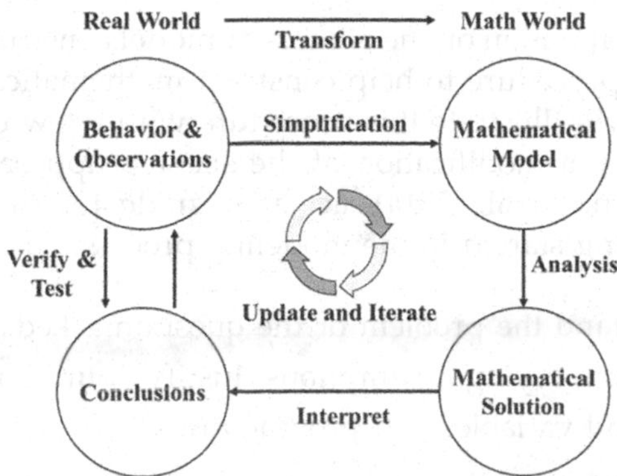


**FIGURE 1.2**
In reaching conclusions about a real-world behavior, the modeling process is a closed system (adapted from Giordano et al., 2014).

**Step 1**. Observe the system, identify the key factors involved in the real-world behavior, simplify initially, and refine later as necessary.

**Step 2**. Conjecture or guess the possible relationships or interrelationships among the factors and variables identified in Step 1.

**Step 3**. Solve the model.

**Step 4**. Interpret the mathematical conclusions in terms of the real-world system.

**Step 5**. Test the model conclusions against real-world observations – the common sense rule.

**Step 6**. Perform model testing or sensitivity analysis.

There are various kinds of models that we will introduce as well as methods or techniques to solve these models in the subsequent chapters. An efficient process would be to build a library of models and then be able to recognize various real-world situations to which they apply. Another task is to formulate and analyze new models. Still another task is to learn to solve an equation or system in order to find more revealing or useful expressions relating to the variables. Through these activities we hope to develop a strong sense of the mathematical aspects of the problem, its physical underpinnings, and the powerful interplay between them.

Most models do simplify reality. Generally, models can only approximate real-world behavior. Next, let's summarize a process for formulating a model.

### 1.2.3 Model Construction

We will focus our attention on the process of model construction. An outline is presented as a procedure to help construct mathematical models. In the next section, we will illustrate this procedure with a few examples.

These nine steps, a modification of the six-step approach presented by Giordano (Giordano et al., 2014), act as a guide for thinking about the problem and getting started in the modeling process.

**Step 1**. Understand the problem or the question asked.

**Step 2**. Make simplifying assumptions. Justify your assumptions.

**Step 3**. Define all variables and provide units.

**Step 4**. Construct a model.

**Step 5**. Solve and interpret the model.

**Step 6**. Verify the model.

**Step 7**. Identify the strengths and weaknesses of your model.

**Step 8.** Sensitivity analysis or model testing of the model. Do the results pass the "common sense" test?

**Step 9.** Implement and maintain the model for future use.

We will start with a discussion of each step in more depth.

**Step 1. Understand the problem or the question asked:**
Identifying the problem to study is usually difficult. In real life, no one walks up to you and hands you an equation to be solved. Usually, it is a comment like, "We need to make more money," or "We need to improve our efficiency." We need to be precise in our formulation of the mathematics to describe the situation.

**Step 2. Make simplifying assumptions:**
Start by brainstorming the situation. Make a list of as many factors, or variables, as you can. Realize we usually cannot capture all these factors influencing a problem. The task is simplified by reducing the number of factors under consideration. We do this by making simplifying assumptions about the factors, such as holding certain factors as constants. We might then examine to see if relationships exist between the remaining factors (or variables). Assuming simple relationships might reduce the complexity of the problem. Once you have a shorter list of variables, classify them as independent variables, dependent variables, or neither.

**Step 3. Define all variables:**
It is critical to define all your variables and provide the mathematical notation to be used for each.

**Step 4. Select or build the modeling approach and formulate the model:**
Using the tools in this text and your own creativity, build a model that describes the situation and whose solution helps to answer important questions.

**Step 5. Solve and interpret the model:**
We take the model we constructed in Steps 1–4 and solve it. Often this model might be too complex or unwieldy so we cannot solve it or interpret it. If this happens, we return to Steps 2–4 and simplify the model further.

**Step 6. Verify the model:**
Before we use the model, we should test it out. There are several questions we must ask. Does the model directly answer the question or does the model allow for the questions to be answered? Is the model useable in a practical sense (can we

obtain data to use the model)? Does the model pass the common sense test?

We like to say that we corroborate the reasonableness of our model rather than verify or validate the model.

**Step 7. Identify the strengths and weaknesses:**
No model is complete with self-reflection of the modeling process. We need to consider not only what we did correctly, but what we did that might be suspect as well as what we could do better. This reflection also helps in refining models.

**Step 8. Sensitivity analysis and model testing:**
A modeler wants to know how the inputs affect the ultimate output for any system. Passing the common sense test is essential. One of the authors once had a class model Hooke's law with springs and weights. He asked them all to use their model to see how far the spring would stretch using their weight. They all provided the numerical answers, but none said that the spring would break under their weight.

**Step 9. Refine, implement, and maintain the model:**
A model is pointless if we do not use it. The more user-friendly the model, the more it will be used. Sometimes the ease of obtaining data for the model can dictate its success or failure. The model must also remain current. Often this entails updating parameters used in the model.

## 1.3 Illustrative Examples

We now demonstrate the modeling process that was presented in the previous section. Emphasis is placed on problem identification and choosing appropriate (useable) variables. We do not build the models as these modeling examples are repeated later in the book and the models are completed and discussed there.

**Example 1.1:** The Size of Prehistoric Creatures

Scenario. *Titanus walleri* really lived about 2 million years ago on the oak and grass savannahs of what is now Florida. Dr. Bob Chandler, one of the world's experts on fossil birds, from Georgia College and State University, has dredged up a number of bits and pieces of this bizarre predatory bird from the Santa Fe River near Gainesville, Florida. Here are some facts about the terror bird from Chandler (1994) (adapted from http://discovermagazine.com/1997/jun/terrortaketwo1149):

- Giant, flightless predatory bird.
- Lived in South America 30 million years before the Interchange.
- Fossils have been found of the terror bird, *Titanus walleri*, in Florida.
- Suspected to be a fierce hunter who would lie in wait and ambush its prey and attack from the tall grasslands.
- Suspected to pin down its prey with its beak.
- Suspected to use 4- to 5-inch inner toe claw with its beak to shred its prey.
- Another unique feature was this bird had arms, not wings (more powerful than the arms of the velociraptor).

We realize that bones provide a good indication of size (height) but not reliable body weight (weight is not easily fossilized). We want to model its weight based upon the information that can be determined from the fossils.

**Understanding the Problem:** What is the relationship between the bones found and the original "grown" size of the creature? This might be too general, so let's restrict our problem. The fossil is a femur bone that is intact. We want to know if we can find a relationship between the weight of an animal and the size of its femur (measured as circumference). If we can find a relationship, then we can use the model to predict the size (in weight measures) of the terror bird as a function of the circumference of the femur.

**Assumptions:** We assume that the thickness (measured by circumference) of the femur is important to the size of the animal. We might conjecture that thicker bones support more weight. We know that the terror bird is an ancestor of modern birds, but should birds be used to help build the model? We might assume that bird data is sufficient and use only that data to build a model. Or should we rely on species that lived about when the terror bird roamed the earth, like the dinosaurs, and use their data to help build the model? We could assume that dinosaur data is more appropriate to use to build a model. We might want to try both approaches and compare results. If bird data is available, we might address the appropriateness of using modern bird data to build a model for a bird that lived 2 million years ago and did not even fly. If more terror bird fossils are discovered, we might attempt to model their size and get a sense of differing terror bird sizes.

Thus, the model that we build is affected by the assumptions we used and data available.

**Example 1.2:** Prescribed Drug Dosage

Scenario. Consider a patient that needs to take a newly marketed prescribed drug. To prescribe a safe and effective regimen for treating the disease, one must maintain a blood concentration above some effective level and below an unsafe level.

**Understanding the Problem:** Our goal is a mathematical model that relates dosage and time between dosages to the level of the drug in the bloodstream. What is the relationship between the amount of drug taken and the amount in the blood after time, $t$? By answering this question, we are empowered to examine other facets of the problem of taking a prescribed drug.

**Assumptions:** We should choose or know the disease in question and the type (name) of the drug that is to be taken. We will assume the drug is rythmol, a drug taken to control the heart rate and which is called an anti-arrhythmic. We need to know or to find decaying rate of rythmol in the blood stream. This might be found from data that has been previously collected. We need to find the safe and unsafe levels of rythmol based upon the drug's "effects" within the body. This will serve as bounds for our model. Initially, we might assume that the patient size and weight have no effect on the drug's decay rate. We might assume that all patients are about the same size and weight. All are in good health, and no one takes other drugs that affect the prescribed drug. We assume all internal organs are functionally properly. We might assume that we can model this using a discrete time period, even though the absorption rate is a continuous function. These assumptions help simplify the model.

**Example 1.3:** Determining Heart Weight

We can start by assuming that we are interested in building a model that shows heart weight as a function of the size of the heart. We might have access to data that relates, for the following seven mammals, their heart weight in grams and their diameter of the left ventricle of the heart measured in millimeters (mm) shown in Table 1.1.

**Problem Identification:** Find a relationship between heart weight and the diameter of the left ventricle of the heart.

**TABLE 1.1**
Data for Mammals Heart Sizes (Adapted from Special Projects in MA 381 at West Point)

| Animal | Heart Weight (g) | Diameter (mm) |
|---|---|---|
| Mouse | 0.13 | 0.55 |
| Rat | 0.64 | 1.00 |
| Rabbit | 5.80 | 2.20 |
| Dog | 102.00 | 4.00 |
| Sheep | 210.00 | 6.50 |
| Ox | 2,030.00 | 12.00 |
| Horse | 3,900.00 | 16.00 |

**Assumptions:** We will assume that the heart weight is typical for a healthy animal. We further assume that the diameter is measured the same way for each heart. We might assume that all mammals are scale models of other mammals. We assume that all mammals are in good health. We assume that the data were collected the same way for each mammal.

**Example 1.4:** Bridge Too Far

Consider an engineering design for a truss bridge, as shown in Figure 1.3. Trusses are lightweight structures capable of carrying heavy loads. In civil engineering bridge design, the individual members of the truss are connected with rotatable pin joints that permit forces to be transferred from one member of the truss to another. The accompanying Figure 1.3 shows a truss that is held stationary at the lower left endpoint #1, is permitted to move horizontally at the lower right endpoint #4, and has pin joints at #1, #2, #3, and #4, as shown. A load of 10 kilo newtons (kN) is placed at joint #3, and the forces on the members of the truss have magnitude given by f1, f2, f3, f4, and f5, as shown.

The stationary support member has both a horizontal force F1 and a vertical force F2, but the movable support member has only the vertical force F3.

If the truss is in static equilibrium, the forces at each joint must add to the zero vector, so the sum of the horizontal and vertical components of the forces at each joint must be zero. This produces the system of linear equations.

**Problem Identification:** Determine the forces on each joint as a function of the angles between joints.

**Assumptions:** We assume the bridge is sturdy and will not falter or collapse. We assume the motion of the bridge due to normal motion is negligible. We assume a design such as the one provided. We might assume the bridge is not in an earthquake zone.

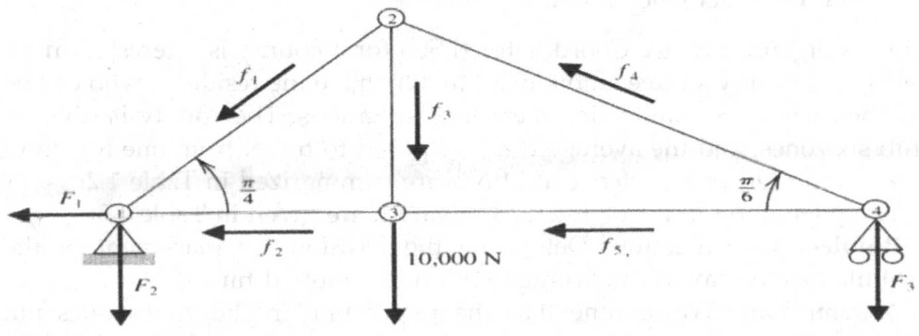


**FIGURE 1.3**
Bridge truss (from Burden and Faires, page 423).

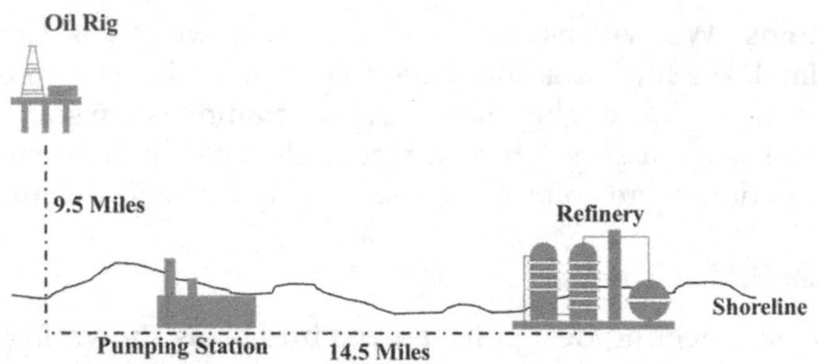


**FIGURE 1.4**
Oil rig, pumping station, and refinery.

**Example 1.5:** Oil Rig and Pumping Station Location

Consider an oil-drilling rig that is 8.5 miles offshore. The drilling rig is to be connected by underwater pipe to a pumping station. The pumping station is connected by land-based pipe to a refinery, which is 14.7 miles down the shoreline from the drilling rig (Figure 1.4). The underwater pipe costs $31,575 per mile, and the land-based pipe costs $13,342 per mile. You are to determine where to place the pumping station to minimize the cost of the pipe.

**Problem Identification:** Build a model to minimize the cost of a pipe from the oil rig to the refinery.

**Assumptions:** We assume that the costs based upon solid estimates and will not fluctuate during the building of the pipeline. We assume no weather delays or any other natural or unnatural delays to building. All workers are competent in their jobs. The materials are not flawed in any way. We assume no natural disaster or storms that might disrupt work. We assume that when built the pipeline works successfully. A schematic is shown in Figure 1.4.

**Example 1.6:** Emergency Medical Response

The Emergency Service Coordinator (ESC) for a county is interested in locating the county's three ambulances to maximize the residents who can be reached within 8 minutes in emergency situations. The county is divided into six zones, and the average times required to travel from one region to the next under semi-perfect conditions are summarized in Table 1.2.

The population in zones 1, 2, 3, 4, 5, and 6 are given in Table 1.3:

**Problem Identification:** Determine the location for placement of the ambulances to maximize coverage within the allotted time.

**Assumptions:** We assume that the population in the zones does not change in the short run. We also assume traffic patterns and time of day do not affect the times to travel from zone i to zone j.

**TABLE 1.2**

Average Travel Times from Zone i to Zone j in Perfect Conditions

| Zones | 1 | 2 | 3 | 4 | 5 | 6 |
|---|---|---|---|---|---|---|
| 1 | 1 | 8 | 12 | 14 | 10 | 16 |
| 2 | 8 | 1 | 6 | 18 | 16 | 16 |
| 3 | 12 | 18 | 1.5 | 12 | 6 | 4 |
| 4 | 16 | 14 | 4 | 1 | 16 | 12 |
| 5 | 18 | 16 | 10 | 4 | 2 | 2 |
| 6 | 16 | 18 | 4 | 12 | 2 | 2 |

**TABLE 1.3**

Population in Each Zone

| Zone | Population |
|---|---|
| 1 | 50,000 |
| 2 | 80,000 |
| 3 | 30,000 |
| 4 | 55,000 |
| 5 | 35,000 |
| 6 | 20,000 |
| Total | 270,000 |

**Example 1.7:** Bank Service Problem

The bank manager is trying to improve customer satisfaction by offering better service. The management wants the average customer wait to be less than 2 minutes and the average length of the queue (length of the line waiting) to be two or fewer. The bank estimates about 150 customers per day. The existing arrival and service times are given in Tables 1.4 and 1.5.

**Problem Identification:** Build a mathematical model to determine whether the bank is meeting its goals. Determine if the current customer service

**TABLE 1.4**

Arrival Times

| Time between Arrival (min) | Probability |
|---|---|
| 0 | 0.10 |
| 1 | 0.15 |
| 2 | 0.10 |
| 3 | 0.35 |
| 4 | 0.25 |
| 5 | 0.05 |

**TABLE 1.5**
Service Times

| Service Time (min) | Probability |
|---|---|
| 1 | 0.25 |
| 2 | 0.20 |
| 3 | 0.40 |
| 4 | 0.15 |

is satisfactory according to the manager guidelines. If not, determine through modeling the minimal changes for servers required to accomplish the manager's goal.

**Assumptions:** We assume customer demands remain consistent over time.

## 1.4 Technology

The mathematical modeling necessary to address the examples above is enhanced with the use of technology. In addition, technology allows you to address problems that are much more robust and difficult than the example problems. This is one of the reasons this chapter is called "Perfect Partners". The partnering of technology with modeling is both key and essential to good modeling principles and practices. In this book, we illustrate three different technologies, Excel©, Maple©, and R©, as we look at and solve many real-world related problems. The intent is to not make the reader an expert in any of the technologies but to demonstrate the capabilities of these technology platforms to address relevant problems.

Although Excel might not be the "go-to" technology for many mathematicians or academicians, it is a common tool used by modeling practitioners in the real world. The goal of illustrating Excel in this book is to demonstrate that it has the capability of solving many different types of problems and to empower students in math, science, and engineering to use Excel© properly for solving future problems.

MAPLE© is an excellent technology for mathematics and operations research majors. Its power and graphical interface in 2D and 3D make it an excellent tool.

R© is a free software programming language and a software environment for statistical computing and graphics. The R language is widely used among statisticians and data miners for developing statistical software and data analysis. R© is a language and environment for statistical computing and graphics. R© provides a wide variety of statistical techniques (linear and nonlinear modeling, classical statistical tests, time-series analysis, classification, clustering, ...) and graphical techniques, and is highly extensible. The S language is often the vehicle of choice for research in statistical methodology, and R© provides an Open-Source route to participation in that activity.

One of R©'s strengths is the ease with which well-designed publication-quality plots can be produced, including mathematical symbols and formulae where needed. Great care has been taken over the defaults for the minor design choices in graphics, but the user retains full control.

R© is available as Free Software under the terms of the Free Software Foundation's GNU General Public License in source code form. It compiles and runs on a wide variety of UNIX platforms and similar systems, Windows, and MacOS.

### 1.4.1 The R© Environment

R© is an integrated suite of software facilities for data manipulation, calculation, and graphical display. It includes:

- an effective data handling and storage facility;
- a suite of operators for calculations on arrays, in particular matrices;
- a large, coherent, integrated collection of intermediate tools for data analysis;
- graphical facilities for data analysis and display either on-screen or on hardcopy; and
- a well-developed, simple, and effective programming language that includes conditionals, loops, user-defined recursive functions, and input and output facilities.

The term "environment" is intended to characterize it as a fully planned and coherent system, rather than an incremental accretion of very specific and inflexible tools, as is frequently the case with other data analysis software.

Many users think of R© as a statistics system. We prefer to think of it as an environment within which statistical techniques are implemented. R© can be extended (easily) via packages. There are about eight packages supplied with the R© distribution, and many more are available through the CRAN family of Internet sites covering a very wide range of modern statistics.

## 1.5 Exercises

1. How would you approach a problem concerning a drug dosage? Do you always assume the doctor is right?

2. In modeling the size of any prehistoric creature, what information would you like to be able to obtain? What additional assumptions might be required?
3. In the oil rig problem (Example 1.5), what other factors might be critical in obtaining a "good" model that predicts reasonably well? What variables could be important that were not considered?
4. For the model in Example 1.3, are the assumptions about the data reasonable? How would you collect data to build the model? What other variables would you consider?
5. In Example 1.4, what is the impact of the location of the bridge?
6. In Example 1.7 on the bank queue problem, discuss the criticality of the assumptions. Do you feel that more training is as valuable as adding another server?

## 1.6 Projects

1. Is Michael Jordan the greatest basketball player of the century? What variables and factors need to be considered?
2. What kind of car should you buy when you graduate from college? What factors should be in your decision? Are car companies modeling your needs?
3. Consider domestic decaffeinated coffee brewing. Suggest some objectives that could be used if you wanted to market your new brew. What variables and data would be useful?
4. Replacing a coaching legend at a school is a difficult task. How would you model this? What factors and data would you consider? Would you equally weigh all factors?
5. How would go about building a model for the "best pro football player of all time"?
6. Rumors abound in major league baseball about steroid use. How would you go about creating a model that could imply the use of steroids? Relate baseball's steroids rules to the Yankee's Alex Rodriquez case.
7. Since 2013, the America League has won 7 straight All-star games and 20 of the last 23 games dating back to 1997. Help the National League prepare to win by designing a model for players or a line-up that could help them change their outcome.

## References and Suggested Further Reading

Albright, B. (2010). *Mathematical Modeling with Excel*, Jones and Bartlett, Burlington, MA.

Albright, B., and W. Fox (2020). *Mathematical Modeling with Excel*, 2nd ed. Taylor & Francis Publishers, Boca Raton, FL.

Burden, R., and D. Faires (1997). *Numerical Analysis*, Brooks-Cole, Pacific Grove, CA.

Chandler, R. M. (1994). The wing of Titanis walleri (Aves: Phorusrhacidae) from the Late Blancan of Florida. *Bulletin of the Florida Museum of Natural History, Biological Sciences* 36: 175–180.

COMAP, Modeling Competition Sites found at www.comap.com/contests

Fox, W. (2018). *Mathematical Modeling for Business Analytics*, Taylor & Francis Publishers, Boca Raton, FL.

Giordano, F., W. Fox, and S. Horton (2014). *A First Course in Mathematical Modeling*, 5th ed. Cengage Publishers, Boston, MA.

# 2

# Modeling Change: Discrete Dynamical Systems (DDS) and Modeling Systems of DDS

**OBJECTIVES**

1. Review of linear and nonlinear discrete dynamical systems.
2. Review use of technology with Excel, Maple, or R.
3. Define, model, and solve systems of discrete dynamical systems.
4. Analyze the long-term behavior of systems of discrete dynamical systems.
5. Understand the concepts of equilibrium and stability in systems.
6. Model both linear and nonlinear systems of discrete dynamical systems.

## 2.1 Introduction and Review of Modeling with Discrete Dynamical Systems

The impact of supply and demand, due to COVID in 2021, drove up car prices and made considering the options of buying new, buying used, or leasing interesting decisions for consumers. Decisions now included how do you go about financing a new car and is it better to buy new, buy used, or lease? Once you have looked at the makes and models to determine what type of car you like, it is time to consider the "costs" and finance packages that lure potential buyers into the car dealerships. This is a perfect process for mathematical modeling as a dynamical system. In most cases, fixed payment amounts are made at the end of each month. The amount of the loan is predetermined, as we will see later in the chapter, and the interest rates are fixed.

DOI: 10.1201/9781003298762-2

These types of buying decisions can easily extend to other sectors, such as helping your company decide about buying or leasing new computers or other major equipment for the company. How could this decision be analyzed? Discrete dynamical system modeling is one classic technique to model and analyze the possible range or set of decisions.

A *discrete dynamical system* (DDS) is a system that changes over time. In this case, the time interval is discrete and can represent any units of time, such a minutes, hours, days, weeks, or years. A DDS is easy to model but may be difficult or even impossible to solve in closed form. In this chapter, we present some example models with easy closed-form solutions, to allow us to concentrate more on building the models and obtaining numerical and graphical solutions.

A simple example will help develop a better understanding of the major terms and elements in a DDS. Let's start by defining the dynamical system $A_n$ to be the amount of antibiotic in our blood stream after $n$ time periods. The domain is non-negative integers representing the time periods from 0, 1, 2, …, which will be the inputs to the function. Since the domain is discrete, our function is a discrete function. The range is the values of $A_n$ determined for each value of the domain. Thus, $A_n$ also represents the dependent variable. For each input value of the domain from 0, 1, 2, …, the result is one and only one $A_n$; thus, $A_n$ is a function.

There are three components to dynamical systems: The first is an equation for the sequence representing $A_n$, the second is a defined time period $n$, and the last element is at least one starting value, commonly referred to as the **initial condition**. Let's assume in our example, that we start with no antibiotic in our system. Then, $A_0 = 0$ mg is our initial condition. But if we started after we took an initial 200 mg tablet, then $A_0 = 200$ mg would represent our initial condition. An example of a DDS with its initial condition would be:

$$A_{n+1} = 0.5A_0, \quad A_0 = 500.$$

We are interested in modeling discrete *change* and modeling with DDSs employs a method to explain certain discrete behaviors or make long-term predictions. A powerful paradigm that we can use to model with DDSs is:

***future value* = *present value* + *change*.**

The DDSs that you encounter may seem to differ in appearance and composition, but we will be able to solve a large class of these "seemingly" different dynamical systems using similar methods. In this chapter, we will use iteration and graphical methods to demonstrate the properties of DDSs.

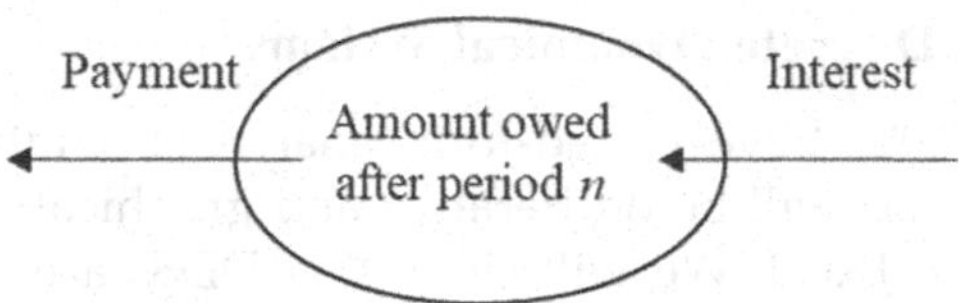


**FIGURE 2.1**
Flow diagram for financing a car.

One common approach to define and visualize the problem is to use flow diagrams. Flow diagrams will allow you to visualize how the dependent variable changes over time. These flow diagrams help illustrate the paradigm and put it into mathematical terms. Let us revisit our earlier discussion concerning cars and examine a simple example of looking to finance a new Ford Truck. The out-the-door cost of the truck is $32,000, and you decide to use $4,000 as a down payment. This means you will need to finance $28,000. The dealership offers 2.5% financing for 72 months. Consider the flow diagram for financing your new tuck, depicted in Figure 2.1.

This flow diagram helps to visualize the elements of the discrete dynamical model and is designed to capture all the changes to the system. First, define, $A_n$ = the amount owed after $n$ months. Notice that the arrow in Figure 2.1 pointing into the circle is the interest to the unpaid balance. This increases your debt or the amount you will owe over time. The arrow pointing out of the circle is your monthly payment that decreases your debt.

We define the following variables:

$A_{n+1}$ = the amount owed in the future;
$A_n$ = amount currently owed;

The element of change, as depicted in the flow diagram, is $iA_n - P$, so the model is

$$A_{n+1} = A_n + iA_n - P,$$

where
$i$ is the monthly interest rate and
$P$ is the monthly payment.

We model dynamical systems that have only **constant coefficients**. A dynamical system with constant coefficients may be written in the form

$$A_{n+3} = b_2 A_{n+2} + b_1 A_{n+1} + b_0 A_n$$

where $b_0$, $b_1$, and $b_2$ are constants.

### 2.1.1 Solutions to Discrete Dynamical Systems

Although some DDSs have closed-form analytical solution that we will discuss, our emphasis will be on iterative and graphical solutions that we will illustrate using Excel. We will show that DDSs are relatively easy to model. The solutions may always be obtained by iterative and graphical methods. Not all DDSs have closed-form analytical solutions.

Let us begin with a simple DDS model that has a closed-form solution.

**Theorem 2.1:** The solution to a linear DDS $A_{n+1} = r\ A_n$ for $r \neq 0$ is

$$A_k = b^k A_n, \tag{2.1}$$

where $A_0$ is the initial condition of the system at time period 0 and $k$ is a generic time period.

Technology is an integral part of models with DDSs. Every DDS has a numerical and graphical solution, which can easily be attained with technology.

**Example 2.1:** A Drug Dosage Example

Suppose that a doctor prescribes that their patient takes a pill containing 100 mg of a certain drug every hour. Assume that the drug is immediately ingested into the bloodstream once taken. Also, assume that every hour the patient's body eliminates 25% of the drug that is in their bloodstream. Suppose that the patient had 0 mg of the drug in their bloodstream prior to taking the first pill. How much of the drug will be in their bloodstream after 72 hours?

Problem Statement: Determine the relationship between the amount of drug in the bloodstream and time.

Assumptions: The patient is of normal size and health. There are no other drugs being taken that will affect the prescribed drug. We assume that there are no internal or external factors that will affect the drug absorption rate. The patient always takes the prescribed dosage at the correct time. Figure 2.2 shows the change diagram for this problem.

Variables:

Define a(n) to be the amount of drug in the bloodstream after period $n$, $n$ = 0, 1, 2… hours.

Change Diagram:

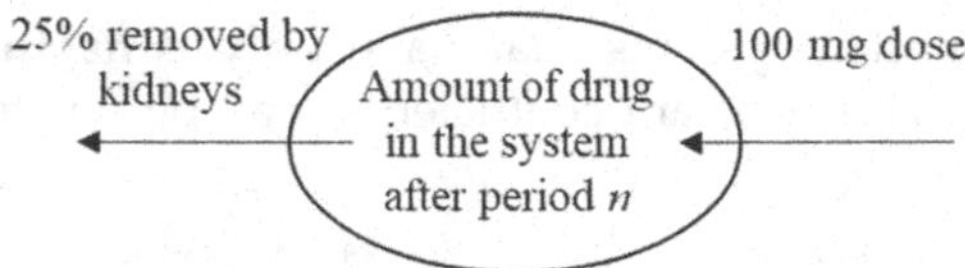


**FIGURE 2.2**
Change diagram for drugs in system.

Model Construction:
We first need to define the following variables:

$a_{n+1}$ = amount of drug in the system on daynin the future;
$a_n$ = amount currently in system.

In this case, we will define change as follows: change = dose – loss in system

$$\text{change} = 100 - 0.25\, a(n);$$

so, Future = Present + Change is

$$a(n + 1) = a(n) - 0.25\, a(n) + 100$$

or

$$a(n + 1) = 0.75\, a(n) + 100.$$

We note that this is not in the form as Theorem 2.1, so we will introduce Theorem 2.2.

**Theorem 2.2:** The solution to a linear DDS $a_{n+1} = ra_n + d$ for $r \neq 0$ and $d \neq 0$ is

$$a(k) = b^k C + (d/(1 - r)), \tag{2.2}$$

where $d/(1 - r)$ is the equilibrium value of the system, C is the initial condition of the system at time period 0, and $k$ is a generic time period that provides $a(0) = V$.

We can employ Theorem 2.2 and find

$$A(k) = 0.75\, k\, C + d/(1 - r)$$

$$D = 100$$

$$1 - 7 = 0.25, \ \text{so}\ d/(1 - r) = 400.$$

Now,

$$A(0) = 0$$

$$\text{So, } 0 = C(0.75)0 + 400$$

$$C = -400.$$

The model, in general form, is

$$A(k) = 0.75^k(-400) + 400.$$

We can also iterate the system numerically. If we let $a(n)$ (we say "$a$ at $n$" or "$a$ of $n$") be the number of milligrams of drug in the bloodstream after $n$ hours, and the initial amount in the bloodstream is 0 mg, then

$$\begin{aligned} a(0) &= 0, \\ a(1) &= 0.75(0) + 100 = 100, \\ a(2) &= 0.75(100) + 100 = 175, \\ a(3) &= 0.75(175) + 100 = 231.25 \text{ mg}. \end{aligned}$$

We could write these equations where we do not substitute the numerical values that we calculated:

$$\begin{aligned} a(0) &= 0, \\ a(1) &= 0.75\,a(0) + 100, \\ a(2) &= 0.75\,a(1) + 100, \\ a(3) &= 0.75\,a(2) + 100. \end{aligned}$$

Here, we see that the amount of drug in the bloodstream is related to the amount of drug in the bloodstream after the previous time period. Specifically, the amount of drug in the bloodstream after any of the first three time periods is 0.75 times the amount of drug in the bloodstream after the previous time period plus an additional 100 mg that is injected every hour.

We see that a pattern has developed that describes the amount of drug in the bloodstream. We are now prepared to conjecture (make an educated guess) about the amount of drug in the bloodstream after any hour. Mathematically, we say that the amount of drug in the bloodstream after $n$ hours is

$$a(n + 1) = 0.75\,a(n) + 100.$$

The relationship above describes the amount of drug in the bloodstream after $n$ hours. With this, we can see the change that occurs every hour within this "system" (amount of drug in the bloodstream), and state of the system, after any hour, is dependent on the state of the system after the previous hour. This is a DSS.

**Analyzing the DDS:** We want to find the value of *a(72)*. First, we can apply Theorem 2.2, as before.

$$a_e = 100/(0.25) = 400$$

$$a(k) = c(0.75)^k + 400$$

Since $a(0) = 0$

$$0 = c + 400$$

$$c = -400.$$

The solution is $a(k) = -400(0.75)^k + 400$, as illustrated with Theorem 2.2 before.

$$\text{For } a(72) = -400(0.75)^{72} + 400 = 399.9996.$$

**Interpretation of Results:** The DDS shows that the drug reaches a value where change stops, and eventually the concentration in the bloodstream levels at 400 mg, as shown in Figure 2.3. If 400 mg is both a safe and effective dosage level, then this dosage schedule is acceptable. We discuss this concept of change stopping (equilibrium) later in this chapter.

**Example 2.2:** A Simple Mortgage Example

Five years ago, your parents purchased a home by financing $80,000 for 20 years, paying monthly payments of $880.87 with a monthly interest of 1%. They have made 60 payments and wish to know what they actually owe on the house at this time. They can use this information to decide whether or not they should refinance their house at a lower interest rate for the next 15 or 20 years. The change in the amount owed each period increases by the amount of the interest and decreases by the amount of the payment. The flow diagram is presented in Figure 2.4.

Problem Identification: Build a model that relates the time with the amount owed on a mortgage for a home.

Assumptions: Initial interest was 12%. Payments are made on time each month.

Variables:

Let $b(n)$ = amount owed on the home after $n$ months

Flow Diagram:

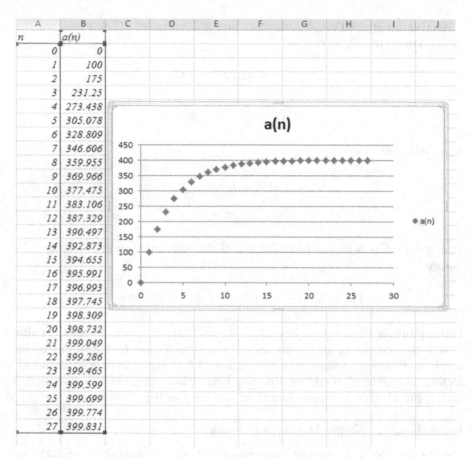

| n | a(n) |
|---|---|
| 0 | 0 |
| 1 | 100 |
| 2 | 175 |
| 3 | 231.25 |
| 4 | 273.438 |
| 5 | 305.078 |
| 6 | 328.809 |
| 7 | 346.606 |
| 8 | 359.955 |
| 9 | 369.966 |
| 10 | 377.475 |
| 11 | 383.106 |
| 12 | 387.329 |
| 13 | 390.497 |
| 14 | 392.873 |
| 15 | 394.655 |
| 16 | 395.991 |
| 17 | 396.993 |
| 18 | 397.745 |
| 19 | 398.309 |
| 20 | 398.732 |
| 21 | 399.049 |
| 22 | 399.286 |
| 23 | 399.465 |
| 24 | 399.599 |
| 25 | 399.699 |
| 26 | 399.774 |
| 27 | 399.831 |

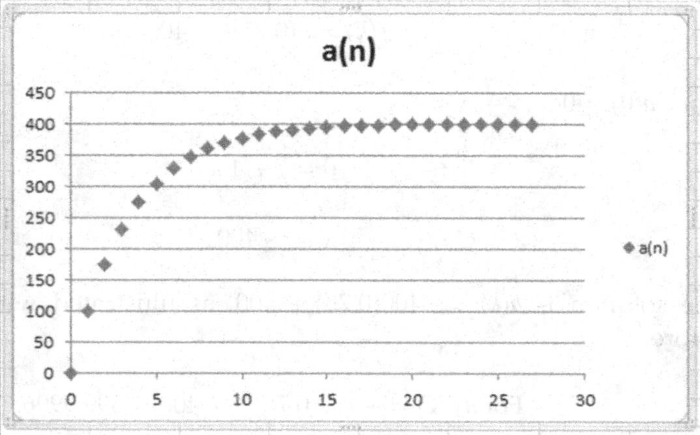

**FIGURE 2.3**
Behavior of drugs in our systems.

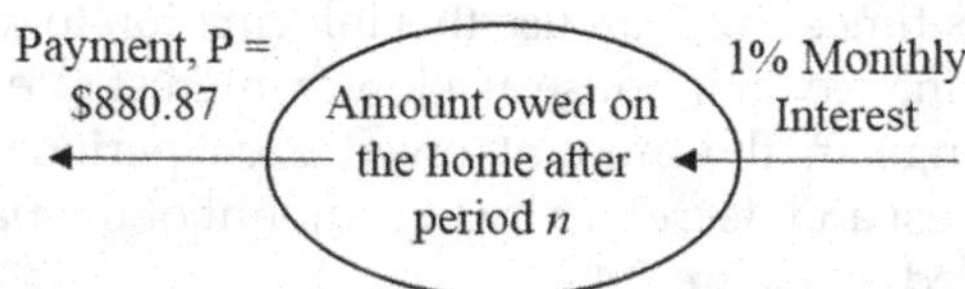


**FIGURE 2.4**
Flow diagram for mortgage example.

**Model Construction:**

$$b(n+1) = b(n) + 0.12/12\, b(n) - 880.87,\ b(0) = 80{,}000$$

$$b(n+1) = 1.01\, b(n) - 880.87,\ b(0) = 80{,}000$$

Model Solution:

$$\text{Mortgage Owed}(n) = -8087(1.01)n + 88087.$$

We can iterate this over the entire 20 years (240 months) in Excel (Figure 2.5):

After paying for 60 months, your parents still owe $73,395.37 out of the original $80,000.

In addition, they have paid a total of $52,852.20, and only $6,605 went toward the principal payment of the home. The rest of the money went toward paying only the interest. If the family continues with this loan, then they will make 240 payments of $880.87, or $211,400.80 total in payments. This is $133,400.80 in interest. They have already paid $46,647.20 in interest. They would pay an additional $86,753.60 in interest over the next 15 years. What should they do? Perhaps an alternative scheme is available, such as refinance.

## 2.2 Equilibrium and Stability Values and Long-Term Behavior

We previously mentioned Theorem 2.1 concerning equilibrium values for linear DDS. We will take a moment to further discuss these equilibriums.

Equilibrium Values

We can use our original paradigm to help find the equilibrium value,

$$Future = Present + Change.$$

Change in the system will stop when flows into and out of the flow diagram are equal. At this point, change equals zero and future values equal the present value. The value for which this happens, if any, is the equilibrium value. This gives us a context for the concept of the equilibrium value.

Models of the Form $a(n + 1) = r\,a(n) + b$, where $r$ and $b$ are constants

We will return to our drug dosage problem (Example 2.1) and consider adding a constant dosage each time period (time periods might be 4 hours). Our model becomes

$$a(n + 1) = 0.5\,a(n) + 16\text{ mg}.$$

We will also assume that there is an initial dosage applied prior to beginning the regime. We will let these initial values be as follows and graphic:

$$a(0) = 20$$
$$a(0) = 10$$
$$a(0) = 40.$$

| n | b(n) | | | int |
|---|---|---|---|---|
| 0 | 80000 | 37 | 76400.67 | |
| 1 | 79919.13 | 38 | 76283.8 | |
| 2 | 79837.45 | 39 | 76165.77 | |
| 3 | 79754.96 | 40 | 76046.56 | |
| 4 | 79671.64 | 41 | 75926.15 | |
| 5 | 79587.48 | 42 | 75804.55 | |
| 6 | 79502.49 | 43 | 75681.72 | |
| 7 | 79416.64 | 44 | 75557.67 | |
| 8 | 79329.94 | 45 | 75432.38 | |
| 9 | 79242.37 | 46 | 75305.83 | |
| 10 | 79153.92 | 47 | 75178.02 | |
| 11 | 79064.59 | 48 | 75048.93 | |
| 12 | 78974.37 | 49 | 74918.55 | |
| 13 | 78883.24 | 50 | 74786.86 | |
| 14 | 78791.2 | 51 | 74653.86 | |
| 15 | 78698.24 | 52 | 74519.53 | |
| 16 | 78604.36 | 53 | 74383.85 | |
| 17 | 78509.53 | 54 | 74246.82 | |
| 18 | 78413.76 | 55 | 74108.42 | |
| 19 | 78317.02 | 56 | 73968.64 | |
| 20 | 78219.32 | 57 | 73827.45 | |
| 21 | 78120.65 | 58 | 73684.86 | |
| 22 | 78020.98 | 59 | 73540.84 | |
| 23 | 77920.32 | 60 | 73395.37 | |
| 24 | 77818.66 | | | |
| 25 | 77715.97 | | | |
| 26 | 77612.26 | | | |
| 27 | 77507.51 | | | |
| 28 | 77401.72 | | | |
| 29 | 77294.87 | | | |
| 30 | 77186.95 | | | |
| 31 | 77077.95 | | | |
| 32 | 76967.85 | | | |
| 33 | 76856.66 | | | |
| 34 | 76744.36 | | | |
| 35 | 76630.93 | | | |
| 36 | 76516.37 | | | |

**FIGURE 2.5**
Screenshot of mortgage iterations in Excel.

Regardless of the starting value, the future terms of $a(n)$ approach 32, as depicted in Figure 2.6. Thus, 32 is the equilibrium value (*ev*). We could have solved for this algebraically as well.

$$a(n+1) = 0.5\,a(n) + 16$$
$$ev = 0.5\,ev + 16$$
$$0.5\,ev = 16$$
$$ev = 32.$$

Another method of finding the equilibrium values involves solving the equation $a = ra + b$, and solving for $a$ (where $a$ is the equilibrium value), we find:

$$a = \frac{b}{1-r}, \quad \text{if } r \neq 1.$$

Using this formula in our previous example, the equilibrium value is

$$a = 16/(1-0.5) = 32.$$

### 2.2.1 Stability and Long-Term Behavior

For a dynamical system, $a(n+1)$ with a specific initial condition, $a(0) = a_0$, we have shown that we can compute $a(1)$, $a(2)$, and so forth. Often these values are not as important as the long-term behavior. By long-term behavior, we refer to what will eventually happen to $a(n)$ for larger values of $n$. There are many types of long-term behavior that can occur with DDS; we will only discuss a few here.

If the $a(n)$ values for a DDS eventually get close to the equilibrium value, *ev*, no matter the initial condition, then the equilibrium value is called a **stable equilibrium value** or **an attracting fixed point**.

Often, we characterize the long-term behavior of the system in terms of its stability. If a DDS has an equilibrium value and if the DDS tends to the equilibrium value from starting values near the equilibrium value, then the DDS is said to be stable (Table 2.1).

Thus, for the dynamical system $a(n+1) = r\,a(n) + b$, where $b \neq 0$:

### 2.2.2 Relationship to Analytical Solutions

If a DDS has an *ev* value, we can use the *ev* value to find the analytical solution.

Recall the mortgage example (Example 2.2) from Section 2.1,

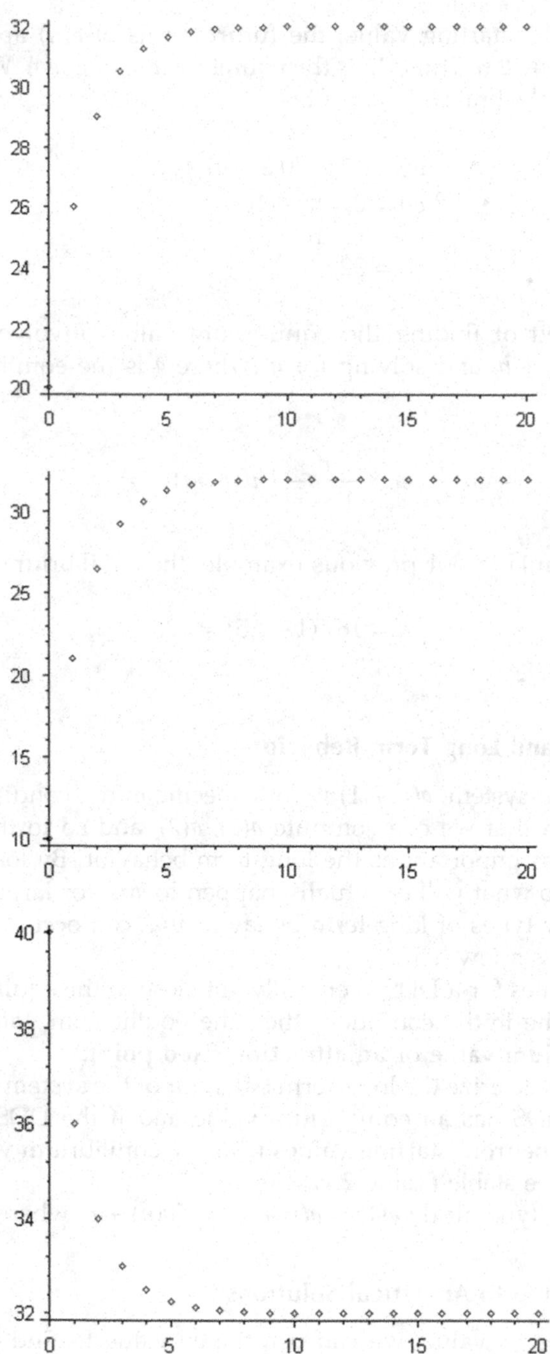


**FIGURE 2.6**
Plot of drugs in our systems with different initial starting conditions.

**TABLE 2.1**

Stability for Linear Functions

| Value of r | DDS Form | Equilibrium | Stability of Solution | Long-Term Behavior |
|---|---|---|---|---|
| $r = 0$ | $a(n + 1) = b$ | $b$ | Stable | Stable equilibrium |
| $r = 1$ | $a(n + 1) = a(n) + b$ | *None* | Unstable | |
| $r < 0$ | $a(n + 1) = r^*a(n) + b$ | $b/(1 - r)$ | Depends on $\|r\|$ | Oscillations |
| $\|r\| < 1$ | $a(n + 1) = r^*a(n) + b$ | $b/(1 - r)$ | Stable | Approaches $b/(1 - r)$ |
| $\|r\| > 1$ | $a(n + 1) = r^*a(n) + b$ | $b/(1 - r)$ | Unstable | Unbounded |

$$B(n + 1) = 1.005\, B(n) - 639.34,\; B(0) = 73,395.$$

The *ev* value is found as 118,031.93.

The analytical solution may be found using the following form:

$$B(k) = (1.005^k)C \; + D \text{ where } D \text{ is the } ev.$$

$$B(k) = (1.005^k)C + 118,031.93,\; B(0) = 73,395.$$

Since $B(0) = 73,395 = 1.005^0\ (C) + 118,031.93.$

$$C = -44,636.93.$$

Thus,

$$B(k) = -\,44,636.93\ (1.005^k) + 118,031.93.$$

Assume we did not know the payment was 639.34 month. We could use the analytical solutions to help find the payment.

$$B(k) = (1.005^k)C + D.$$

We build a system of two equations and two unknowns.

$$B(K) = (1.0057^k)C + D$$

$$B(0) = 73395 = C + D$$

$$B(180) = 0 = 1.00541667^{180}C + D$$

$$C = -44638.70,\; D = 118033.7$$

$$B(K) = -44638.70(1.005)^k + 118033.7.$$

$D$ represents the equilibrium value, and we accepted some round-off error. From our model form:

$$B(n+1) = 1.005\,B(n) - P, \text{ we can find } P.$$

Solving analytically for the equilibrium value,

$$X - 1.005X = -P$$
$$X = P/0.005$$

$$X \text{ is } 118{,}033.70 \text{ so}$$

$$18{,}033.70 = P/0.00541667$$

$$P = 639.34.$$

**Example 2.3:** Growth of a Bacteria Population

We often model population growth by assuming that the change in population is directly proportional to the current size of the given population. This produces a simple, first-order DDS similar to those seen earlier. It might appear reasonable at first examination, but the long-term behavior of growth without bound is disturbing. Why would growth without bound of a yeast culture in a jar (or controlled space) be alarming?

There are certain factors that affect population growth. Things include resources (food, oxygen, space, etc.). These resources can support some maximum population. As this number is approached, the change (or growth rate) should decrease, and the population should never exceed its resource supported amount.

Problem Identification: Predict the growth of yeast in a controlled environment as a function of the resources available and the current population.

Assumptions and Variables: We assume that the population size is best described by the weight of the biomass of the culture. We define *y(n)* as the population size of the yeast culture after period n. There exists a maximum carrying capacity, M, that is sustainable by the resources available. The yeast culture is growing under the conditions established.

Model:

$y(n+1) = y(n) + k\,y(n)\,(M - y(n))$ where
$y(n)$ is the population size after period $n$,
$n$ is the time period measured in hours,
$k$ is the constant of proportionality,
$M$ is the carrying capacity of our system.

**TABLE 2.2**

Bacteria Growth in a Petri Dish

| $n$ | 0 | 1 | 2 | 3 | 4 | 5 | 6 | 7 | 8 | 9 |
|---|---|---|---|---|---|---|---|---|---|---|
| $y$ | 10.3 | 17.2 | 27 | 45.3 | 80.2 | 125.3 | 176.2 | 256.6 | 330.8 | 390.4 |
| $n$ | 10 | 11 | 12 | 13 | 14 | 15 | 16 | 17 | 18 | 19 |
| $y$ | 440 | 520.4 | 560.4 | 600.5 | 610.8 | 614.5 | 618.3 | 619.5 | 620 | 621 |

We have data shown in Table 2.2 for the growth of bacteria in a Petri dish. The variable, $y(n)$, is the number of bacteria at the end of period $n$.

It is often convenient to think about the way the variables change between time periods. We compute $\Delta y(n) = y(n+1) - y(n)$. The values are provided in Table 2.3 and used to find the proportionality constant.

We find the constant slope is 0.0008. We also find the *ev* of 621.

In our experiment, we first plot $y(n)$ versus $n$ and find a stable equilibrium value of approximately 621. Next, we plot $y(n+1) - y(n)$ versus $y(n)$ $(621 - y(n))$ to find the slope, $k$, is approximately 0.0008, with $k = 0.0008$ and the carrying capacity in biomass is 621. This model is

$$y(n+1) = y(n) + .0008\, y(n)\,(621 - y(n)).$$

Again, this is nonlinear because of the $y^2(n)$ term. There is no closed form analytical solution for this equation; however, we may obtain a solution through iteration and graphing the iterated values in Excel.

The model shows stability in that the population (biomass) of the yeast culture approaches 621 as $n$ gets large. Thus, the population is eventually stable at approximately 621 units, as shown in Figure 2.7.

**Example 2.4:** Spread of a Contagious Disease

There are 5,000 students in college dormitories, and some students have been diagnosed with COVID-19, a highly contagious disease. The health center wants to build a model to determine how fast the disease will spread.

**TABLE 2.3**

Bacteria Growth in a Petri Dish Solution Values

| $n$ | 0 | 1 | 2 | 3 | 4 | 5 | 6 | 7 | 8 |
|---|---|---|---|---|---|---|---|---|---|
| $y(n)$ | 10.3 | 17.2 | 27 | 45.3 | 80.2 | 125.3 | 176.2 | 256.6 | 330.8 |
| $\Delta y(n)$ | 6.9 | 9.8 | 18.3 | 34.9 | 45.1 | 50.9 | 79.4 | 74.2 | 59.6 |
| $n$ | 9 | 10 | 11 | 12 | 13 | 14 | 15 | 16 | 17 |
| $y(n)$ | 390.4 | 440 | 520.4 | 560.4 | 600.5 | 614.5 | 618.3 | 619.5 | 620 |
| $\Delta y(n)$ | 46.4 | 79.4 | 40 | 40.1 | 14 | 3.9 | 1.2 | 0.5 | 1 |

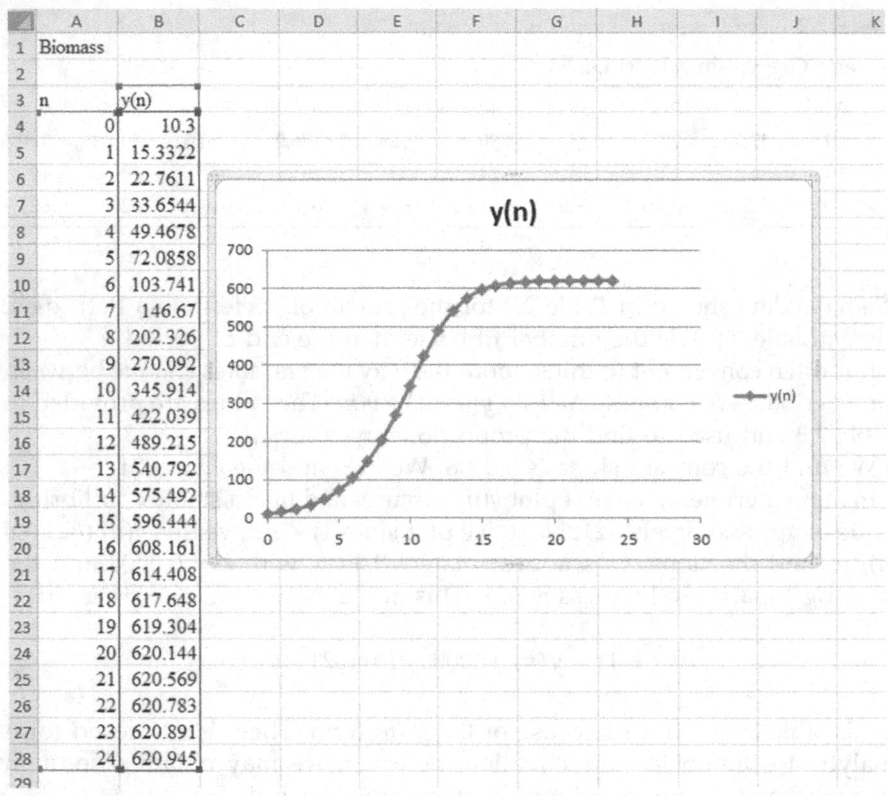

Biomass

| n | y(n) |
|---|---|
| 0 | 10.3 |
| 1 | 15.3322 |
| 2 | 22.7611 |
| 3 | 33.6544 |
| 4 | 49.4678 |
| 5 | 72.0858 |
| 6 | 103.741 |
| 7 | 146.67 |
| 8 | 202.326 |
| 9 | 270.092 |
| 10 | 345.914 |
| 11 | 422.039 |
| 12 | 489.215 |
| 13 | 540.792 |
| 14 | 575.492 |
| 15 | 596.444 |
| 16 | 608.161 |
| 17 | 614.408 |
| 18 | 617.648 |
| 19 | 619.304 |
| 20 | 620.144 |
| 21 | 620.569 |
| 22 | 620.783 |
| 23 | 620.891 |
| 24 | 620.945 |



**FIGURE 2.7**
Screenshot of DDS for bacteria growth in a Petri dish.

**Problem Identification:** Predict the number of students affected with COVID-19 as a function of time.

**Assumptions and Variables:** Let $m(n)$ be the number of students affected with COVID-19 after n days. We assume all students are susceptible to the disease. The possible interactions of infected and susceptible students are proportional to their product (as an interaction term).

The model is,

$$m(n + 1) - m(n) = k\, m(n)\,(5{,}000 - m(n)) \text{ or}$$
$$m(n + 1) = m(n) + k\, m(n)\,(5{,}000 - m(n)).$$

Two students returned from spring break with COVID-19, so $m(0) = 2$. The rate of spreading per day is characterized by $k = 0.00090$. It is assumed that there is no vaccine that can be introduced to slow the spread.

Interpretation: The results show that most students will be affected within 2 weeks. Since only about 10% will be affected within 1 week, every effort

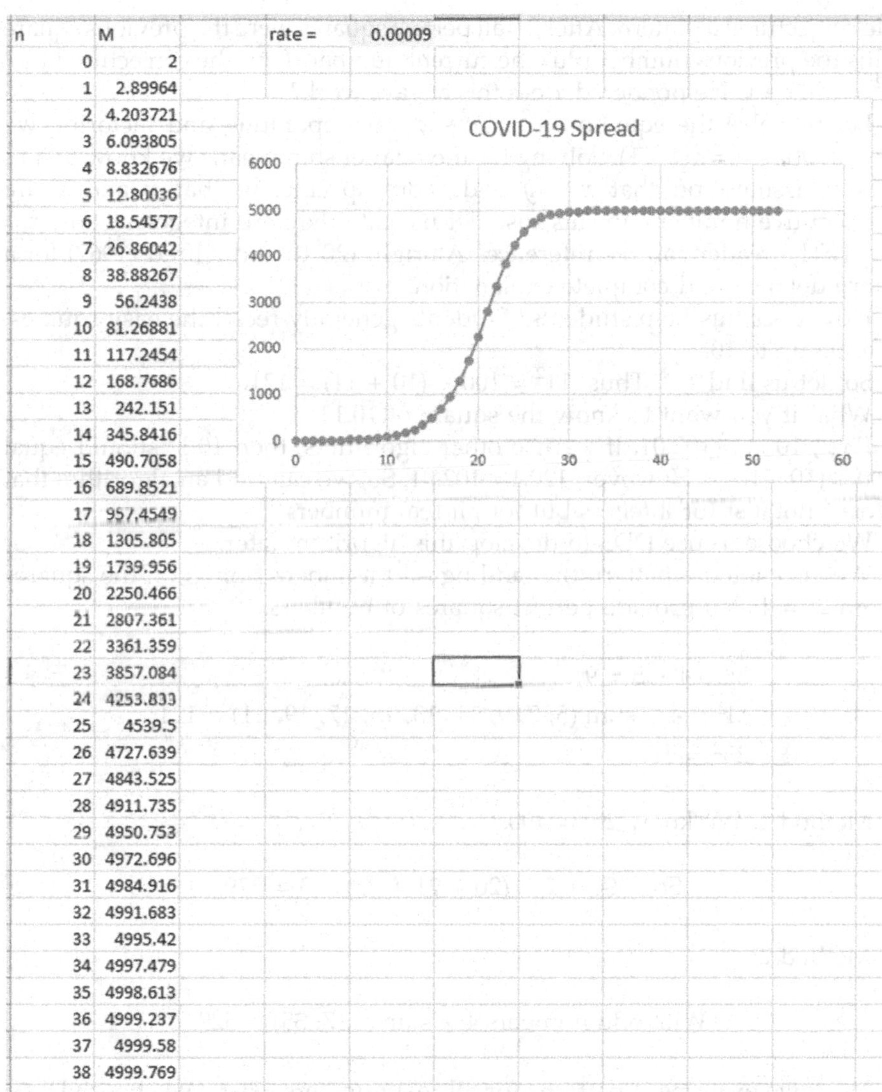

| n | M | rate = | 0.00009 |
|---|---|---|---|
| 0 | 2 | | |
| 1 | 2.89964 | | |
| 2 | 4.203721 | | |
| 3 | 6.093805 | | |
| 4 | 8.832676 | | |
| 5 | 12.80036 | | |
| 6 | 18.54577 | | |
| 7 | 26.86042 | | |
| 8 | 38.88267 | | |
| 9 | 56.2438 | | |
| 10 | 81.26881 | | |
| 11 | 117.2454 | | |
| 12 | 168.7686 | | |
| 13 | 242.151 | | |
| 14 | 345.8416 | | |
| 15 | 490.7058 | | |
| 16 | 689.8521 | | |
| 17 | 957.4549 | | |
| 18 | 1305.805 | | |
| 19 | 1739.956 | | |
| 20 | 2250.466 | | |
| 21 | 2807.361 | | |
| 22 | 3361.359 | | |
| 23 | 3857.084 | | |
| 24 | 4253.833 | | |
| 25 | 4539.5 | | |
| 26 | 4727.639 | | |
| 27 | 4843.525 | | |
| 28 | 4911.735 | | |
| 29 | 4950.753 | | |
| 30 | 4972.696 | | |
| 31 | 4984.916 | | |
| 32 | 4991.683 | | |
| 33 | 4995.42 | | |
| 34 | 4997.479 | | |
| 35 | 4998.613 | | |
| 36 | 4999.237 | | |
| 37 | 4999.58 | | |
| 38 | 4999.769 | | |



**FIGURE 2.8**
Screenshot of plot and iteration for the spread of meningitis.

must be made to get the vaccination at the school and get the students vaccinated within 1 week. This is illustrated graphically in Figure 2.8.

**Example 2.5:** Modeling for Number Theory

Introduction Square Integers

My friend was teaching his young granddaughter about squares of numbers. He recently went blind, so he was using mental mathematics to help.

He conjectured a pattern. After $2^2$, all perfect squares were the previous square plus the previous number plus the current number. Thus, he conjectured that $y^2 = x^2 + x + y$. He pondered, does this always work?

Let us solve the equation, $y^2 = x^2 + x + y$. Separating and factoring we obtain $y(y - 1) = x(x - 1)$. Solving for the relationship means we know to add a key assumption that $x \neq y$ and more specifically that $x$ and $y$ are consecutive numbers. In this case, let us state they are integers. Thus, $x = y - 1$. The reader can see references Albright (2010) and Alfred (1967) for a more detailed and complete explanation.

How does this help students? Students generally recall the easy squares, maybe up to 10.

So, let us find $11^2$. Thus, $11^2 = 100 + (10 + 11) = 121$.

What if you want to know the square of 10.1?

First, $10.1^2 = 102.01$. If we use other algorithms, then $10.1^2$ should equal $100 + (10 + 10.1)$. However, $120.1 \neq 102.01$. So, we can find an algorithm that works not just for integers but for all real numbers.

We choose to use DDS to develop this algorithm later.

We also noted a pattern that adding odd numbers from 5 on the squares from $2^2 = 4$ also provide perfect squares of numbers.

$$3^2 = 4 + 5 = 9,$$
$$11^2 = 4 + \text{sum}\,(5, 7, 9, 11, 13, 15, 17, 19, 21) = 121,$$
$$23^2 = 529.$$

Method 1: We know $20^2 = 400$.

$$\text{So, } 400 + 2 * (20 + 21 + 22) + 3 = 529.$$

Method 2:

$$\text{With odd numbers } 4 + \text{sum}\,(5, 7, 35) = 529.$$

Next, we examine if patterns for all real numbers exist. We choose DDSs to model and look for patterns.

For example, let $A(n)$ = square of $n$. We define our initial condition to be $A(2) = 4$. Now, $A(n+1) = A(n) + \Delta$, where we define $\Delta$ to be the sum of $n - 1 + n$.

We illustrate in Excel to iterate from $n = 2$ to 20 (Table 2.4).

We note that in our modeling change there is a multiplier of 1 for $(n + (n - 1))$ for the change term. This is important as we shall see in the next section.

**TABLE 2.4**
Example Data for Pattern Recognition

| n | A(n) |
|---|---|
| 2 | 4 |
| 3 | 9 |
| 4 | 16 |
| 5 | 25 |
| 6 | 36 |
| 7 | 49 |
| 8 | 64 |
| 9 | 81 |
| 10 | 100 |
| 11 | 121 |
| 12 | 144 |
| 13 | 169 |
| 14 | 196 |
| 15 | 225 |
| 16 | 256 |
| 17 | 289 |
| 18 | 324 |
| 19 | 361 |
| 20 | 400 |

### 2.2.3 Extending to Squares of Real Numbers Other than Integers

Although dealing with integers is fun, we conjectured that there must be something similar for squares of decimals. Next, we built a DDS to apply to decimals. Depending on the number of decimal places, we find a multiplier must be used in the addition of the numbers to get them into the proper decimal equivalents. The multiplier that works is the size of the step used to iterate from $n$ to $n$ + *step size*, where $n$ and the step size are not necessarily an integer.

$$A(n+1) = A(n) + \textit{step size} * (n - 1 + n).$$

We will iterate three examples to illustrate this process.

1. $B\ \ (n+1) = B(n) + 0.01(n + n - 1)$.
   Initial Condition: $B(1) = 1$.
2. $C\ \ (n+1) = C(n) + 0.2(n + n - 1)$.
   Initial Condition: $C(10) = 100$.
3. $D\ \ (n+1) = D(n) + 0.5(n + n - 1)$.
   Initial Condition: $D(10) = 100$.

| n | B(n) | | n | C(n) | | n | D(n) |
|---|---|---|---|---|---|---|---|
| | stepsize | | | stepsize | | | stepsize |
| | 0.01 | | | 0.2 | | | 0.5 |
| n | B(n) | | n | C(n) | | n | D(n) |
| 1 | 1 | | 10 | 100 | | 10 | 100 |
| 1.01 | 1.0201 | | 10.2 | 104.04 | | 10.5 | 110.25 |
| 1.02 | 1.0404 | | 10.4 | 108.16 | | 11 | 121 |
| 1.03 | 1.0609 | | 10.6 | 112.36 | | 11.5 | 132.25 |
| 1.04 | 1.0816 | | 10.8 | 116.64 | | 12 | 144 |
| 1.05 | 1.1025 | | 11 | 121 | | 12.5 | 156.25 |
| 1.06 | 1.1236 | | 11.2 | 125.44 | | 13 | 169 |
| 1.07 | 1.1449 | | 11.4 | 129.96 | | 13.5 | 182.25 |
| 1.08 | 1.1664 | | 11.6 | 134.56 | | 14 | 196 |
| 1.09 | 1.1881 | | 11.8 | 139.24 | | 14.5 | 210.25 |
| 1.1 | 1.21 | | 12 | 144 | | 15 | 225 |
| | | | 12.2 | 148.84 | | 15.5 | 240.25 |
| | | | 12.4 | 153.76 | | 16 | 256 |
| | | | 12.6 | 158.76 | | 16.5 | 272.25 |
| | | | 12.8 | 163.84 | | 17 | 289 |
| | | | 13 | 169 | | 17.5 | 306.25 |
| | | | 13.2 | 174.24 | | 18 | 324 |
| | | | 13.4 | 179.56 | | 18.5 | 342.25 |
| | | | 13.6 | 184.96 | | 19 | 361 |
| | | | | | | 19.5 | 380.25 |
| | | | | | | 20 | 400 |

**FIGURE 2.9**
Screenshot of DDS iterations for our three examples in Excel.

We can iterate these problems in Excel to find the values. These values are displayed in Figure 2.9.

### 2.2.4 Finding Patterns of Cubes with Discrete Dynamical Systems

After experimenting with squares and discovering the use of DDS to model the merging patterns to obtain a formula for all real squares, we can turn toward cubes. In most literature, cubic numbers are defined only by multiplication. The result is using a whole number in a multiplication three times.

For example, 3 × 3 × 3 = 27, so 27 is a cube number.

Here are the first few cube numbers:

| | |
|---|---|
| 1 | (= 1 × 1 × 1) |
| 8 | (= 2 × 2 × 2) |
| 27 | (= 3 × 3 × 3) |
| 64 | (= 4 × 4 × 4) |
| 125 | (= 5 × 5 × 5) |
| … etc. | |

We begin by assuming that there has to be more as we search for a distinguishable pattern.

Since we were able to use DDS to explore squares of numbers, it makes sense to start with DDS to explore cubes.

### 2.2.5 Cubes and DDS with Cubes

A cube, written with exponents as $x^3$, is also known as a product of $x * x * x$. For example, $6^3 = 6 * 6 * 6 = 216$.

So, we created a spreadsheet in Excel of the numbers 1, 2, 3, … n. The next column we cubed those numbers to get, 1, 8, 27, 64, …. We then created columns subtracting the differences in consecutive cubes obtaining, 7, 19, 37, 61, 91, …. T first these look like prime numbers but 91 is not prime but it is the product of 2 primes, 13 and 7. We took the differences of these odd numbers to obtain, 12, 18, 24, 30, 36, … We note a pattern that the next difference is always 6. Thus, we can work backward to obtain a DDS model (Table 2.5).

We define $A(n)$ to be the cube of the $n$th number.

The conjectured pattern appears to be a second order DDS of the form:

$$A(n+2) = (n-1) * 6 + A(n+1) + [A(n+1) - A(n)].$$

This simplifies to $A(n+2) = (n-1) * 6 + 2 * A(n+1) - A(n)$ with initial conditions $A(1) = a_1$ and $A(0) = a_0$.

For example, given $A(1) = 8$, $A(0) = 1$, determine $A(7)$.

We iterate our DDS to obtain, $A(7) = 343 = 7^3$.

| n | A(n) |
|---|---|
| 1 | 1 |
| 2 | 8 |
| 3 | 27 |
| 4 | 64 |
| 5 | 125 |
| 6 | 216 |
| 7 | 343 |

**TABLE 2.5**

Calculating Differences

| N | a | 1st diff | 2nd diff | 3rd diff |
|---|---|---|---|---|
| 1 | 1 | | | |
| 2 | 8 | 7 | | |
| 3 | 27 | 19 | 12 | |
| 4 | 64 | 37 | 18 | 6 |
| 5 | 125 | 61 | 24 | 6 |
| 6 | 216 | 91 | 30 | 6 |

**TABLE 2.6**
Iterated Sequence

| N | A(n) |
|---|---|
| 1.00 | 1.00 |
| 1.10 | 1.331 |
| 1.20 | 1.728 |
| 1.30 | 2.197 |
| 1.40 | 2.744 |
| 1.50 | 3.375 |
| 1.60 | 4.096 |
| 1.70 | 4.913 |
| 1.80 | 5.832 |
| 1.90 | 6.859 |
| 2.00 | 8.000 |
| 2.10 | 9.261 |
| 2.20 | 10.648 |
| 2.30 | 12.167 |

Further, for any real consecutive cube, we have:

$$A(n+2) = 2 * A(n+1) - A(n) + step - size\ (n - \ step - size) * (step - size * 6).$$

For example, consider wanting the cube of 1.2 through 2.3 given $1^3 = 1, 1.1^3 = 1.331$. We use our DDS and iterate to obtain the sequence: (Table 2.6)

For example, let us assume we want to start at 1, and get cubes by a step size of 2. Thus, we want cubes for 1, 3, 5, 7. We need two initial conditions so we know $A(0) = 1$ and $A(1) = 27$. We use our DDS formula,

$$A(n+2) = 2 * A(n+1) - A(n) + step\ size\,(n - step\ size) * (step\ size * 6),$$

and iterate to obtain the cubes for 5, 7, 9, etc.

| n | Cubes |
|---|---|
| 5 | 125 |
| 7 | 343 |
| 9 | 729 |

## 2.3 Introduction to Systems of Discrete Dynamical Systems

In the previous sections, we reviewed linear and nonlinear DDS models. Now, we extend the discussion to systems of systems, but we still use our paradigm:

$$\textit{future} = \textit{present} + \textit{change}.$$

Consider wanting to retire on a lake that you stock with bass and trout for endless fishing. Will it be endless? Can the species co-exist in your lake? How often do you need to restock the lake? This is an example of a competitive hunter model where both species compete for the same resources.

First, we define a system of DDS.

$$A(n) = f(A(n),\ B(n)),$$
$$B(n) = g(A(n),\ B(n)).$$

As before, simple linear systems of DDS have a closed-form analytical solution; however, most systems do not. We will analyze all those DDS through iteration and graphs.

For a selected set of initial conditions, we build numerical solutions to get a sense of long-term behavior for the system. For the systems that we will study, we will find their equilibrium values.

We then explore starting values near the equilibrium values to see if by starting close to an equilibrium value, the system:

a. will remain close;
b. approaches the equilibrium value;
c. does not remain close.

What happens near these values gives great insight concerning the long-term behavior of the system. We can study the resulting pattern of the numerical solutions and the resulting plots.

### 2.3.1 Simple Linear Systems and Analytical Solutions

As an example, consider that there are students in a public school (PS) and a private magnet (PM) school that now have access to school vouchers. Assume that a pre-survey of families used as historical records determined that 75% of the magnet school remain while 25% preferred to transfer. We found the 65% of the public school want to remain but 35% preferred to transfer. We can now build a model to determine the long-term behavior of

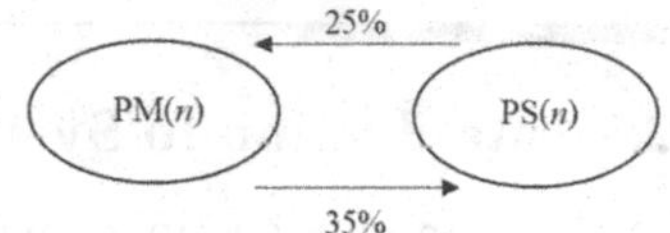


**FIGURE 2.10**
Change diagram for the school vouchers.

these students based upon this historical data. The change diagram is shown in Figure 2.10.

**Problem Identification:** Determine the number of students who will select the public or private school.

**Assumptions and Variables:** Let $n$ represent the number of student months. We define

$PS(n)$ = the number of students in public school at the end of $n$ months;

$PM(n)$ = the number of students in the magnet school at the end of $n$ months.

We assume that no other incentives are given to the students for either staying or moving.

The Model:

Mathematically, the system model is written as:

$$PM(n+1) = 0.75\,PM(n) + 0.35\,PS(n),$$
$$PS(n+1) = 0.25\,PM(n) + 0.65\,PS(n).$$

There are initially 1,500 students in the magnet school and 2,000 students in the public school. We seek to find the long-term behavior of this system.

We rewrite the model as a system of DDS:

$$PM(n+1) = 0.75\,PM(n) + 0.35\,PS(n),$$
$$PS(n+1) = 0.25\,PM(n) + 0.65\,PS(n).$$
$$PM(0) = 1{,}500 \text{ and } PS(0) = 1{,}000, \text{ respectively.}$$

This is a simple linear system. We may use the initial conditions to iterate this system DDS, as shown in Figure 2.11. We see from the solution plot that there are stable equilibria at about 2,042 and 1,458 in the magnet and public schools, respectively.

### 2.3.2 Analytical Solutions

Analytical solutions assume knowledge of linear algebras through eigenvalues and eigenvectors.

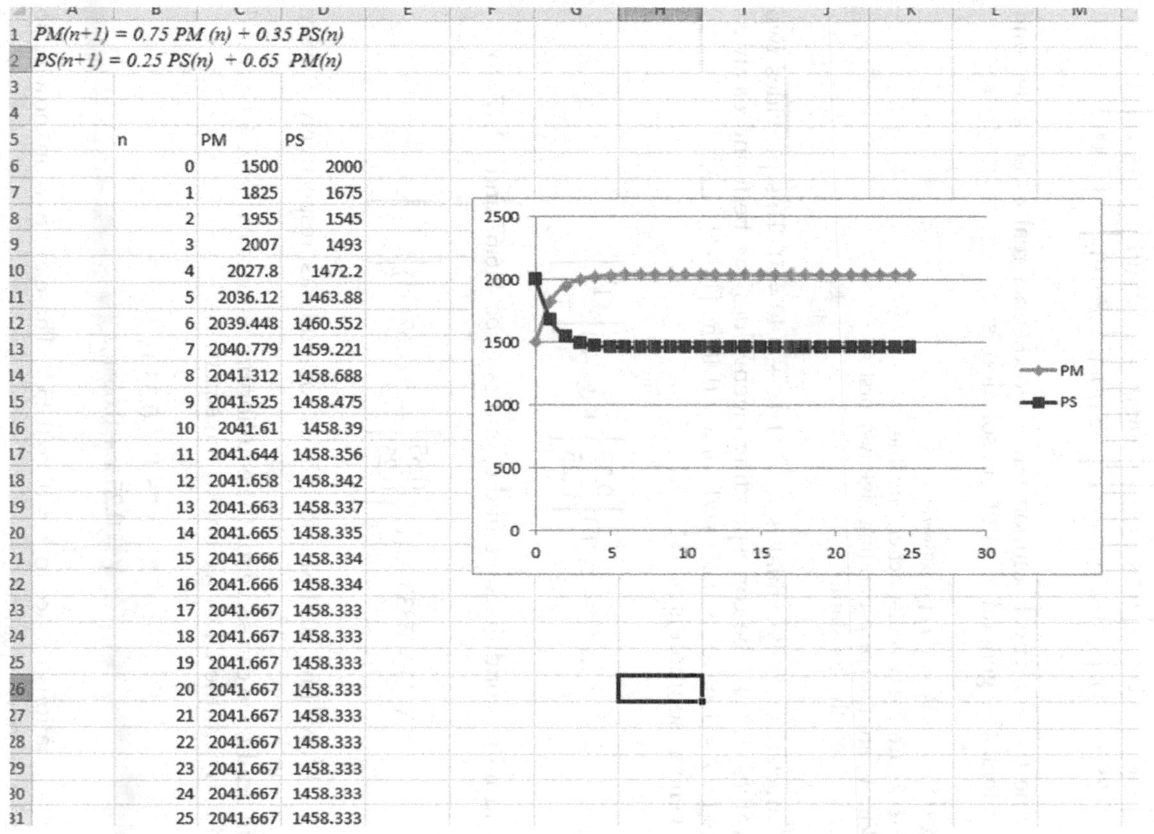

PM(n+1) = 0.75 PM (n) + 0.35 PS(n)
PS(n+1) = 0.25 PS(n) + 0.65 PM(n)

| n | PM | PS |
|---|---|---|
| 0 | 1500 | 2000 |
| 1 | 1825 | 1675 |
| 2 | 1955 | 1545 |
| 3 | 2007 | 1493 |
| 4 | 2027.8 | 1472.2 |
| 5 | 2036.12 | 1463.88 |
| 6 | 2039.448 | 1460.552 |
| 7 | 2040.779 | 1459.221 |
| 8 | 2041.312 | 1458.688 |
| 9 | 2041.525 | 1458.475 |
| 10 | 2041.61 | 1458.39 |
| 11 | 2041.644 | 1458.356 |
| 12 | 2041.658 | 1458.342 |
| 13 | 2041.663 | 1458.337 |
| 14 | 2041.665 | 1458.335 |
| 15 | 2041.666 | 1458.334 |
| 16 | 2041.666 | 1458.334 |
| 17 | 2041.667 | 1458.333 |
| 18 | 2041.667 | 1458.333 |
| 19 | 2041.667 | 1458.333 |
| 20 | 2041.667 | 1458.333 |
| 21 | 2041.667 | 1458.333 |
| 22 | 2041.667 | 1458.333 |
| 23 | 2041.667 | 1458.333 |
| 24 | 2041.667 | 1458.333 |
| 25 | 2041.667 | 1458.333 |



**FIGURE 2.11**
Screenshot Iterative solution and plot for voucher students.

We rewrite the DDS in matrix form:

$$X(n+1) = MX(n),\ X(0) = B.$$

$$\begin{bmatrix} PM(n+1) \\ PS(n+1) \end{bmatrix} = \begin{bmatrix} 0.75 & 0.35 \\ 0.25 & 0.65 \end{bmatrix} \begin{bmatrix} PM(n) \\ PS(n) \end{bmatrix}, \begin{bmatrix} PM(0) \\ PS(0) \end{bmatrix} = \begin{bmatrix} 1500 \\ 2000 \end{bmatrix}.$$

We define the analytical solution with two distinct real eigenvalues and eigenvectors. The general form of the solution is

$\mathbf{X}(k) = \lambda_1^k c_1 V_1 + \lambda_2^k c_2 V_2$, where
$\lambda_1$& $\lambda_2$ are the two distinct eigenvalues,
$\mathbf{V}_1$& $\mathbf{V}_2$ are the corresponding eigenvectors.
$c_1$ & $c_2$ are the constant.

The characteristic polynomial is $\lambda^2 - 1.4\lambda + 0.40 = 0$. This provides two eigenvalues: 1, 0.4. The corresponding eigenvalues can be found easily as vectors for $\lambda = 1$ of [0.35, 0.25] and for $\lambda = 0.40$ as [1, −1].

The general solution is

$$X(k) = c_1(1^k)\begin{bmatrix} 0.35 \\ 0.25 \end{bmatrix} + c_2(.4^k)\begin{bmatrix} 1 \\ -1 \end{bmatrix}.$$

With our initial conditions we find $c_1$ and $c_2$ to be 416.67 and −4.17, so

$$X(k) = 5833.33(1^k)\begin{bmatrix} 0.35 \\ 0.25 \end{bmatrix} - 541.67(.4^k)\begin{bmatrix} 1 \\ -1 \end{bmatrix}.$$

When $k = 10$, we find 2041.61 and 1458.31 for students, respectively. We see this again in Figure 2.12.

Analytically, we can solve for the equilibrium values. We let $X = D(n)$ and $Y = M(n)$. From the DDS, we obtain the equations:

$$X = 0.75\,X + 0.35\,Y,$$
$$Y = 0.25\,x + 0.65\,Y,$$

and both equations reduce to $X = 0.35/0.25\,Y$. There are two unknowns, so we need a second equation.

From the initial conditions, we know that $X + Y = 3500$. We can use the equations

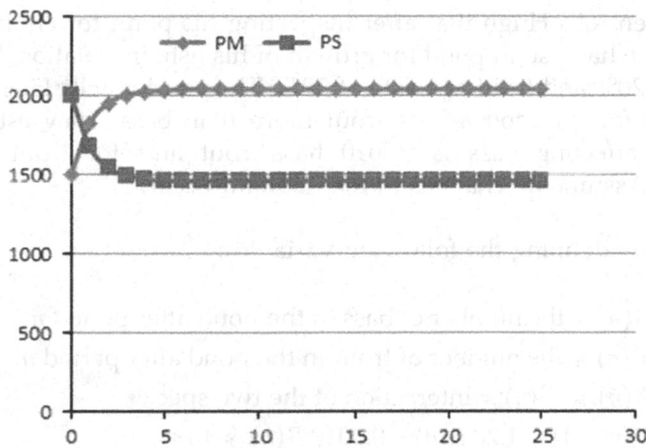


**FIGURE 2.12**
Plot of DDS for student's relocation example.

$X + Y = 3,500$ and $X = 0.35/0.25\ Y$ to find the equilibrium values:
$X = 2041.67$ and $Y = 1458.33$.

We previously iterated the solution, and now we start with initial conditions near to those equilibrium values and we find the sequences tend toward those values. We conclude the system has *stable* equilibrium values.

You should go back and change the initial conditions and see what behavior follows.

**Interpretation:** The long-term behavior shows that eventually (without other influences) of the 3,500 students about 1,458 remain in public school and 2,042 go to the magnet school. We might want to try to attract students with advertising and perhaps add incentives.

## 2.4 Iteration and Graphical Solution

**Example 2.6:** Competitive Hunter Models

Competitive hunter models involve species vying for the same resources (e.g. food or living space) in the habitat. The effect of the presence of a second species diminishes the growth rate of the first species. We now consider a specific example concerning trout and bass in a small pond. Hugh Ketum owns a small pond that he uses to stock fish and eventually allows fishing. He has decided to stock both bass and trout. The fish and game warden tells Hugh that after inspecting his pond for environmental conditions he has a solid pond for growth of his fish. In isolation, bass grow at a rate of 20% and trout at a rate of 30%. The warden tells Hugh that the interactions for the food affects trout more than bass. They estimate the interaction affecting bass is 0.0010 bass*trout and for trout is 0.0020 bass*trout. Assume no changes in the habitant occur.

Model:

We start by defining the following variables:

$$B(n) = \text{the number of bass in the pond after period } n.$$
$$T(n) = \text{the number of trout in the pond after period } n.$$
$$B(n) * T(n) = \text{interaction of the two species.}$$
$$B(n+1) = 1.20\,B(n) - 0.0010\,B(n) * T(n).$$
$$T(n+1) = 1.30\,T(n) - 0.0020\,B(n) * T(n).$$

The equilibrium values can be found by allowing $X = B(n)$ and $Y = T(n)$ and solving for $X$ and $Y$.

$$X = 1.2X - 0.001\,X * Y,$$
$$Y = 1.3\,Y - 0.0020\,X * Y.$$

We rewrite these equations as

$$0.2\,X - 0.001\,X * Y = 0,$$
$$0.3\,Y - 0.002\,X * Y = 0.$$

We can rewrite the two equations to obtain:

$$X(0.2 - 0.001\,Y) = 0,$$
$$Y(0.3 - 0.002\,X) = 0.$$

Solving we find $X = 0$ or $Y = 2{,}000$ and $Y = 0$ or $X = 1{,}500$.

We want to know the long-term behavior of the system and the stability of the equilibrium points.

Hugh initially considers 151 bass and 199 trout for his pond. The solution is left to the student as an exercise. From Hugh's initial conditions, bass will grow without bound and trout will eventually die out.

We iterated the system and obtained the plot, Figure 2.13, of bass and trout over time. Trout die out at about period 29.

This is certainly not what Hugh had in mind so he must find ways to improve the environment for the fish that alter the parameters from the model.

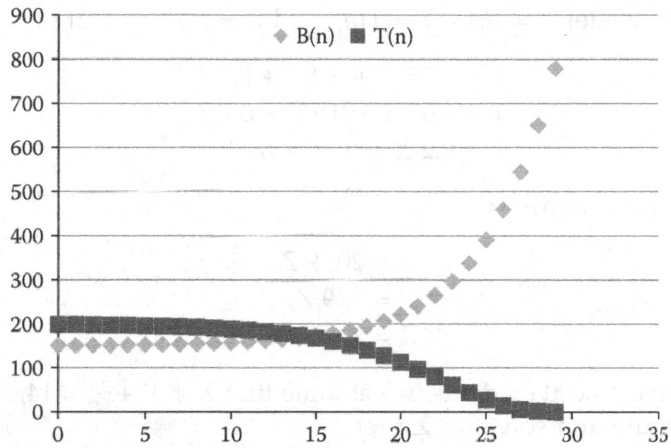


**FIGURE 2.13**
Bass and trout over time.

**Example 2.7:** Fast Food Tendencies

Consider that your student union center desires to have three fast food chains available to students serving: burgers, tacos, and pizza. These chains run a survey of students and find the following information concerning lunch: 75% that ate burgers will eat burgers again at the next lunch, 5% will eat tacos next, and 20% will eat pizza next. Of those who ate tacos last, 20% will eat burgers next, 60% will stay will tacos, and 35% will eat pizza next. Of those who ate pizza, 40% will eat burgers next, 20% tacos, and 40% pizza again.

We formulate the problem as follows:

Let $n$ represent the $n$th day's lunch, so we define the following variables:

$$B(n) = \text{the number of burger eaters in the } n^{\text{th}} \text{ lunch.}$$
$$T(n) = \text{the number of taco eaters in the } n^{\text{th}} \text{ lunch.}$$
$$P(n) = \text{the number of pizza eaters in the } n^{\text{th}} \text{ lunch.}$$

Formulating the system, we have the following dynamical system:

$$B(n+1) = 0.75\,B(n) + 0.20\,T(n) + 0.40\,P(n),$$
$$T(n+1) = 0.05\,B(n) + 0.60\,T(n) + 0.20P(n),$$
$$P(n+1) = 0.20B(n) + 0.20\,T(n) + 0.40\,P(n).$$

Analytically, we let $X = B(n)$, $Y = T(n)$, and $Z = P(n)$ so that

$$X = 0.75\,X + 0.2\,Y + 0.4\,Z,$$
$$Y = 0.05\,X + 0.6\,Y + 0.2\,Z,$$
$$Z = 0.2\,X + 0.2Y + 0.4\,Z.$$

These equations reduce to

$$X = 20/9\,Z,$$
$$Y = 7/9\,Z,$$
$$Z = Z.$$

Since we have 14,000 students, we assume that $X + Y + Z = 14{,}000$

We substitute and solve for $Z$ first.

$$4\,Z = 14{,}000,$$
$$Z = 3{,}500,$$
$$X = 20/9\,Z = 20/9(3{,}500) = 7{,}777.77,$$
$$Y = 7/9\,Z = 7/9(3{,}500) = 2{,}722.22.$$

Suppose the campus has 14,000 students that eat lunch. The graphical results also show that an equilibrium value is reached at a value of about 7,778 burger eaters, 2,722 taco eaters, and 3,500 pizza eaters. This allows the fast food establishments to plan for a projected future. We see this in the iterated table and Figure 2.14. By varying the initial conditions for 14,000 students we find that these are stable equilibrium values.

## 2.5 Modeling of Predator–Prey Model, SIR Model, and Military Models

**Example 2.8:** A Predator–Prey Model: Foxes and Rabbits

In the study of the dynamics of a single population, we typically take into consideration such factors as the "natural" growth rate and the "carrying capacity" of the environment. Mathematical ecology requires the study of populations that interact, thereby affecting each other's growth rates. In this module, we study a very special case of such an interaction, in which there are exactly two species, one of which the predators eat the prey. Such pairs exist throughout nature, such as lions and gazelles, birds and insects, pandas and eucalyptus trees, and Venus fly traps and flies.

To keep our model simple, we will make some assumptions that would be unrealistic in most of these predator–prey situations. Specifically, we will assume that

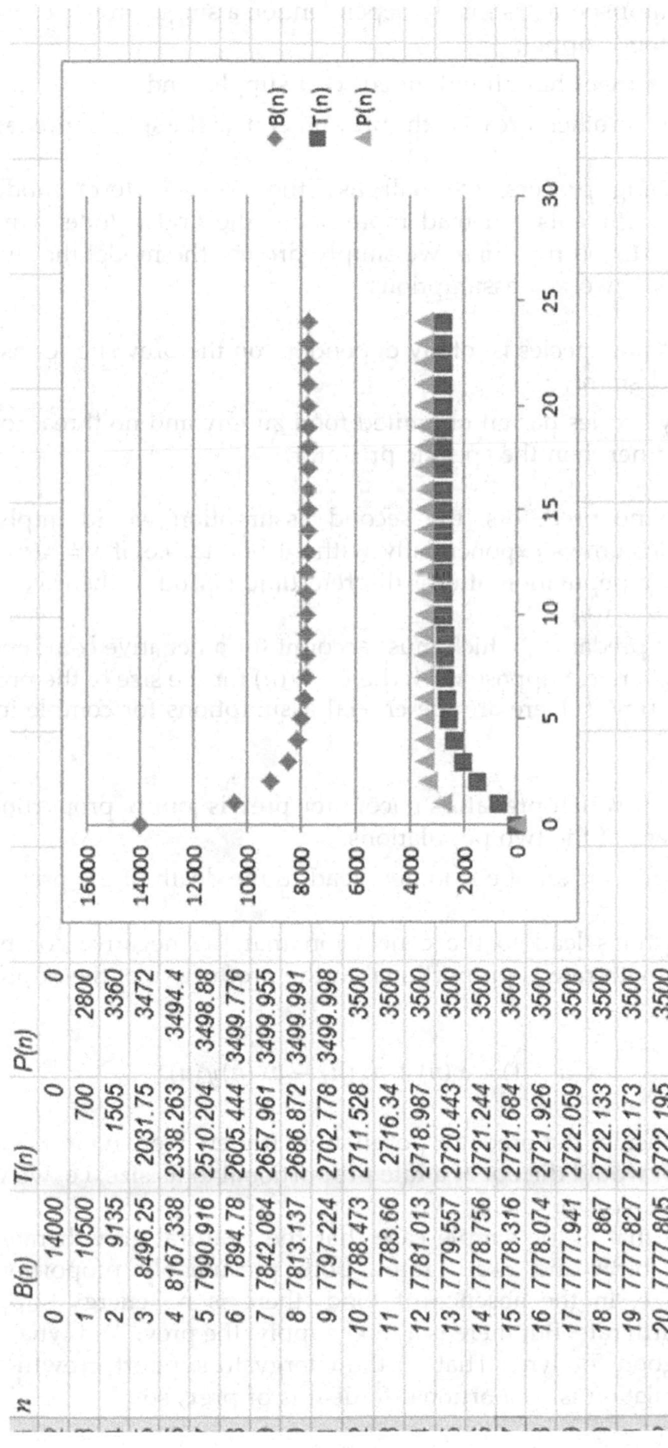

| n | B(n) | T(n) | P(n) |
|---|---|---|---|
| 0 | 14000 | 0 | 0 |
| 1 | 10500 | 700 | 2800 |
| 2 | 9135 | 1505 | 3360 |
| 3 | 8496.25 | 2031.75 | 3472 |
| 4 | 8167.338 | 2338.263 | 3494.4 |
| 5 | 7990.916 | 2510.204 | 3498.88 |
| 6 | 7894.78 | 2605.444 | 3499.776 |
| 7 | 7842.084 | 2657.961 | 3499.955 |
| 8 | 7813.137 | 2686.872 | 3499.991 |
| 9 | 7797.224 | 2702.778 | 3499.998 |
| 10 | 7788.473 | 2711.528 | 3500 |
| 11 | 7783.66 | 2716.34 | 3500 |
| 12 | 7781.013 | 2718.987 | 3500 |
| 13 | 7779.557 | 2720.443 | 3500 |
| 14 | 7778.756 | 2721.244 | 3500 |
| 15 | 7778.316 | 2721.684 | 3500 |
| 16 | 7778.074 | 2721.926 | 3500 |
| 17 | 7777.941 | 2722.059 | 3500 |
| 18 | 7777.867 | 2722.133 | 3500 |
| 19 | 7777.827 | 2722.173 | 3500 |
| 20 | 7777.805 | 2722.195 | 3500 |



**FIGURE 2.14**
Screenshot Excel iterated solution and plot for fast food on campus.

- the predator species is totally dependent on a single prey species as its only food supply;
- the prey species has an unlimited food supply; and
- there exist no other threats to the prey other than the specific predator.

In this modeling process, we will use the Lotka-Volterra model for predator–prey. Students can read more about the Lotka-Volterra models in the suggested readings. Here we simply present the model that we use.

We repeat our two key assumptions:

- The predator species is totally dependent on the prey species as its only food supply.
- The prey species has an unlimited food supply and no threat to its growth other than the specific predator.

If there were no predators, the second assumption would imply that the prey species grows exponentially without bound, i.e. if $x = x(n)$ is the size of the prey population after a discrete time period $n$, then we would have $x(n + 1) = a\ x(n)$.

But there *are* predators, which must account for a negative component in the prey growth rate. Suppose we write $y = y(n)$ for the size of the predator population at time $t$. Here are the crucial assumptions for completing the model:

- The rate at which predators encounter prey is jointly proportional to the sizes of the two populations.
- A fixed proportion of encounters lead to the death of the prey.

These assumptions lead to the conclusion that the negative component of the prey growth rate is proportional to the product $xy$ of the population sizes, i.e.

$$x(n + 1) = x(n) + ax(n) - bx(n)y(n).$$

Now, we consider the predator population. If there were no food supply, the population would die out at a rate proportional to its size, i.e. we would find $y(n + 1) = -cy(n)$.

We assume that is the simple case that the "natural growth rate" is a composite of birth and death rates, both presumably proportional to population size. In the absence of food, there is no energy supply to support the birth rate. But there is a food supply: the prey. And what is bad for hares is good for lynx. That is, the energy to support growth of the predator population is proportional to deaths of prey, so

$$y(n+1) = y(n) - cy(n) + px(n)y(n).$$

This discussion leads to the discrete version of the Lotka–Volterra Predator–Prey Model:

$$\begin{aligned} x(n+1) &= (1+a)x(n) - bx(n)y(n) \\ y(n+1) &= (-c)y(n) + px(n)y(n) \\ n &= 0, 1, 2, \ldots \end{aligned}$$

where $a$, $b$, $c$, and $p$ are positive constants.

The Lotka–Volterra model consists of a system of linked dynamical system equations that cannot be separated from each other and that cannot be solved in closed form. Nevertheless, they can be solved numerically and graphed in order to obtain insights about the scenario being studied.

Let us return to our foxes and hares scenario. We can assume this discrete model is as explained above. Further, data investigation yields the following estimates for the parameters that we require: {a, b, c, p} = {0.04, 0.0004, 0.09, 0.001} We will further assume that initially there are 600 rabbits and 125 foxes.

We iterate and plot the results for rabbits and foxes versus time and then plot rabbits versus foxes, shown in Figures 2.15 and 2.16.

If we ran this model for many more iterations, we would find the plot of foxes versus rabbits spiral in a similar fashion as above. We conclude that the model appears reasonable. We could find the equilibrium values for the system. There is set of feasible equilibrium points for rabbits and foxes at

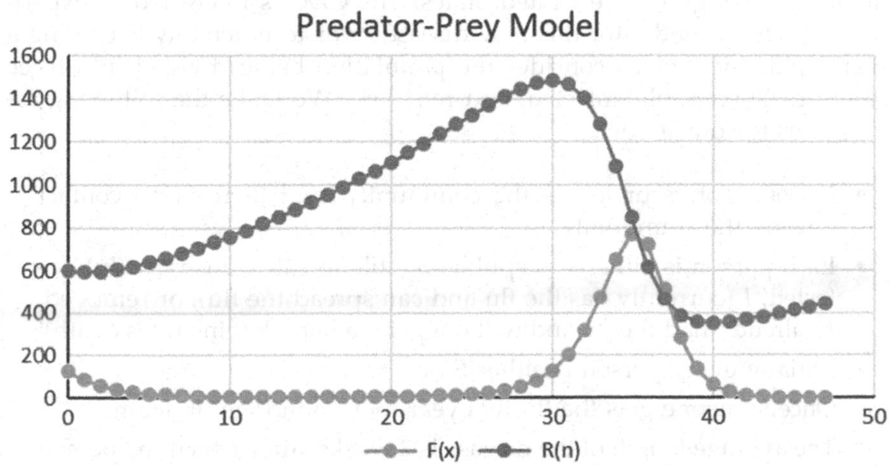


**FIGURE 2.15**
Foxes and rabbits over time.

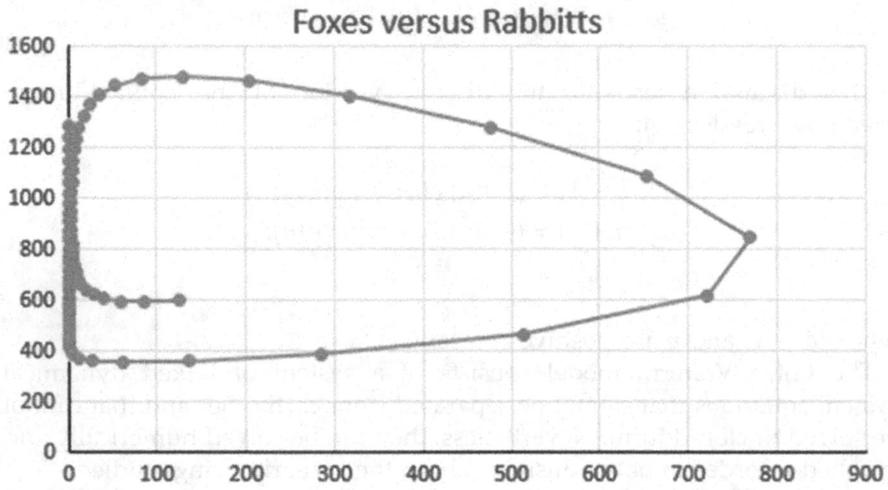


**FIGURE 2.16**
Foxes versus rabbits in a spiral motion.

(0,0) and at (2725,960). The orbits of the spiral indicate that the system is moving away from both (0,0) and (2725,960), so we conclude the system is not stable.

In models of predator–preys, it is often that "managers" of the ecological system must intervene in some way to keep both species flourishing.

**Example 2.9:** Discrete SIR Model of Epidemics

Consider a generic contagious disease, such as a new flu strand, that is spreading throughout the United States. The CDC is interested in experimenting with a model for this new disease prior to it actually becoming a "real" epidemic. Let us consider the population being divided into three categories: susceptible, infected, and removed. We make the following assumptions for our model:

- No one enters or leaves the community, and there is no contact outside the community.
- Each person is either susceptible, $S$ (able to catch this new flu); infected, $I$ (currently has the flu and can spread the flu); or removed, $R$ (already had the flu and will not get it again; that includes death).
- Initially every person is either $S$ or $I$.
- Once someone gets the flu this year they cannot get it again.
- The average length of the disease is 2 weeks, after which the person is deemed infected and can spread the disease.
- Our time period for the model will be per week.

The model we will consider is the SIR model (Allman & Rhodes, 2004).

We will start by defining our variables.

$$S(n) = \text{number in the population susceptible after period } n.$$
$$I(n) = \text{number infected after period } n.$$
$$R(n) = \text{number removed after period } n.$$

We will start our modeling process with $R(n)$. Our assumption for the length of time someone has the flu is 2 weeks. Thus, half the infected people will be removed each week:

$$R(n+1) = R(n) + 0.5\,I(n).$$

The value, 0.5, is called the removal rate per week. It represents the proportion of the infected persons who are removed from infection each week. If real data is available, then we could do "data analysis" to obtain the removal rate.

$I(n)$ will have terms that both increase and decrease its amount over time. It is decreased by the number that are removed each week, 0.5 * $I(n)$. It is increased by the numbers of susceptible that come in contact with an infected person and catch the disease, $aS(n)I(n)$. We define the rate, $a$, as the rate in which the disease is spread or the transmission coefficient. We realize this is a probabilistic coefficient. We will assume, initially, that this rate is a constant value that can be found from initial conditions.

Assume we have a population of 1,000 students in the dorms. Our nurse found 3 students reporting to the infirmary initially. The next week, 5 students came into the infirmary with flu-like symptoms. $I(0) = 3$, $S(0) = 997$. In week 1, the number of newly infected is 30.

$$5 = a\,I(n)S(n) = a(3) * (995),$$
$$a = 0.00167.$$

We now consider $S(n)$. This number is decreased only by the number that becomes infected. We may use the same rate, $a$, as before to obtain the model:

$$S(n+1) = S(n) - aS(n)I(n).$$

Our coupled SIR model is:

$$R(n+1) = R(n) + 0.5I(n)$$
$$I(n+1) = I(n) - 0.5I(n) + 0.00167S(n)I(n)$$
$$S(n+1) = S(n) - 0.00167S(n)I(n)$$
$$I(0) = 3,\ S(0) = 997,\ R(0) = 0.$$

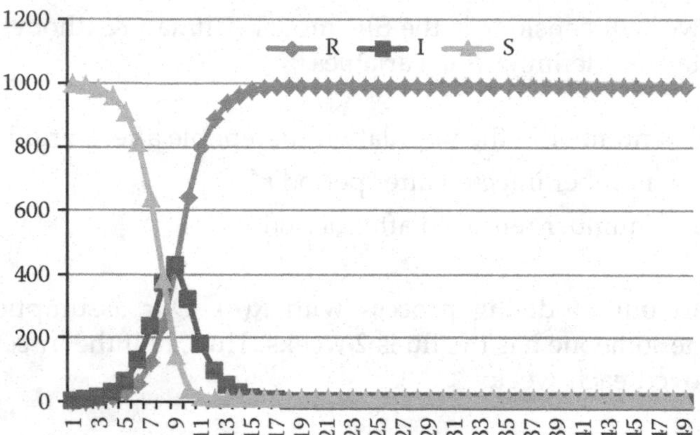


**FIGURE 2.17**
Plot of SIR model over time.

The SIR model can be solved iteratively and viewed graphically. We can iterate the solution and obtain the graph, Figure 2.17, to observe the behavior to obtain some insights.

The worse of the flu epidemic occurs around week 8, at the maximum of the infected graph. The maximum number is slightly larger than 400, from the table it is 427. After 25 weeks, slightly more than nine persons never get the flu. You will be asked to check for sensitivity to the coefficient in the exercise set.

## 2.6 Technology Examples for Discrete Dynamical Systems

Using DDS is an interesting and productive approach. The use of computer technology is essential to the methods described in this chapter. It allows for interactions between instructors and students. It provides a means to use technology in a nonstandard way. It provides another way to educated students concerning squares and cubes using discrete mathematics. In this section, we present technology and examples of solving linear and nonlinear discrete dynamical systems.

### 2.6.1 Excel for Linear and Nonlinear DDS

We will start with a discrete dynamical system such as $a(n + 1) = 0.5a(n)$, with $a(0) = 100$.

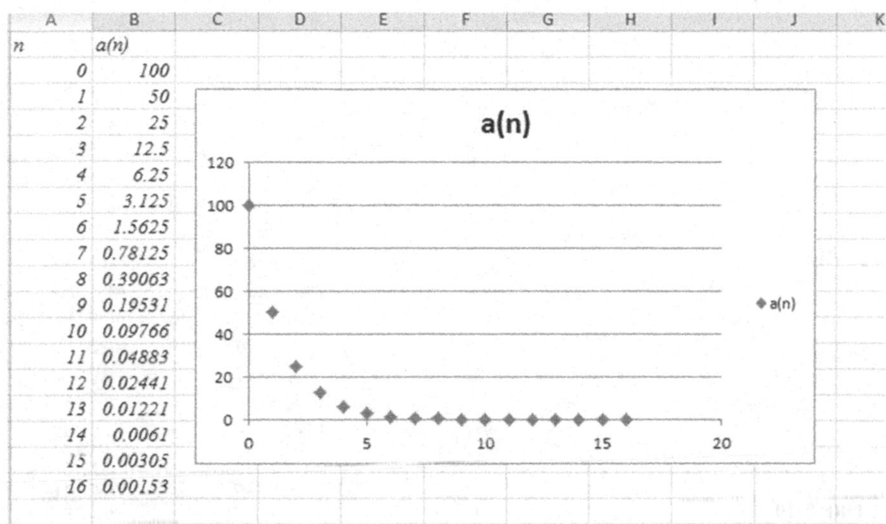

| *n* | *a(n)* |
|---|---|
| 0 | 100 |
| 1 | 50 |
| 2 | 25 |
| 3 | 12.5 |
| 4 | 6.25 |
| 5 | 3.125 |
| 6 | 1.5625 |
| 7 | 0.78125 |
| 8 | 0.39063 |
| 9 | 0.19531 |
| 10 | 0.09766 |
| 11 | 0.04883 |
| 12 | 0.02441 |
| 13 | 0.01221 |
| 14 | 0.0061 |
| 15 | 0.00305 |
| 16 | 0.00153 |



**FIGURE 2.18**
Screenshot of Excel's solution.

Steps to iterate and graph in Excel are as follows:

Step 1. Open a new worksheet and name it DDS or some appropriate name.

Step 2. Label the following columns as $n$ and $a(n)$ in cell *a1* and *b1*.

Step 3. In cells *a2* and *b2* input the initial condition by putting in 0 in cell *a2* and 100 in cell *b2*.

Step 4. In cell *a3* type = *1 + cell a2*.

Step 5. In cell *b3* type = *0.5 * cell b2*.

Step 6. Highlight cells *a3* and *b3* and drag the curser down to fill in cells as far as desired or needed, in this case to about a4:b16.

Step 7. Highlight cells a1:b16, INSERT scatterplot to obtain the graph.

Step 8. Interpret the results.

A screenshot of the model and results is provided in Figure 2.18, where we see that our DDS tends to zero over time.

In Figure 2.19, we show the appropriate formulas used.

### 2.6.2 Maple for Linear and Nonlinear DDS

In Maple, DDS are referred to as recursion equations. One might obtain closed-form solutions, if they exist. One might also iterate and graph the

| n | a(n) |
|---|---|
| 0 | 100 |
| =1+A2 | =0.5*B2 |
| =1+A3 | =0.5*B3 |
| =1+A4 | =0.5*B4 |
| =1+A5 | =0.5*B5 |
| =1+A6 | =0.5*B6 |
| =1+A7 | =0.5*B7 |
| =1+A8 | =0.5*B8 |
| =1+A9 | =0.5*B9 |
| =1+A10 | =0.5*B10 |
| =1+A11 | =0.5*B11 |
| =1+A12 | =0.5*B12 |
| =1+A13 | =0.5*B13 |
| =1+A14 | =0.5*B14 |
| =1+A15 | =0.5*B15 |
| =1+A16 | =0.5*B16 |
| =1+A17 | =0.5*B17 |

**FIGURE 2.19**
Screenshot of Excel formulas used in our example.

behavior of the recursion equation. We will use both commands and libraries from Maple such as *with(plots)* and we will add some commands to our Maple toolbox, **rsolve** and **seq**.

rsolve – **recurrence equation solver**

Calling Sequence
rsolve(**eqns**, **fcns**)
rsolve(**eqns**, **fcns**, 'genfunc'(**z**))
rsolve(**eqns**, **fcns**, 'makeproc')

Parameters
eqns – single equation or a set of equations,
fcns – function name or set of function names,
z – name, the generating function variable.

seq – **create a sequence**

Calling Sequence
seq(**f**, **i** = **m**..**n**)
seq(**f**, **i** = x)

Parameters
f – any expression,
I – name,
m, n – numerical values,
x – expression.

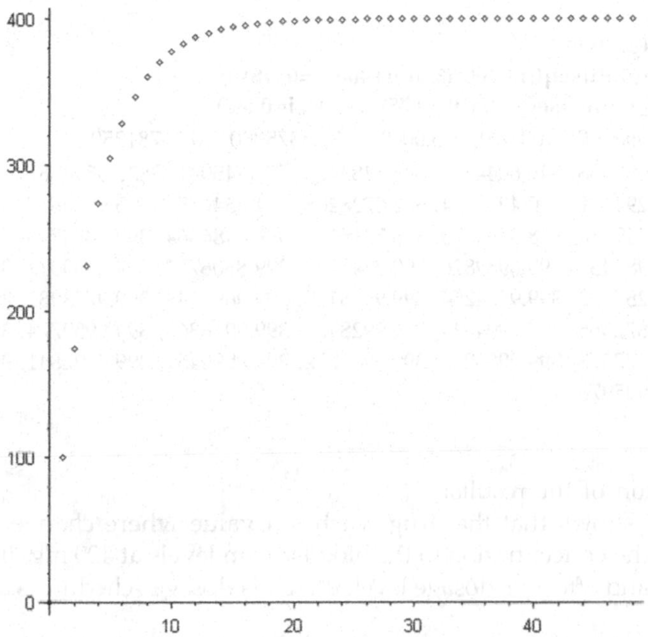


**FIGURE 2.20**
Behavior of drugs in our systems.

Many of the models that we will solve have closed-form solutions so that we can use the command **rsolve** to obtain the closed solution, and then we can use the sequence command (**seq**) to obtain the numerical values in the solution. Many dynamical systems do not have closed-form analytical solutions so we cannot use the rsolve and seq commands to obtain solutions. When this occurs, we will write a small program using PROC to obtain the numerical solutions. To plot the solution to the dynamical systems, we will use plot commands to plot the sequential data pairs. We will illustrate all these commands in our example.

**Example 2.10:** Solve the DDS $a(n+1) = 0.75\, a(n) + 100$, $a(0) = 0$. Determine the value of $a(72)$.

We can illustrate the iterative technique for analyzing a DDS in Maple. Figure 2.20 shows the graphical representation of the solution.

We type the following commands:

**>restart;**
**>drug:=rsolve({a(n+1)=.75*a(n)+100,a(0)=0},a(n));**
$drug{:=}400 - 400\left(\frac{3}{4}\right)^n$
**>L:=limit(drug,n=infinity);**

*L*:=400.
**>with(plots):**
**>pointplot({seq([i,–400*(3/4)^i+400],i=0..48)});**
**>drug_table:=seq(–400.0*(0.75)^i+400.,i=0..48);**
*drug_table*:=0. ,100.000, 175.00000, 231.4375000, 305.0781250,
328.8085938, 346.604453, 359.9548340, 377.4745941, 383.1059456,
387.3294592, 390.4970944, 392.8728208, 394.6546156, 395.9909617, 396.9932213
397.7449160, 398.3086870, 398.7315152, 399.0486364, 399.2864773, 399.4648580,
399.5986435, 399.6989826, 399.7742370, 399.8306777, 399.8730083, 399.9047562,
399.9285672, 399.9464254, 399.9598190, 399.9698643, 399.9773982, 399.9830487,
399.9872865, 399.9904649, 399.9928487, 399.9946365, 399.9959774, 399.9969830,
399.9977373, 399.9983030, 399.9987272, 399.9990454, 399.9992841, 399.9994630,
399.9995973

Interpretation of the results:

The DDS shows that the drug reaches a value where change stops and eventually the concentration in the bloodstream levels at 400 mg. If 400 mg is both a safe and effective dosage level, then this dosage schedule is acceptable.

### *2.6.2.1 Using Maple for a System of DDS*

Again, we use Systems of DDS with rsolve, numerical, and plotting for the problem $PS(n) = 0.65\ PS(n-1) + 0.25\ PM(n-1)$, (Figures 2.21–2.23)

$$PM(n) = 0.35\,PS(n-1) + 0.75\,PM(n-1),$$

$$PS(0) = 2{,}000,$$
$$PM(0) = 1{,}500.$$

*The following is output from MAPLE:*

> dds:=rsolve({PS(n)=.65*PS(n–1)+.25*PM(n–1),PM(n)=.35*PS(n–1)
+.75*PM(n–1),
PS(0) = 2000, PM(0) = 1500}, {PM, PS});

$$dds := \left\{ PM(n) = -\frac{1625}{3}\left(\frac{2}{5}\right)^n + \frac{6125}{3},\ PS(n) = \frac{1625}{3}\left(\frac{2}{5}\right)^n + \frac{4375}{3} \right\}$$

> plot({–(1625/3)*(2/5)^n+6125/3, (1625/3)*(2/5)^n+4375/3},n=0..15,
thickness=3, title='Student Vouchers');
> public:=n-> if n=0 then 2000 else.65*public(n-1)+.25*magnent(n-1)
end if;

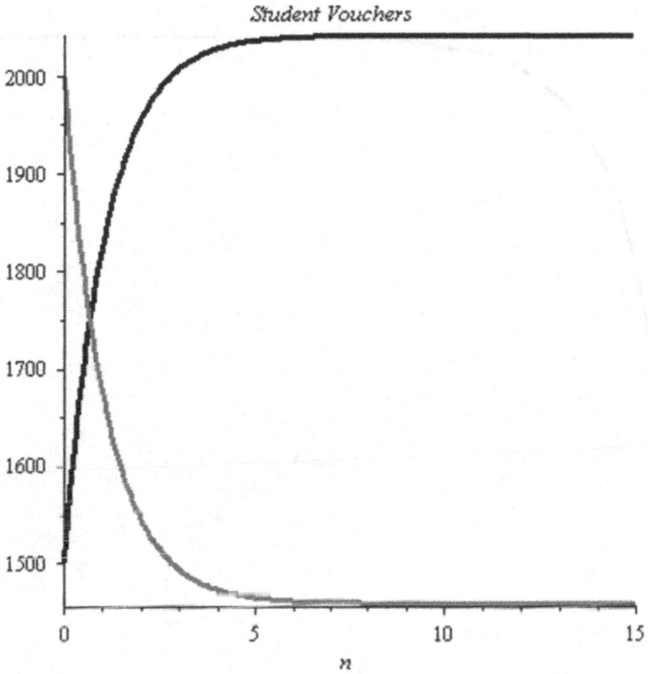


**FIGURE 2.21**
Student vouchers.

$$public{:=}n \rightarrow \textbf{if } n = 0 \textbf{ then } 2000 \textbf{ else } 0.65\, public(n-1) + 0.25\, magnent(n-1) \textbf{ end if}$$

```
> magnent:=n-> if n=0 then 1500 else 0.35*public(n-1)+.75*magnent(n-1)
end if;
```

$$magnent{:=}n \rightarrow \textbf{if } n = 0 \textbf{ then } 1500 \textbf{ else } 0.35\, public(n-1) + 0.75\, magnent(n-1) \textbf{ end if}$$

```
> seq([public(n),magnent(n)],n=0..10);
```

[2000, 1500], [1675.00, 1825.00], [1545.0000, 1955.0000],
[1493.000000, 2007.000000], [1472.200000, 2027.800000],
[1463.880000, 2036.120000], [1460.552000, 2039, 448000],
[1459.220800, 2040.779200], [1458.688320, 2041.311680],
[1458.475328, 2041.524672], [1458.390131, 2041.609869]

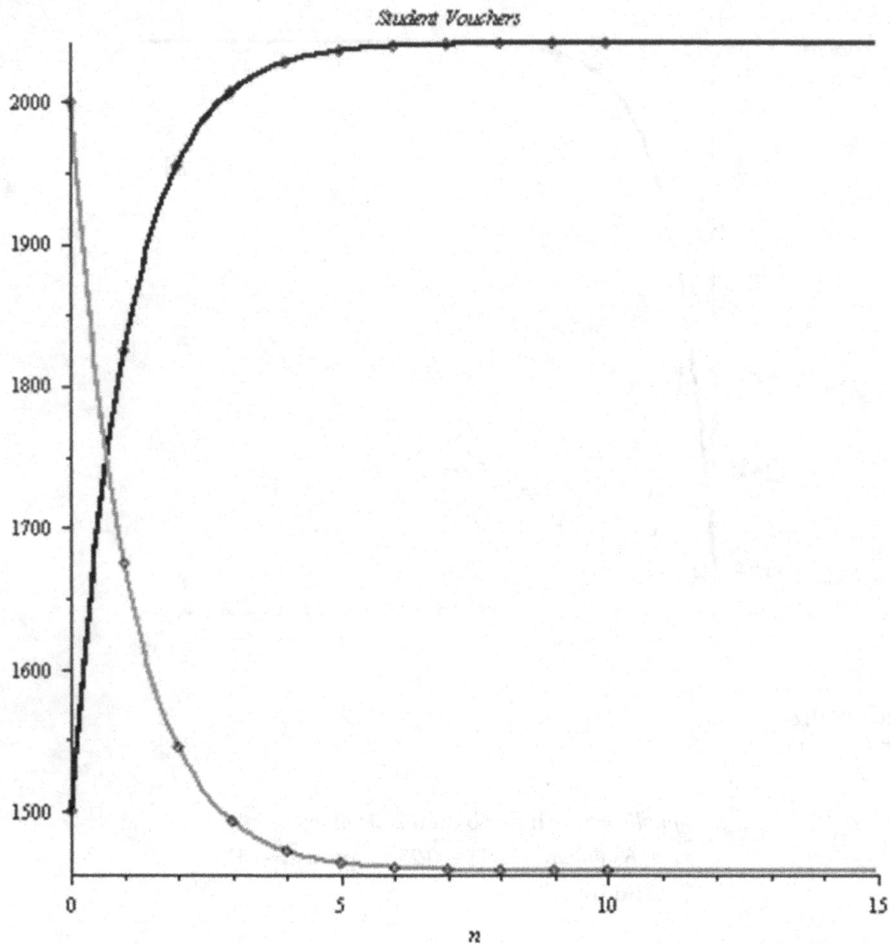


**FIGURE 2.22**
Student vouchers – update.

```
> u:=seq(public(n),n=0..10);
```

*u*:=2000, 1675.00, 1545.0000.1493, 000000, 1472.200000, 1463.880000, 1460.552000, 1459.220800, 1458.688320, 1458.475328, 1458.390131

```
> w-=seq(magnent(n),n=0..10);
```

*w*:=1500, 1825.00, 1955.0000, 2007.000000, 2027.800000, 2036.120000, 2039.448000, 2040.779200, 2041.311680, 2041.524672, 2041.609869

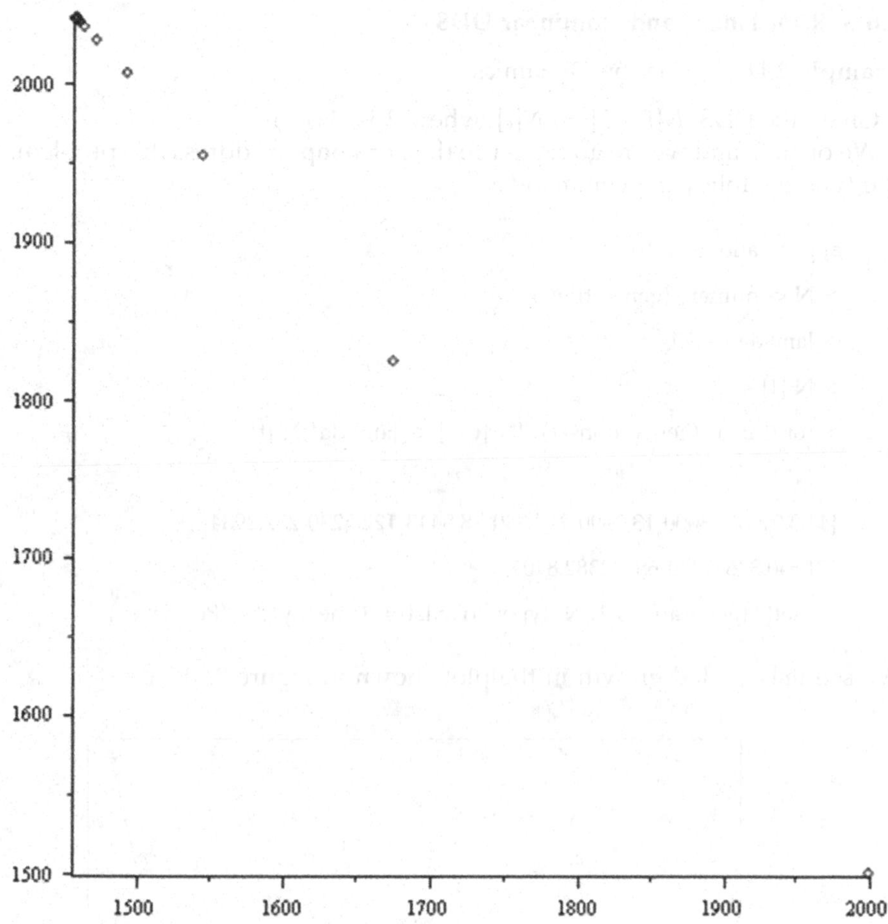


**FIGURE 2.23**
Student vouchers – plot points.

```
> with(plots):
> a:=plot({-(1625/3)*(2/5)^n+6125/3, (1625/3)*(2/5)^n+4375/3},n=0.15,
thickness=3, title='Student Vouchers'):
> b:=pointplot({seq([n,public(n)],n=0..10)}):
> c:=pointplot({seq([n,magnent(n)],n=0..10)}):
> display(a,b,c);
> pointplot({seq([public(n),magnent(n)],n=0..20)});
```

### 2.6.3 R for Linear and Nonlinear DDS

**Example 2.11:** Population Dynamics

Given the DDS, $N[t + 1] = \lambda N[t]$, where $\lambda$ is $(1 + r)$.

We open R. and we are going to use the **For Loop** to address this problem. We type the following commands:

```
> generations <- 10
> N <- numeric(generations)
> lambda <- 2.1
> N [1] <- 3
> for (t in 1: (generations-1)) {N [t+1] <- lambda* N [t]}
> N
[1] 3.0000 6.3000 13.2300 27.7830 58.3443 122.5230 257.2984
[8] 540.32661134.68582382.8401
> plot(0:(generations-1),N, type="o",xlab="Time", ylab="Pop Size")
```

We see unbounded growth in the plot shown in Figure 2.24.

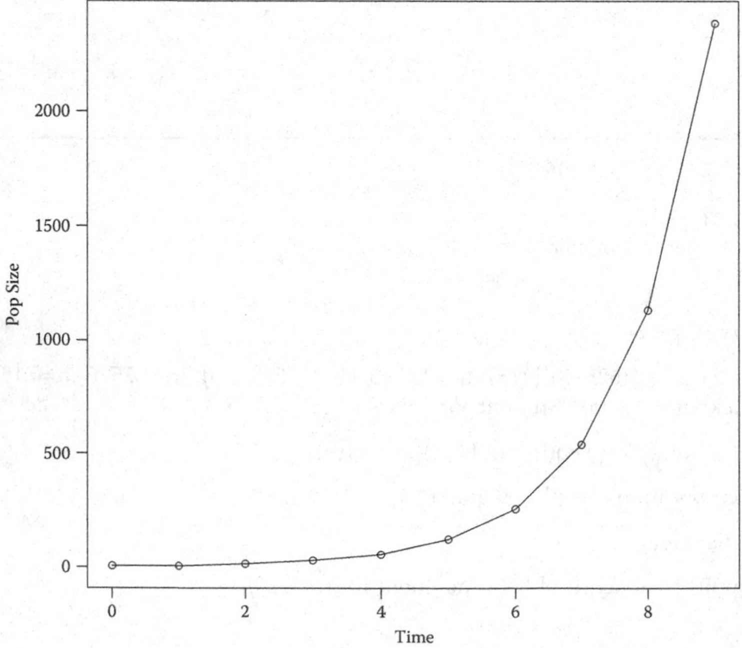


**FIGURE 2.24**
Screenshot from R.

**Example 2.12:** Repeat Example 2.10 $a(n + 1) = 0.75\, a(n) + 100$, $a(0) = 0$ using *R*.

We type the following commands:

```
> gener <- 20
> D <- numeric(gener)
> lam <- 0.75
> D[1] <- 0
> for (t in 1:(gener-1)) {D[t+1] <- lam*D[t]+100}
> D
[1] 0.0000 100.0000 175.0000 231.2500 273.4375 305.0781 328.8086
346.6064
[9] 359.9548 369.9661 377.4746 383.1059 387.3295 390.4971 392.8728
394.6546
[17] 395.9910 396.9932 397.7449 398.3087
> plot(0:(gener-1),D,type="o",xlab="Time",ylab="Drug_in_Sys")
```

In R, we can see that the drug becomes stable at approximately 400 units, as shown in Figure 2.25.

We now present a drug dosage model analytically.

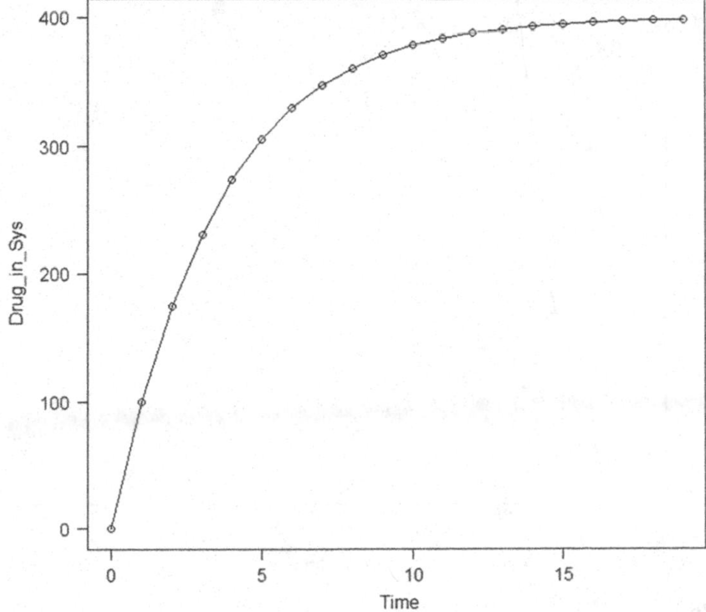


**FIGURE 2.25**
Screenshot from R solution graph for $a(n + 1) = 0.75\, a(n) + 100$, $a(0) = 0$.

#### 2.6.3.1 Logistics Growth

We can modify this model to a nonlinear model using a logistics growth DDS in R.

Given the DDS, N[$t$ + 1] = $N[t] + r\,\lambda\, N[t]\,(1 - N[t]/K)$, we have the following R script using the **function** command.

```
DDSL <- function(K,r,N0,generations)
+ {N <- c(N0,numeric(generations-1))
+ for (t in 1:(generations-1)) N [t+1] <- {N [t] + r*N [t]* (1– (N [t]/K))}
+ return(N)}
> Output <-DDSL(K=1000,r=1.5,N0=10,generations=30)
> generations <-30
> plot(0:(generations-1),Output, type='o', xlab="time",ylab="Population")
```

The plot (Figure 2.26) shows the exponential growth at the beginning and then leveling off to the carrying capacity, $K$ = 1,000.

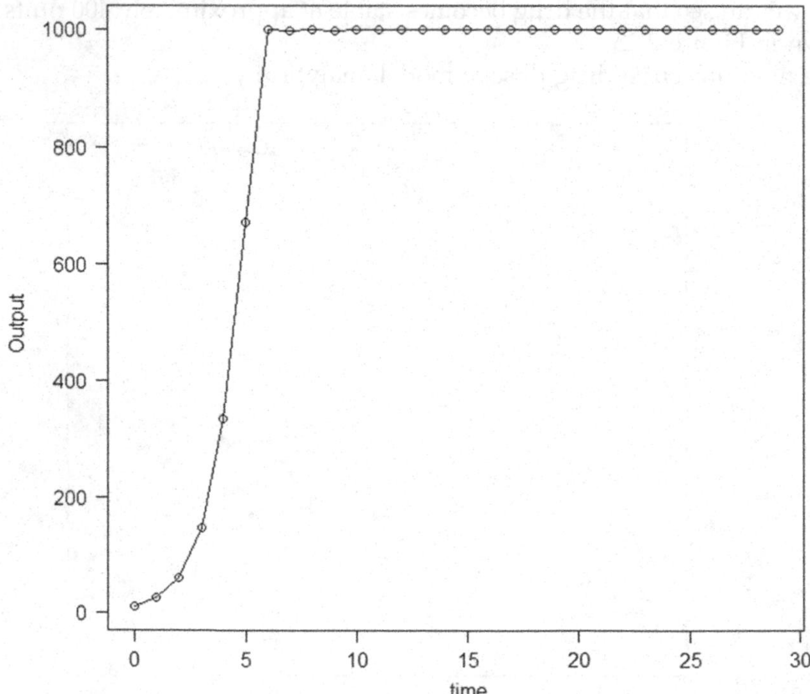


**FIGURE 2.26**
Logistics growth.

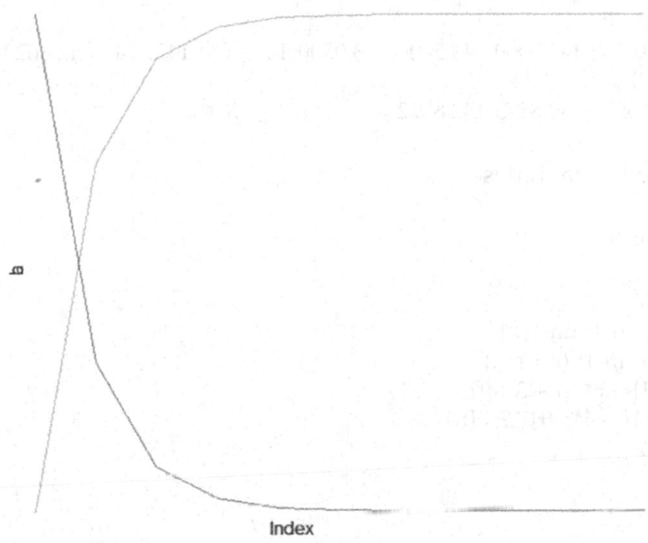


**FIGURE 2.27**
DDS system in R.

Next, we present a system of DDS using *R*.
DDSs using a multiple **for** loop command. (Figure 2.27)

```
nn <-11
a[1] <-100
b[1]<-150
for (t in 1: (nn−1)) {
for (tt in 1: (nn−1)){

>plot(a, type="l",col="green")

>par(new=TRUE)

> plot(b, type="l",col="red", axes=FALSE)

> for (t in 1:(nn−1)) {
+ for (tt in 1: (nn−1)) {
+ a[t+1] <-.6*a[t]+.3*b[t]
+ b[tt+1] <-.4 *a[tt]+.7*b[tt]
+}
+}

> a
[1] 100.0000 105.0000 106.5000 106.9500 107.0850 107.1255
107.1376107.1413
[9] 107.1424 107.1427 107.1428
```

```
> b
[1] 150.0000 145.0000 143.5000 143.0500 142.9150 142.8745 142.8623
142.8587
[9] 142.8576 142.8573 142.8572
```

Steady state probabilities

```
> a[1] <- 1

> b[1] <-0

> for (t in 1:(nn-1)) {
+ for (tt in 1: (nn-1)) {
+ a[t+1]<-.6*a[t]+.3*b[t]
+ b[tt+1]<-.4*a[tt]+.7*b[tt]
+}
+}
> a

[1] 1.0000000 0.6000000 0.4800000 0.4440000 0.4332000 0.4299600
0.4289880
[8] 0.4286964 0.4286089 0.4285827 0.4285748

> b
[1] 0.0000000 0.4000000 0.5200000 0.5560000 0.5668000 0.5700400
0.5710120
[8] 0.5713036 0.5713911 0.5714173 0.5714252
0.42857 and 0.57143 respectively.
```

The matrix product yields the same solution as before.

Getting multiple plots: here are some suggested R commands:

```
plot.new()
plot.window(xlim=range(x1),ylim=range(y1))
lines(x1,y1)
axis(1); axis(2); box()
plot.window(xlim=range(x2),ylim=range(y2))
lines(x2,y2)
axis(1); axis(2); box()
## Using the 'deSolve' package
library(deSolve)
## Time
t <- seq(0, 100, 1)
## Initial population
N0 <- 10
## Parameter values
params <- list(r=0.1, K=1000)
## The logistic equation
fn <- function(t, N, params) with (params, list(r * N * (1 – N/K)))
```

```
## Solving and plotin the solution numerically
out <- ode(N0, t, fn, params)
plot(out, lwd=2, main="Logistic equation\nr=0.1, K=1000, N0=10")
## Plotting the analytical solution
with(params, lines(t, K * N0 * exp(r * t)/ (K + N0 * (exp(r * t) - 1)), col=2,
lwd=2))
```

## 2.7 Exercises

Consider the model $a(n+1) = r\, a(n)\, (1 - a(n))$. Let $a(0) = 0.2$. Determine the numerical and graphical solution for the following values of $r$. Find the pattern in the solution.

1. $r = 2$.
2. $r = 3$.
3. $r = 3.6$.
4. $r = 3.7$.

For problems 5–8, find the equilibrium value by iteration and determine if it is stable or unstable.

5. $a(n+1) = 1.7\, a(n) - 0.14\, a(n)^2$.
6. $a(n+1) = 0.8\, a(n) + 0.1\, a(n)^2$.
7. $a(n+1) = 0.2\, a(n) - 0.2\, a(n)^3$.
8. $a(n+1) = 0.1\, a(n)^2 + 0.9\, a(n) - 0.2$.
9. Consider spreading a rumor through a company of 1,000 employees all working in the same building. We assume that the spread of a rumor is similar to the spread of a contagious disease in that the number of people hearing the rumor each day is proportional to the product of the number hearing the rumor and the number who have not heard the rumor. This is given by the formula:

$$r(n + 1) = r(n) + 1000k\, r(n) - k\, r(n)^2,$$

where k is the parameter that depends on how fast the rumor spreads. Assume k = 0.001 and further assume that four people initially know the rumor. How soon will everyone know the rumor?

10. Determine the equilibrium values of the bass and trout model presented in Section 2.2. Can these levels ever be achieved and maintained? Explain.

11. Test the fast food models with different starting conditions summing to 14,000 students. What happens? Obtain a graphical output and analyze the graph in terms of long-term behavior.
    a. Find the equilibrium values for the Predator–Prey Model presented in Section 2.5.
    b. In the Predator–Prey Model, presented in Section 2.5, determine the outcomes with the following sets of parameters.
        i. Initial foxes are 200, and initial rabbits are 400.
        ii. Initial foxes are 2,000, and initial rabbits are 10,000.
        iii. Birth rate of rabbits increases to 0.1.
    c. In the SIR model, presented in Section 2.5, determine the outcome with the following parameters changing: The flu lasts 1 week.
        i. Initially 5 are sick and 10 the next week.
        ii. The flu lasts 4 weeks.
        iii. There are 4,000 students in the dorm, and 5 are initially infected and 30 more the next week.

## 2.8 Projects

1. Consider the contagious disease as the Ebola virus. Use the internet to find out how deadly this virus actually is. Now consider an animal research laboratory in Restin, VA., a suburb of Washington, DC, with population 856,900 people. A monkey with the Ebola virus has escaped its captivity and infected one employee (unknown at the time) during its escape. This employee reports to University hospital later with Ebola symptoms. The Infectious Disease Center (IDC) in Atlanta gets a call and begins to model the spread of the disease. Build a model for the IDC with the following growth rates to determine the number infected after 2 weeks:
    a. $k = 0.00025$
    b. $k = 0.000025$
    c. $k = 0.00005$
    d. $k = 0.000009$

    List some ways of controlling the spread of the virus.

2. Consider the spread of a rumor concerning termination among 1,000 employees of a major company. Assume that the spreading of a rumor

is similar to the spread of contagious disease in that the number hearing the rumor each day is proportional to the product of those who have heard the rumor and those who have not heard the rumor. Build a model for the company with the following rumor growth rates to determine the number having heard the rumor after 1 week:

a. $k = 0.25$
b. $k = 0.025$
c. $k = 0.0025$
d. $k = 0.00025$

List some ways of controlling the spread of the rumor.

3. Lions and spotted hyena: Predict the number of lions and spotted hyena in the same environment at a function of time

Assumptions:

The variables: $L(n)$ = number of lions at the end of period $n$.

$H(n)$ = number of hyenas at the end of period $n$.

Assume the Model:

$$L(n+1) = 1.2\,L(n) - 0.002\,L(n)H(n).$$
$$H(n+1) = 1.3\,H(n) - 0.001\,H(n)L(n).$$

a. Find the equilibrium values of the system.
b. Iterate the system from the following initial conditions and determine what happens to the lions and the spotted hyenas in the long term (Table 2.7).

**TABLE 2.7**

Data for Project 3

| Lions | Spotted Hyena |
|---|---|
| 150 | 200 |
| 151 | 199 |
| 149 | 201 |
| 20 | 20 |
| 200 | 200 |

4. It is getting close to election day. The influence of the new Independent Party is of concern to the Republicans and Democrats. Assume that in the next election that 75% of those who vote

Republican vote Republican again, 5% vote Democratic, and 20% vote Independent. Of those that voted Democratic before, 20% vote Republican, 60% vote Democratic, and 20% vote Independent. Of those that voted Independent, 40% vote Republican, 20% vote Democratic, and 40% vote Independent.

a. Formulate and write the DDS that models this situation.
b. Assume that there are 399,998 voters initially in the system, how many will vote Republican, Democratic, and Independent in the long run? (Hint: you can break down the 399,998 voters in any manner that you desire as initial conditions.)
c. (New scenario) In addition to the above, the community is growing (18-year-olds + new people – deaths – losses to the community, etc.). Republicans predict a gain of 2,000 voters between elections. Democrats estimate a gain of 2,000 voters between elections. The Independents estimate a gain of 1,000 voters between elections. If this rate of growth continues, what will be the long-term distribution of the voters?

Normal 0 false false false false EN-US X-NONE X-NONE

## References

Albright, B. (2010). *Mathematical Modeling with Excel*, Jones and Bartlett, Sudberry, MA.

Alfred, U. (1967). Sums of squares of consecutive odd integers. *Mathematics Magazine* 40(4): 194–199.

Allman, E., and J. Rhodes (2004). *Mathematical Models in Biology: An Introduction*, Cambridge University Press, Cambridge, UK.

Arney, D., F. Giordano, and J. Robertson (2002). *Mathematical Modeling with Discrete Dynamical Systems*, McGraw Hill, Boston, MA.

Fox, W. P. (2010). Discrete combat models: Investigating the solutions to discrete forms of Lanchester's combat models. *International Journal of Operations Research and Information Systems* (IJORIS) 1(1): 16–34.

Fox, W. P. (2012a). Discrete combat models: Investigating the solutions to discrete forms of Lanchester's combat models. In *Innovations in Information Systems for Business Functionality and Operations Management*, IGI Global & SAGE Publishers, pp. 106–122.

Fox, W. P. (2012b). Mathematical modeling of the analytical hierarchy process using discrete dynamical systems in decision analysis. *Computers in Education Journal* 3(3): 27–34.

Fox, W. P. (2012c). *Mathematical Modeling with Maple*, Cengage Publishing, Boston, MA.

Fox, W. P., and Patrick J. Driscoll (2011). Modeling with dynamical systems for decision making and analysis. *Computers in Education Journal* (COED) 2(1): 19–25.

Giordano, F. R., W. P. Fox, and S. Horton (2014). *A First Course in Mathematical Modeling*, 5th Edition. Cengage Publishing, Boston, MA.

Leyendekker, J., and A. Shannon (2015). The odd-number sequence: Squares and sums. *International Journal of Mathematical Education in Science and Technology* 46(8): 1222–1228.

Sandefur, J. (1990). *Discrete Dynamical System: Theory and Applications*, Oxford University Press, New York.

Sandefur, J. (2002). *Elementary Mathematical Model: A Dynamic Approach*, 1st Edition. Brooks-Cole Publishers, Belmont, CA.

Sandefur, J. (2003). *Elementary Mathematical Modeling: A Dynamic Approach*, Thomson Publishing, Belmont, CA.

# 3

# *Statistical and Probabilistic Models*

**OBJECTIVES**

1. Understand the basic statistical displays of data.
2. Understand the common display techniques of quantitative data.
3. Know the basic principles of measures of central tendency.

## 3.1 Introduction

Statistical and probability models provide the bread-and-butter techniques for understanding and gaining inferences from data and for developing predictive models. In probability problems, the properties of the population are assumed to be known and questions regarding a sample are posed and answered. In a statistics problem, characteristics of the sample are available to the experimenter, and this information enables the experimenter to draw conclusions about the population. The relationship between the two disciplines is illustrated in Figure 3.1.

In Chapter 3 and Chapter 4, we will cover multiple probability and statistics topics that will enable us to readily use probability and statistics to analyze data or make decisions about mission and/or operations. These topics include the following:

- **Displays of Data and Statistics (Chapter 3)**
  - Pie and Bar Charts for Qualitative Data
  - Stem and Leaf, Histogram, Boxplot for Quantitative Data
- **Descriptive Statistics (Chapter 3)**
  - Displays
  - Data Types
  - Quantitative and Qualitative (categorical)
  - Mean

DOI: 10.1201/9781003298762-3

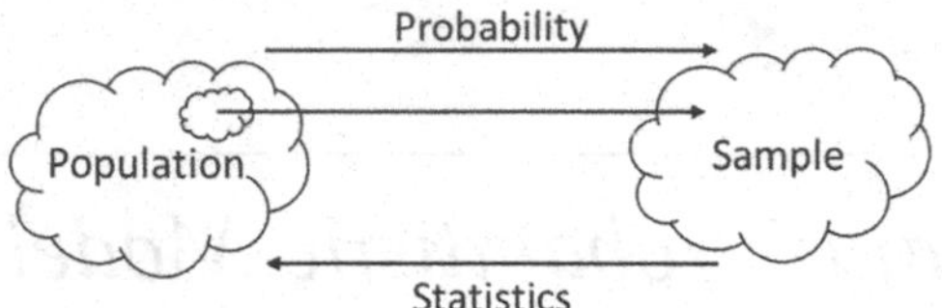


**FIGURE 3.1**
The relationship between probability and inferential statistics.

- Median
- Mode
- Standard Deviation
- Variance
- Coefficient of Skewness

- **Classical Probability** (Chapter 4)
  - Probability Rules
  - Tree Diagrams
  - Conditional Probability
  - Independence
  - *Bayes' Rule*
- **Random Variables** (Chapter 4)
  - PMF and CDF for Discrete Distributions
    - Poisson and Binomial Distributions
  - PDF and CDF Continuous Distributions
    - Exponential and Normal Distributions
    - Expected Value
    - Central Limit Theorem
- **Hypothesis tests** (Chapter 4)

We will start with a simple example to help demonstrate the power of simple probability and statics techniques to gain insights into a problem.

**Example 3.1:** Cruise Ship Disaster

A cruise ship loses power in the Mediterranean Sea, and a disaster ensues. Table 3.1 provides the results of the disaster.

We have all heard the rule of a disaster at sea that woman and children are the first to be rescued or saved. We are interested in discovering if this rule was followed in our Mediterranean Sea disaster (Example 3.1). We will present the formulas later, but a few basic statistical calculations reveal that only 19.3% of the men (340 survived out of a total of 1,760 men) and 69.7%

**TABLE 3.1**

Disaster Results of Cruise Ship

| | Men | Women | Boys | Girls | Total |
|---|---|---|---|---|---|
| **Survived** | 332 | 318 | 29 | 27 | 706 |
| **Died** | 1360 | 104 | 35 | 18 | 1517 |
| **Total** | 1692 | 422 | 64 | 45 | 2223 |

of the woman and children survived (365 out of 524). It does appear that the data support the fact that the rule of woman and children first was followed in this case. These simple calculations are examples of powerful tools in analyzing information and providing insights into results. Later, in Chapter 4, we will explore if these differences are truly significantly different.

## 3.2 Understanding Univariate and Multivariate Data

The most fundamental principle in statistics is that of variability. If the world were perfectly predictable and all our data showed no variability, then there would be no need to study statistics. First, for this discussion, we will consider statistics to be the science of reasoning from data. Therefore, a natural starting point is to understand what is meant by "data". You will also need to discover the notion of a variable and then learn how to classify variables before we can begin a deeper discussion of probability and statistics.

We will refer to any characteristic of a person, item, event, or thing that can be expressed as a number is called a ***variable***. A *value* of that variable is the actual number that describes that person, thing, etc. Think of classic demographic variables that might be used to describe an individual: height, weight, eye color, hair color, income, grade level, political affiliation, gender, and debt. Data can be ***quantitative*** or ***categorical,*** as we will define below.

**Quantitative** is defined as data that are numerical in nature and where the number has relative meaning. Examples could be a list of heights of students in your class, weights of students on a football team, or batting averages of the starting line-up for the 2000 Mets.

Heights of students in a class (Table 3.2):

**TABLE 3.2**

Student Heights

| | | | | | |
|---|---|---|---|---|---|
| 5′10″ | 6′2″ | 5′5″ | 5′2″ | 6′ | 5′9″ |

Weights of students in a class (Table 3.3):

**TABLE 3.3**

Student Weights

| | | | | | | | |
|---|---|---|---|---|---|---|---|
| 135 | 155 | 215 | 192 | 173 | 170 | 165 | 142 |

Batting averages for the San Francisco lineup (Table 3.4):

**TABLE 3.4**

San Francisco Batting Averages

| | | | | | | | | |
|---|---|---|---|---|---|---|---|---|
| 0.276 | 0.320 | 0.345 | 0.354 | 0.269 | 0.275 | 0.300 | 0.254 | 0.309 |

These data elements provide numerical information. We can determine from the data which height is the tallest or smallest or which batting average is the greatest or the smallest. We also have the ability to mathematically compare and contrast these values.

Quantitative data are either discrete (counting data) or continuous in nature. These characteristics become important as we analyze them and use them in models.

**Categorical** data are used to describe the objects, such as recording the people with a particular hair color by: blonde = 1 or brunette = 0. For example, if we had four colors of hair: blonde, brunette, black, and red, we could use codes: brunette = 0, blonde = 1, black = 2, and red = 3. Logically, we cannot have an average hair color from these numbers. Another example is categories by gender: male = 0 and female = 1. In general, it may not make sense, in terms of interpretation, to do arithmetic operations using categorical variables.

In dealing with data, it makes sense to do arithmetic with quantitative data. Once you have learned to distinguish between quantitative and categorical data, we need to move on to a fundamental principle of data analysis – developing a visual display of the data set.

### 3.2.1 Displaying the Data

Why do we want to display data? Visual displays such as bar graphs, pie charts, and histograms are very useful because they provide a quick and efficient way to present the revealing characteristics of the data. These displays allow our eyes to take in the overall pattern and see if there are unusual observations of data elements. Graphs and numbers that we will introduce to describe the data are not ends in themselves, but merely aids to our overall understanding of the data.

Unfortunately, "displaying data badly" is a common problem. We can illustrate and explain some common elements in displaying data badly.

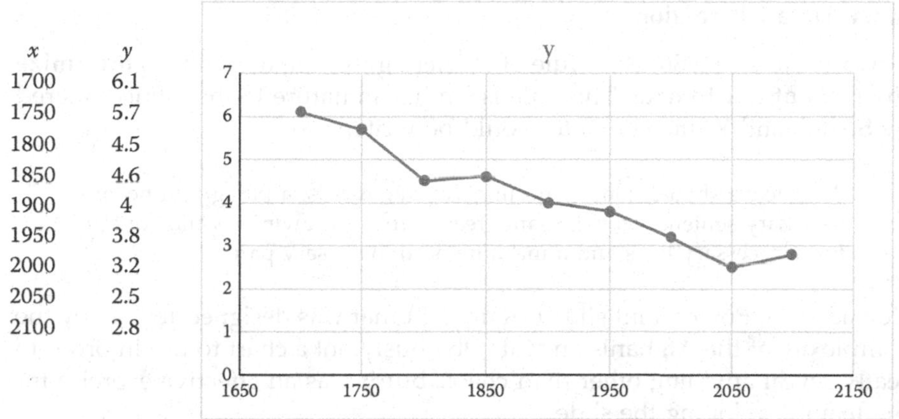

| x | y |
|---|---|
| 1700 | 6.1 |
| 1750 | 5.7 |
| 1800 | 4.5 |
| 1850 | 4.6 |
| 1900 | 4 |
| 1950 | 3.8 |
| 2000 | 3.2 |
| 2050 | 2.5 |
| 2100 | 2.8 |



**FIGURE 3.2**
Data ambiguity example does not provide information as to what y represents.

The three fundamental elements of bad graphical display are these: **data ambiguity, data distortion, and data distraction.**

### 3.2.2 Data Ambiguity

Data ambiguity arises from the failure to precisely define just what the data represent. Every dot on a scatterplot, every point on a time series line, every bar on a bar chart represents a number, and in the case of a scatterplot, two numbers. It is the job of the legend and labels on the chart to provide information to tell us just what each of those numbers represents. If a number represented in a chart is, say, 33½, the text in the graph – in the title, the axis labels, the data labels, the legend, and sometimes the footnote – must answer the question: "Thirty-three and a half what?" We see in Figure 3.2 a plot where we have no clue what information is provided on the "*y*" axis. The bottom line is to label everything for clarity.

### 3.2.3 Data Distortion

Before the development of spreadsheet graphing, the most common graphical mistake was the use of artist-drawn 3D images with the height of 3D objects representing the magnitude of the data points. In these charts, both the height and the width of the drawn object increase proportionate to the magnitude of the data points. The effect is to exaggerate the differences in magnitude as the viewer tends to perceive the area of the figures rather than just the height as representing the magnitude.

### 3.2.4 Data Distraction

Edward Tufte's (1986) basic rule of efficient graphical design is to **minimize the ratio of ink-to-data**. This rule is similar in nature to the advice offered by Strunk and White (1999) to would be writers:

> A sentence should contain no unnecessary words, a paragraph no unnecessary sentences for the same reason that a drawing should contain no unnecessary lines and a machine no unnecessary part.

Consider the PowerPoint slide (Figure 3.3) that was designed to convey the complexity of the Afghanistan war. Obviously not a chart to use in order to really reveal anything other than chaos, but it was an effective exercise for the team developing the slide.

## 3.3 Displays of Data and Statistics

### 3.3.1 Good Displays of the Data

There are a couple of common tools used in developing a visual representation of data to provide a clearer picture, versus looking at raw numbers, of what the data are conveying. We will discuss five methods of displaying data: pie chart, bar chart, stem and leaf (by hand and an Excel template), histogram, and boxplot (by hand and an Excel template). The displays should supply visual information to the viewer without a struggle.

### 3.3.2 Displaying Categorical Data – Pie Chart

The *pie chart* is useful to show the division of a total quantity into component parts. A pie chart, if done correctly, is usually safe from misinterpretation. The total quantity, or 100%, is shown as the entire circle. Each wedge of the circle represents a component part of the total. These parts are usually labeled with *percentages* of the total. Thus, a pie chart helps us see what part of the whole each group forms.

We will start with a review of percentages. Let $a$ represent the partial amount and $b$ represent the total amount. Then, $P$ represents a relative frequency found by calculating $a/b$, which can be made into a percentage by $= a/b\ (100)$.

A percentage is thus a part of a whole. For example, \$0.25 is what part of \$1.00? We let $a = 25$ and $b = 100$. Then, $P = 25/100\ (100) = 25\%$.

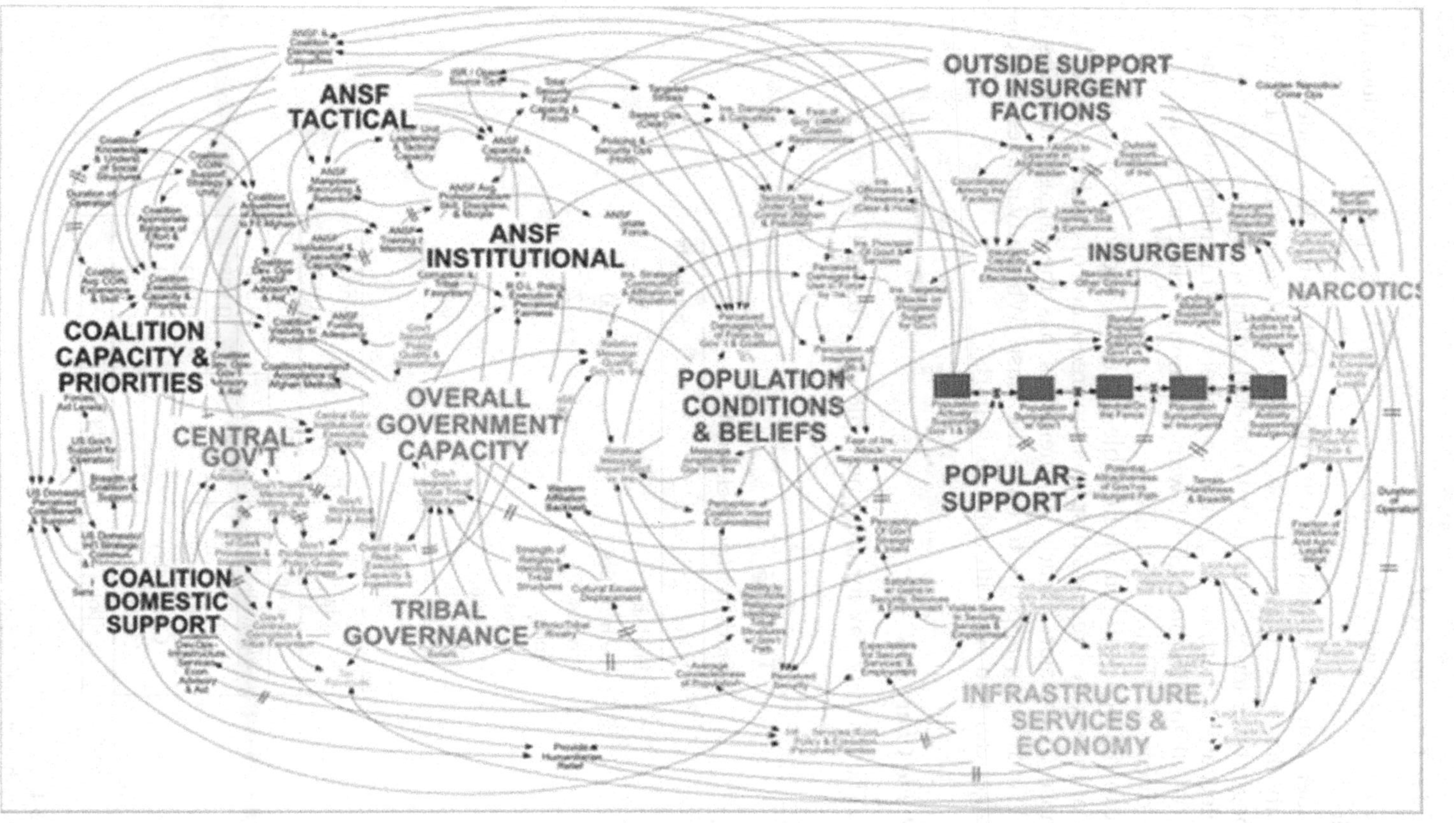


**FIGURE 3.3**
Powerpoint diagram to portray the complexity of Afghanistan strategy (https://www.nytimes.com/2010/04/27/world/27powerpoint.html).

#### *3.3.2.1 Pie Chart in Excel*

Now, we can see how Excel would create a pie chart for us in the following scenario.

Consider students choosing their major. Out of the 388 students in a university that selected a major, the breakdown of selection is in Table 3.5.

We begin by entering the data into labeled columns in Excel (Figure 3.4).

To select a pie chart, we highlight the labels and the numbers for the six majors. We then click on INSERT with the mouse, click on Pie, and follow editing directions (Figure 3.5).

You can right-click to put in the table values as shown above.

Each of the shaded regions displays the frequency of students out of 632 that chose that major. Clearly Mathematics has the largest numbers of major. Which major appears to have the least? What advantages and disadvantages can you see with using pie charts?

#### *3.3.2.2 Pie Chart in R*

Syntax: The basic syntax for creating a pie chart using the R is

**TABLE 3.5**

Student Major Selection

| Major | Students |
|---|---|
| English | 58 |
| Mathematics | 200 |
| Physics | 42 |
| Biology | 37 |
| Chemistry | 21 |
| US History | 30 |
| Total | 388 |

| Major | Frequency | Relative Frequency |
|---|---|---|
| English | 58 | 0.149 |
| Mathematics | 200 | 0.515 |
| Physics | 42 | 0.108 |
| Biology | 37 | 0.095 |
| Chemistry | 21 | 0.054 |
| US History | 30 | 0.077 |
| Total | 388 | |

**FIGURE 3.4**
Excel screenshot of student data.

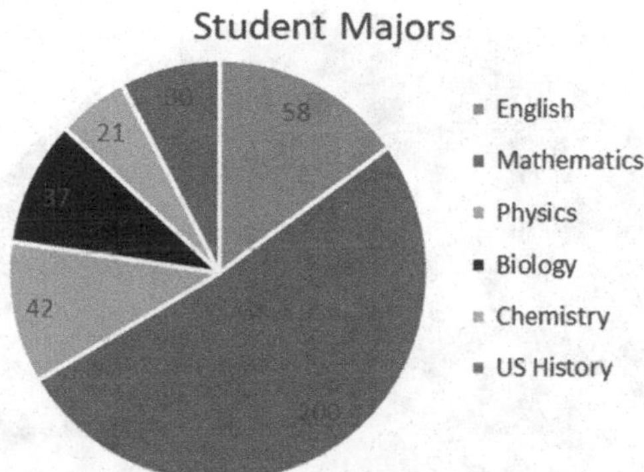


**FIGURE 3.5**
Student major pie chart.

```
pie(x, labels, radius, main, col, clockwise)
```

Following is the description of the parameters used:

- **x** is a vector containing the numeric values used in the pie chart.
- **labels** is used to add a description to the slices.
- **radius** indicates the radius of the circle of the pie chart (value between −1 and +1).
- **main** indicates the title of the chart.
- **col** indicates the color palette.
- **clockwise** is a logical value indicating if the slices are drawn clockwise or counterclockwise.

```
mm <- c(12,10,9,6,3,5)
labels <- c("blue", "red", "green", "yellow", "brown", "orange")
piepercent <- round(100*mm/sum(mm),1)
mm <- c(12,10,9,6,3,5)
labels <- c("blue", "red", "green", "yellow", "brown", "orange")
piepercent<- round(100*mm/sum(mm),1)
pie(mm, labels = piepercent, main = "MM Pie Chart", col = rainbow
(length(mm)))
```

Our recommendation is to enter your data (with names), then cut, paste, and then edit the names in these bar or pie chart commands, as necessary. Check your output to ensure it works properly.

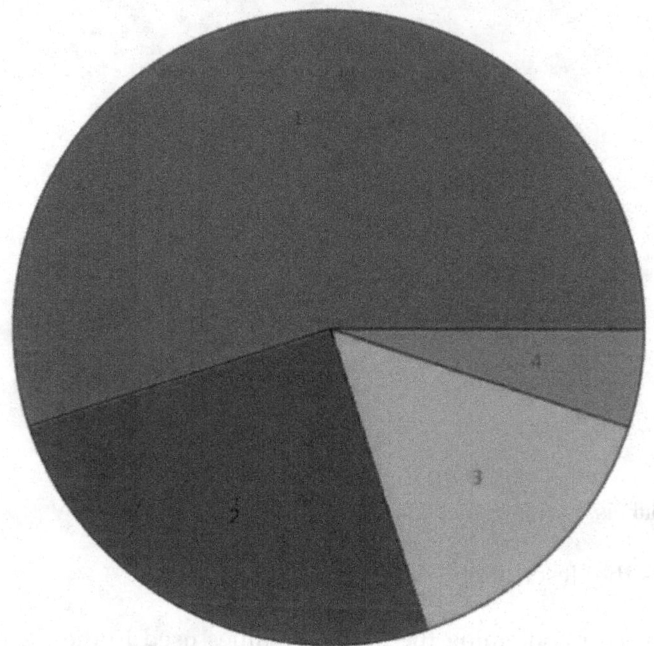


**FIGURE 3.6**
Maple pie chart.

#### *3.3.2.3 Pie Charts in Maple*

Syntax: The basic syntax for creating a pie-chart using Maple is (Figure 3.6)

```
> with(Statistics);
> PieChart([1, 1, 1, 1, 1, 1, 1, 1, 1, 1, 1, 2, 2, 2, 2, 2, 3, 3, 3, 4]);
```

The Help menu in Maple provides additional information that is helpful to view the display.

### 3.3.3 Bar Charts for Qualitative Data or Discrete Quantitative Data

Next, we can review how bar charts present data in a different manner from pie charts (Figure 3.7):

Is the bar chart any clearer in making your point than the pie chart? In this case, we think it does make it clearer.

### 3.3.4 Displaying Categorical Data – Bar Chart

Bar charts are useful when comparing relative sizes of data groups, especially when they come from **categorical** variables. For example, consider the eye color from patients visiting the local eye clinic last year (Table 3.6).

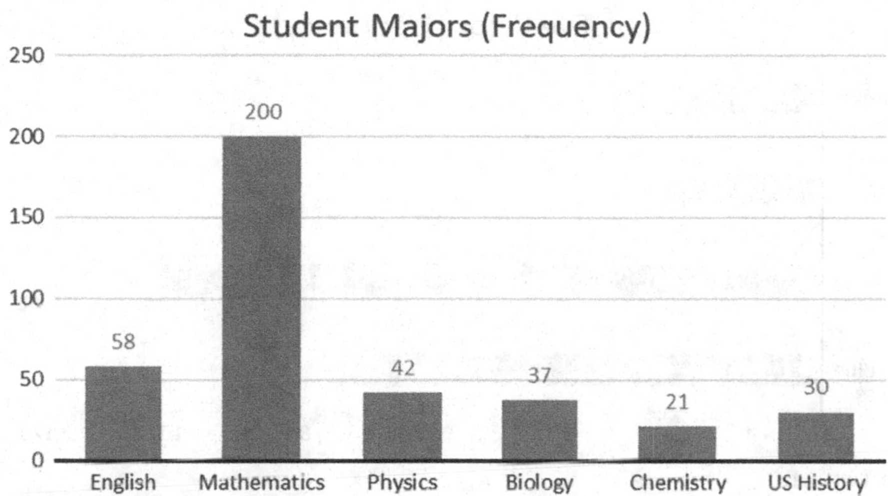


**FIGURE 3.7**
Student major bar chart.

**TABLE 3.6**
Patient Eye Color

| Eye Color | Count | Percent | Green | Yellow | Brown | Orange |
|---|---|---|---|---|---|---|
| **Blue** | 75 | 0.339 | 9 | 6 | 3 | 5 |
| **Hazel** | 101 | 0.457 | | | | |
| **Green** | 20 | 0.090 | | | | |
| **Brown** | 25 | 0.113 | | | | |
| **Total** | 221 | | | | | |

In Excel, you can simply click on INSERT and obtain the Bar Chart. Decide whether your display is better as a column or horizontal chart (Figure 3.8).

You can quickly and clearly compare the relative sizes of the color groups. From the bar chart, which eye color occurs the most frequently? Hazel eyes occur the most frequently. It appears twice as large as the next most frequent – brown eyes. Again, bar charts are most useful to display categorical data.

### *3.3.4.1 Bar Charts in R*

The basic syntax to create a bar-chart in R is:

```
>barplot(H, xlab, ylab, main, names. arg, col)
```

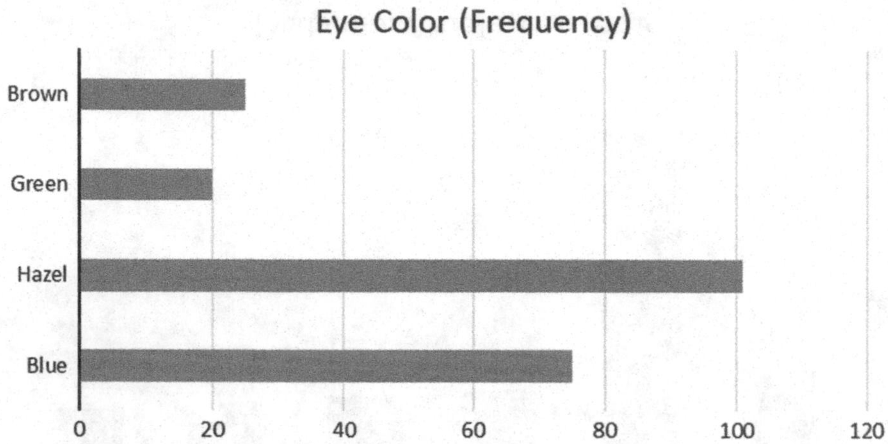


**FIGURE 3.8**
Excel patient eye color bar chart.

Following is the description of the parameters used:

- **H** is a vector or matrix containing numeric values used in the bar chart.
- **xlab** is the label for the x axis.
- **ylab** is the label for the y axis.
- **main** is the title of the bar chart.
- **names.arg** is a vector of names appearing under each bar.
- **col** is used to give colors to the bars in the graph.

Table 3.7 contains a count, by color of M & Ms found in a small bag.

The R commands to generate a bar chart (Figure 3.9) of the data in Table 3.7 is

```
mm <- c(12,10,9,6,3,5)
mc <- c("blue", "red", "green", "yellow", "brown", "orange")
barplot(mm,names.arg=mc,xlab="color",ylab="Frequency",col="blue",
main="MM's chart",border="red")
```

**TABLE 3.7**
M & M Candies Count by Color

| Color | Blue | Red | Green | Yellow | Brown | Orange |
|---|---|---|---|---|---|---|
| Count | 12 | 10 | 9 | 6 | 3 | 5 |

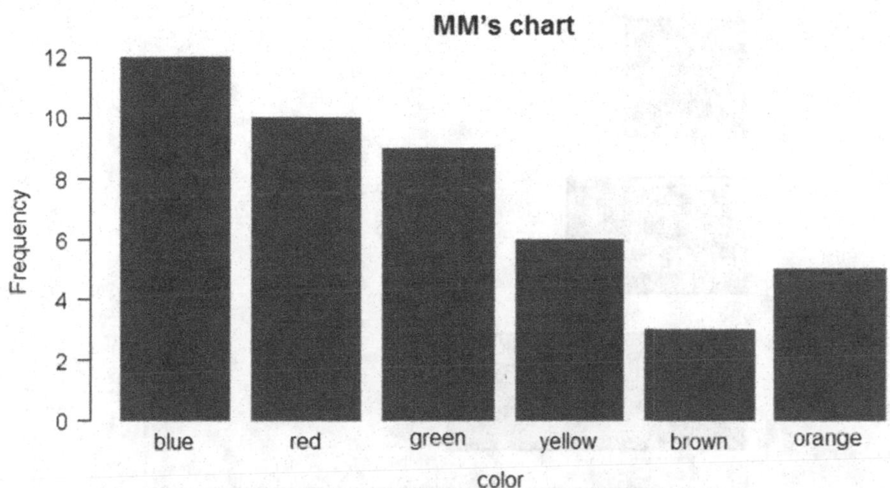


**FIGURE 3.9**
Bar chart in R of M & M counts by color.

```
mm <- c(12,10,9,6,3,5)

mc <- c("blue", "red", "green", "yellow", "brown", "orange")
barplot(mm,names.arg=mc,xlab="color",ylab="Frequency",col="blue",
main="MM's chart",border="red")

> mm
[1] 12 10 9 6 3 5

> mc

[1] "blue" "red" "green" "yellow" "brown" "orange"
```

#### 3.3.4.2 Bar charts in Maple

The basic syntax to create a bar chart (Figure 3.10) in Maple is:

```
with(Statistics);

dataset := [Vector([10, 5, 3, 2])];
                    [[10]]
                    [[ ]]
                    [[5 ]]
        dataset := [[ ]]
                    [[3 ]]
```

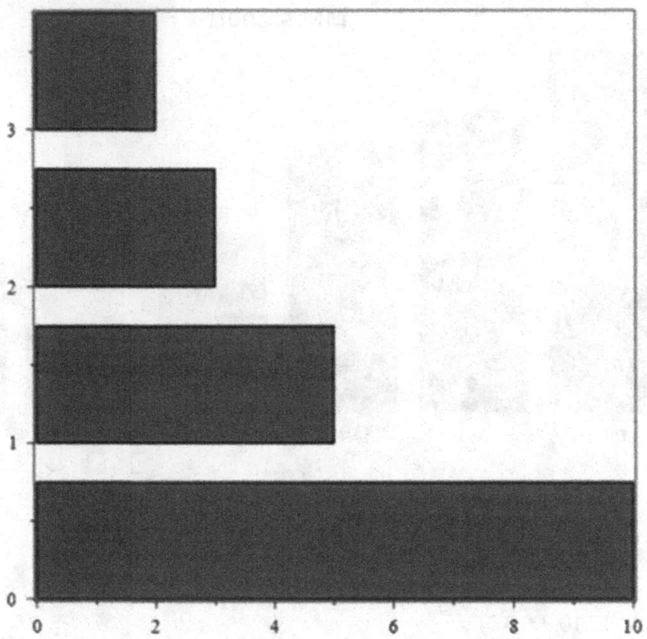


**FIGURE 3.10**
Simple bar chart in Maple.

[[ ]]

[[2 ]]

```
A := Array([1, 1, 1, 1, 1, 1, 1, 1, 1, 1]);
B := Array([2, 2, 2, 2, 2]);
C := Array([3, 3]);
BarChart(dataset);
```

### 3.3.5 Displaying Quantitative Data – Stem and Leaf

In quantitative data, we are concerned with the shape of the data. Shape refers to symmetry of data. "Is it symmetric?" and "Is it skewed?" are questions we ask and answer.

**Step 1**. Order data.

**Step 2**. Separate according to one or more leading digits. List stems in a vertical column.

**Step 3**. Separate the stem from the leaves by a vertical line. Leading digit is the stem and trailing digit is the leaf. For example, 32 is the stem and 2 is the leaf.

**Step 4**. Indicate the units for stems and leaves in the display.

A stem and leaf plot uses the real data points in making a plot. The plot will appear strange because your plot is sideways. The rules are as follows:
You will probably create these plots by hand. **Excel** will not produce a stem and leaf plot unless you have a stem-leaf template built. However, R does.

#### *3.3.5.1 Steam and Leaf Plots in R*

The command is straight-forward. For example, enter your data as

```
grades <-53, 55, 66, 69, 71, 78, 75, 79, 77, 75, 76, 73, 82, 83, 85, 74, 90, 92, 95, 99
stem(grades)
```

```
> grades<-c(53, 55, 66, 69, 71, 78, 75, 79, 77, 75, 76, 73, 82, 83, 85, 74, 90,
92, 95, 99)
> stem(grades)
```

The decimal point is 1 digit to the right of the |

```
5 | 35
6 | 69
7 | 134556789
8 | 235
9 | 0259
```

We will take a moment to demonstrate the process by hand.

**Example 3.2:** Mathematics Course Grades

Consider that you have the grades for 20 students in a mathematics course:

53, 55, 66, 69, 71, 78, 75, 79, 77, 75, 76, 73, 82, 83, 85, 74, 90, 92, 95, 99

Stems are the leading digit:

5

6

7

8

9

Standing for 50s, 60s, 70s, 80s, and 90s.

If there had been a score of 100, then the leading digit is in 100s. So, we would need:

05

06

07

08

09

10

for 50s, 60s, 70s, 80s, 90s, and 100s.

Draw a vertical line after each stem.

5|

6|

7|

8|

9|

Now, add the leaves, which are the trailing digits,

53, 55, 66, 69, 71, 73, 74, 75, 75, 76, 77, 78,79, 82, 83, 85, 90, 92, 95, 99

5| 3, 5

6| 6, 9

7| 1, 3, 4, 5, 5, 6, 7, 8, 9

8| 2, 3, 5

9| 0, 2, 5, 9

We can characterize this shape as almost *symmetric*. Note, how we read the values from the stem and leaf.

For example, we read

5| 3, 5

as data elements 53 and 55.

We have a way to program Excel with commands that give us a stem and leaf plot. It is still up to us to determine the shape. Here is an example with the data we just used. First, we show the result.

Stem and leaf in Excel by hand (Figure 3.11):

We might accept this as symmetric.

Now, how did we obtain this stem and leaf. We first enter the data into column A (Figure 3.11) in any order. Next, we determine the min and max of this data. We do this using the min and max commands over the range of the data (Figure 3.12).

The values are 53 for the min and 99 for the max. These are essential as we manually list our stems from the 5 for 53 to 9 for 99. If our values were 25 and 150 for the min and max, we might use 2, 3, 4, 5, 6, 7, 8, 9, 10, 11, 12 as

| A | B | C | D | E | F |
|---|---|---|---|---|---|
| grades | | | | Minimum | 53 |
| 53 | | | | Maximum | 99 |
| 55 | | | | | |
| 66 | | | | Stems | Leafs |
| 69 | | | | 5 | 35 |
| 71 | | | | 6 | 69 |
| 78 | | | | 7 | 134556789 |
| 75 | | | | 8 | 235 |
| 79 | | | | 9 | 0259 |
| 77 | | | | | |
| 75 | | | | | |
| 76 | | | | | |
| 73 | | | | | |
| 82 | | | | | |
| 83 | | | | | |
| 85 | | | | | |
| 74 | | | | | |
| 90 | | | | | |
| 92 | | | | | |
| 95 | | | | | |
| 99 | | | | | |

**FIGURE 3.11**
Excel screenshot of Example 3.2.

| Minimum | =MIN(A3:A22) |
|---|---|
| Maximum | =MAX(A3:A22) |

**FIGURE 3.12**
Excel screenshot of min and max commands.

our stems. Next is a more complicated command that we start with the first stem, 5. In this case, we want it to repeat the unit values between 0 for 50 and 9 for 59. We type the following using the REPT command.

```
=REPT("0",COUNTIF($A$3:$A$22,E6*10+0))&REPT("1",COUNTIF
($A$3:$A$22,E6*10+1))&REPT("2",COUNTIF($A$3:$A$22,E6*10+2))
&REPT("3",COUNTIF($A$3:$A$22,E6*10+3))&REPT("4",COUNTIF
($A$3:$A$22,E6*10+4))&REPT("5",COUNTIF($A$3:$A$22,E6*10+5))&
REPT("6",COUNTIF($A$3:$A$22,E6*10+6))&REPT("7",COUNTIF
($A$3:$A$22,E6*10+7))&REPT("8",COUNTIF($A$3:$A$22,E6*10+8))
&REPT("9",COUNTIF($A$3:$A$22,E6*10+9))
```

Then, we merely copy down through all stems. Next, we must interpret the shape as symmetric or skewed (left or right).

### 3.3.6 Symmetry Issues with Data

We look at these shapes as symmetric or skewed. Symmetric looks like a bell-shaped curve, whereas skewed means that the plot appears lopsided to one side or the other.

### 3.3.7 Displaying Quantitative Data with Histograms

We begin by stating that there is a difference between bar charts and histograms. Bar charts have discrete values as their horizontal axis. Thus, bars are centered at discrete values. A histogram has continuous values as its horizontal axis. Thus, there are no spaces between the bars unless no data are in that range. Since most of the data that you will use are large, we will go quickly to displays with technology.

Steps:

**Step 1.** Obtain descriptive statistics for the data or order the data smaller to larger.

**Step 2.** Determine the Interval [smallest, largest].

**Step 3.** Calculate the class intervals (largest-smallest)/$n$ where $n$ is the number of intervals desired. The value of $n$ must be between

5 and 20. Start with 5 and go up until a good view of the histogram is obtained.

**Step 4.** List the endpoints as Bin values.

**Step 5.** Go to Data Analysis, Histogram and bring up the dialog box. Put data in data input and endpoints in bins.

**Step 6.** The output is a table.

**Step 7.** Highlight the frequencies of the table and go to insert Bar chart.

**Step 8.** Right-click in bar char (on a bar) and close GAP size to 0.

**Step 9.** Comment on the shape regarding symmetry and skewness.

Histograms of data series can be created using the Analysis ToolPak's Histogram tool. Data are grouped into intervals (known as bins) and the number of observations that fall into each are displayed both in a table and, also graphically, as a bar chart. We must edit the bar chart so that the gap width is 0 to be a true histogram and not just a bar chart.

**Example 3.3:** Revisiting Grades

Let us take another look at our grades example (Example 3.2) but this time as a histogram. Using the Data Analysis ToolPak, we open and highlight histogram to get the dialog box below (Figure 3.13).

We have asked the histogram to take the data from $A$1 to $A$21 including labels and the bin range (that we enter by hand by taking the min and max from before and ensuring that there are at least 5 but less than 20 bins to collect the data). We asked for the Chart Output to be placed in cell A25 (Figure 3.14).

This is the output:

The output fills in the Frequency for the counts in the Bins and plots a Bar Chart. To obtain a true histogram, right-click into the histogram to Format Data Series (Figures 3.15 and 3.16).

We now have a histogram (Figure 3.17), and we can analyze the shape. Sometimes if the shape is not clear, we can create more bins to see if we can make it clearer. We next try increments of 5 for the bins.

In this case, we did make the histogram (Figure 3.18) clearer.

We might determine the distribution is skewed to the right.

Below is the output generated by the Histogram tool for the **weight** data using a step-size of 25 instead of 50, as in the previous graph (Figure 3.19).

We must close the gap width to 0 for a histogram (Figure 3.20):

Our Examination of the data appears to be skewed right.

Histogram Summary in Excel

Using the Histogram tool will allow us to make a histogram and create a frequency distribution chart at the same time.

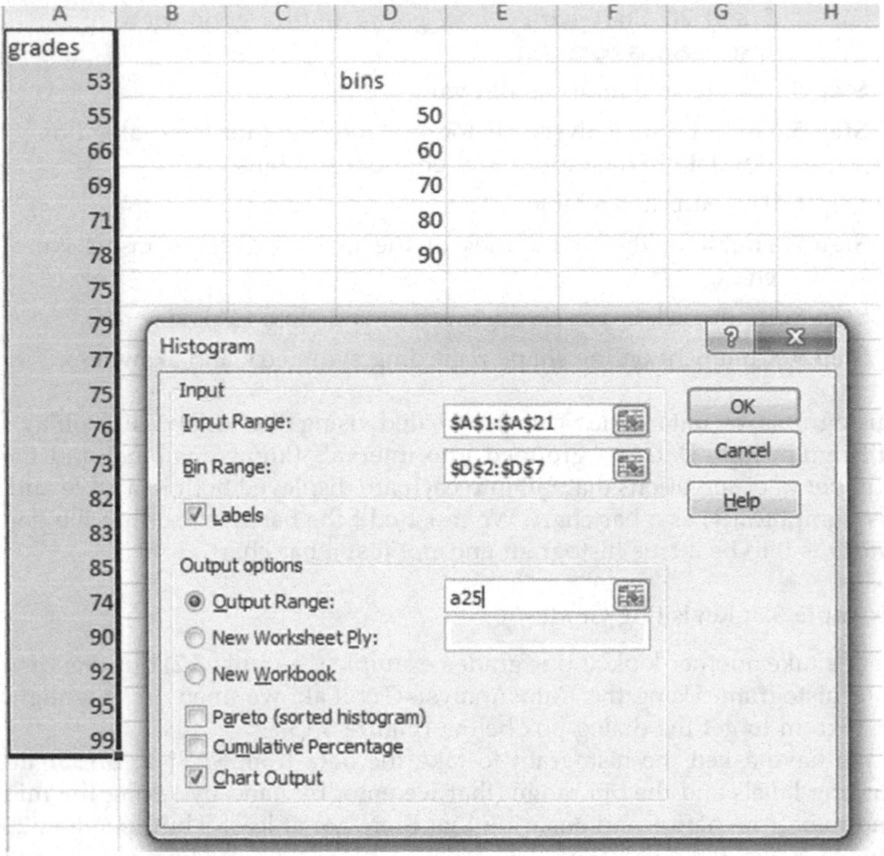

| A | B | C | D | E | F | G | H |
|---|---|---|---|---|---|---|---|
| grades | | | | | | | |
| 53 | | | bins | | | | |
| 55 | | | 50 | | | | |
| 66 | | | 60 | | | | |
| 69 | | | 70 | | | | |
| 71 | | | 80 | | | | |
| 78 | | | 90 | | | | |
| 75 | | | | | | | |
| 79 | | | | | | | |
| 77 | | | | | | | |
| 75 | | | | | | | |
| 76 | | | | | | | |
| 73 | | | | | | | |
| 82 | | | | | | | |
| 83 | | | | | | | |
| 85 | | | | | | | |
| 74 | | | | | | | |
| 90 | | | | | | | |
| 92 | | | | | | | |
| 95 | | | | | | | |
| 99 | | | | | | | |

Histogram

Input

Input Range: $A$1:$A$21

Bin Range: $D$2:$D$7

☑ Labels

Output options

◉ Output Range: a25

○ New Worksheet Ply:

○ New Workbook

☐ Pareto (sorted histogram)

☐ Cumulative Percentage

☑ Chart Output

OK

Cancel

Help

**FIGURE 3.13**
Excel screen shot of Example 3.2.

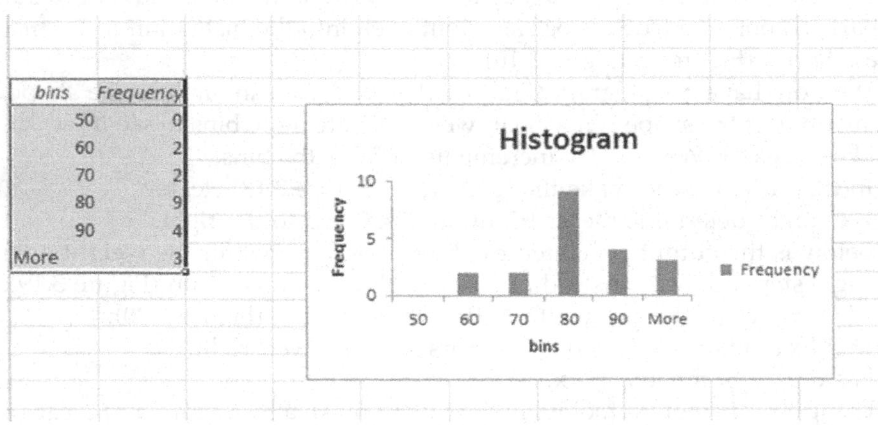

| bins | Frequency |
|---|---|
| 50 | 0 |
| 60 | 2 |
| 70 | 2 |
| 80 | 9 |
| 90 | 4 |
| More | 3 |



**FIGURE 3.14**
Excel screenshot of Example 3.2 histogram.

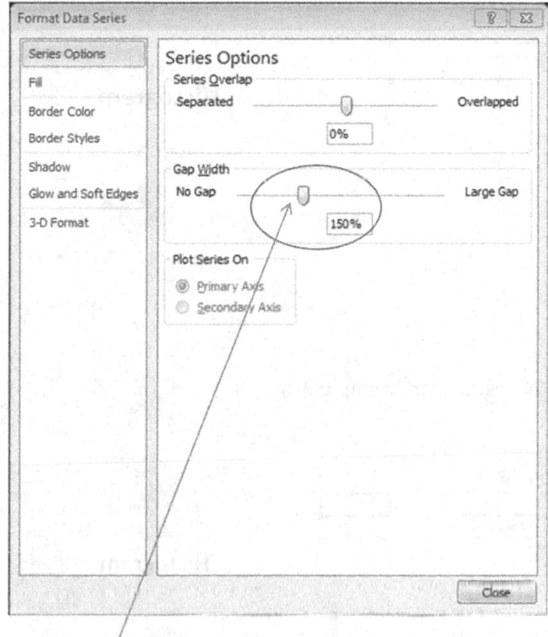


**FIGURE 3.15**
Excel format data series command box.

**FIGURE 3.16**
Excel screenshot – reducing the gap to 0.

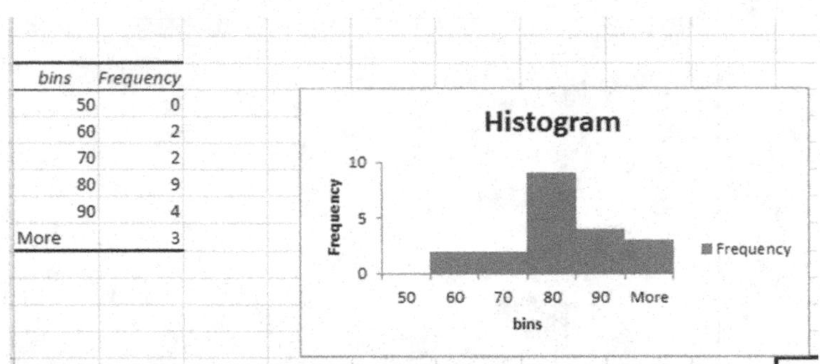

| bins | Frequency |
|---|---|
| 50 | 0 |
| 60 | 2 |
| 70 | 2 |
| 80 | 9 |
| 90 | 4 |
| More | 3 |



**FIGURE 3.17**
Excel Screenshot of histogram for Example 3.2.

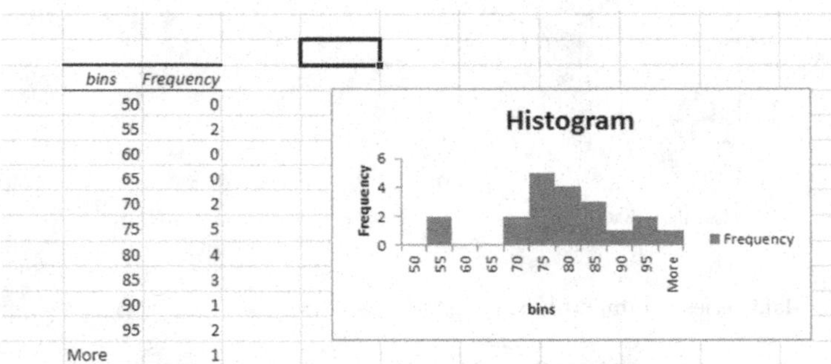

| bins | Frequency |
|---|---|
| 50 | 0 |
| 55 | 2 |
| 60 | 0 |
| 65 | 0 |
| 70 | 2 |
| 75 | 5 |
| 80 | 4 |
| 85 | 3 |
| 90 | 1 |
| 95 | 2 |
| More | 1 |



**FIGURE 3.18**
Excel screenshot of histogram.

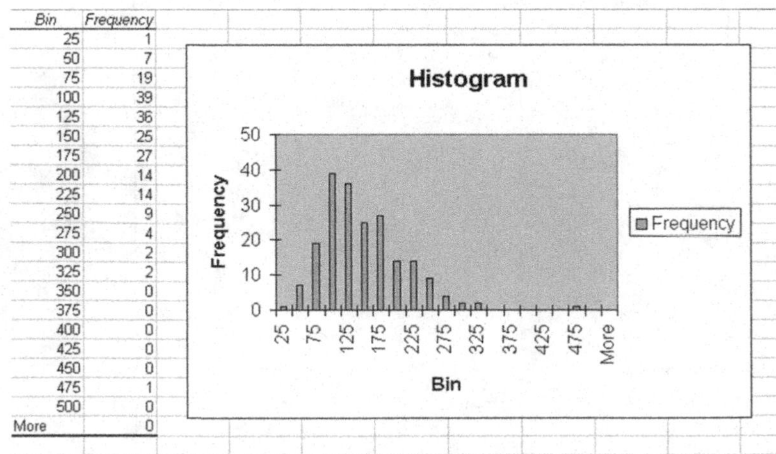

| Bin | Frequency |
|---|---|
| 25 | 1 |
| 50 | 7 |
| 75 | 19 |
| 100 | 39 |
| 125 | 36 |
| 150 | 25 |
| 175 | 27 |
| 200 | 14 |
| 225 | 14 |
| 250 | 9 |
| 275 | 4 |
| 300 | 2 |
| 325 | 2 |
| 350 | 0 |
| 375 | 0 |
| 400 | 0 |
| 425 | 0 |
| 450 | 0 |
| 475 | 1 |
| 500 | 0 |
| More | 0 |



**FIGURE 3.19**
Excel screenshot of histogram of weight data.

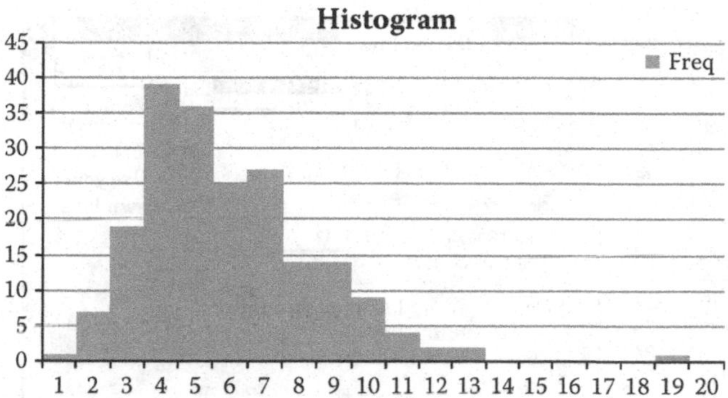


**FIGURE 3.20**
Skewed right histogram.

1. Click Tools > Data Analysis.
2. In the Data Analysis window, select Histogram and click OK.
3. A new window titled Histogram should appear. This window has many options. Below is a brief explanation of each:
   - Input Range is where the data being used to create the histogram go. Simply put your cursor back into the spreadsheet and highlight the variable name and all the data in that column.
   - Click Labels. If a variable name was highlighted in the Input Range, then this needs to be checked.
   - You must select one of the following Output options:
     - Click Output Range if you want the histogram to be placed on the current sheet. Next, simply input the cell where you want the output to be placed.
     - Click New Worksheet Ply if you want the histogram to be placed on a new sheet. Next, type the name of the new sheet where you want the output to be placed.
   - Clicking Cumulative Percentage will list the cumulative percentage for each class and include a cumulative percentage line on your histogram.
   - Click Chart Output under Output options. This step is ***necessary*** to obtain the histogram. If this is not highlighted, you will only receive a frequency distribution chart.
4. Click OK. The histogram and frequency distribution chart should be placed onto your spreadsheet (Figure 3.21).

We will present the information on how to construct a histogram using Excel.

5. Right-click in the chart to remove the gap width.

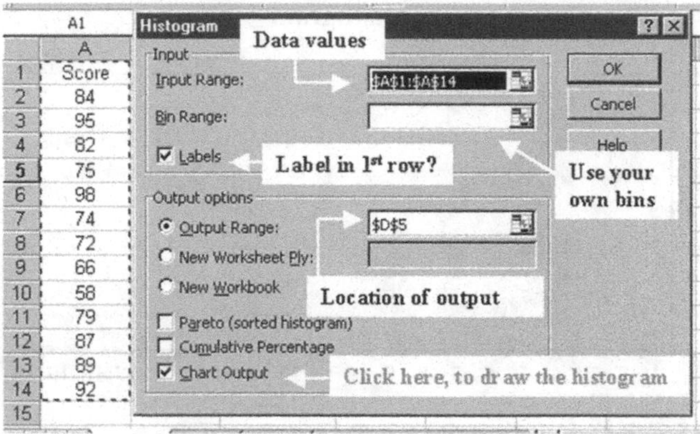


**FIGURE 3.21**
Excel histogram dialogue box.

The following process allows us to build a reasonable histogram of our data.

**Histogram:**

**Step 1.** Determine and select the classes, 5–15 classes. Find the range (lowest to highest value). Classes should be evenly spaced if possible.
**Step 2.** Tally the data in the classes.
**Step 3.** Find the numerical (relative) frequencies from the tallies.
**Step 4.** Find the cumulative frequencies.

*Histogram*: connects class interval as a base and tallies (or relative frequencies) as the height of a rectangle. Rectangle is centered at the mid-point of class interval.

53, 55, 66, 69, 71, 73, 74, 75, 75, 76, 77, 78,79, 82, 83, 85, 90, 92, 95, 99

Possible class intervals:

a. Classes 51–60, 61–70, 71–80, 81–90, 91–100 (5 classes intervals).
b. Classes 50–59, 60–69, 70–79, 80–89, 90–99, 100–109 (6 classes intervals).
c. Classes 51–55, 56–60, 61–65, 66–70, 71–75, 76–80, 81–85, 86–90, 91–95, 96–100 (10 class intervals).

**TABLE 3.8**
Interval Frequency

| Interval | Count | Percentage |
|---|---|---|
| 51–60 | 2 | 0.10 |
| 61–70 | 2 | 0.10 |
| 71–80 | 9 | 0.45 |
| 81–90 | 4 | 0.20 |
| 91–100 | 3 | 0.15 |
| Total | 20 | |

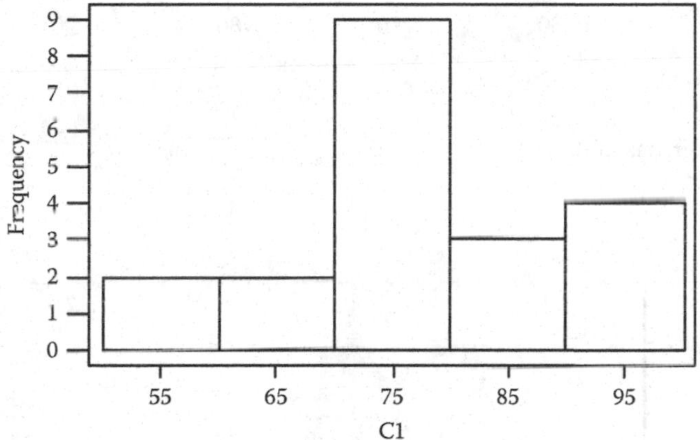


**FIGURE 3.22**
Histogram of data.

We can now look at interval selection (a) (Table 3.8 and Figure 3.22).
We note the data are somewhat symmetric.

### 3.3.7.1 Histogram in R

The syntax command to generate the histogram (Figure 3.23) is

```
>hist(grades)
```

### 3.3.7.2 Histogram in Maple

The syntax command to generate the histogram (Figure 3.24) is
with(Statistics);

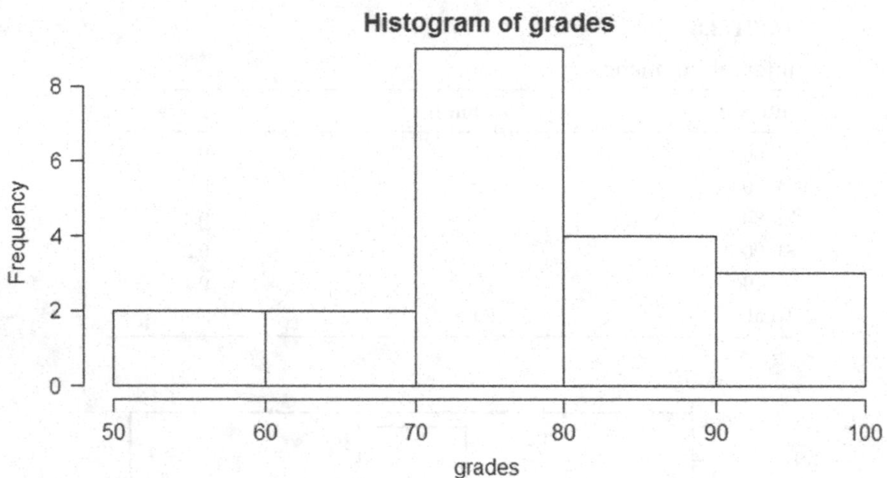


**FIGURE 3.23**
Histogram of grades in R.

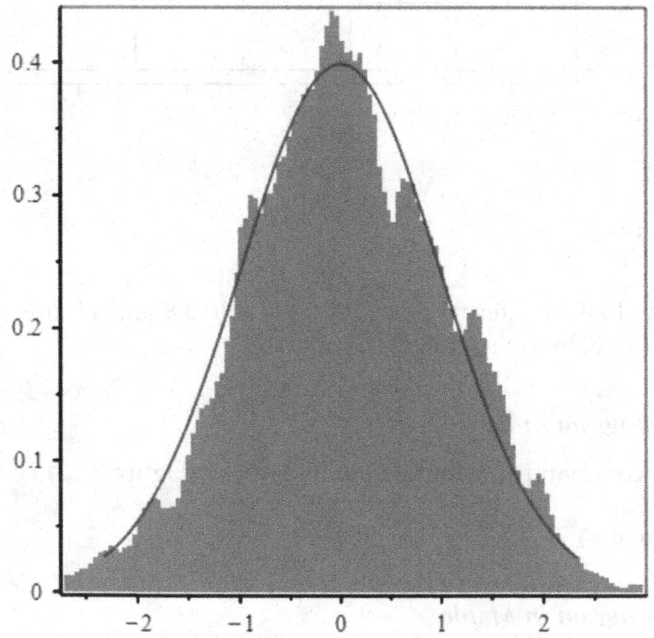


**FIGURE 3.24**
Example histogram of grades in Maple.

```
A := Sample(Normal(0, 1), 1000);
P := DensityPlot(Normal(0, 1), color = "Niagara Red");
Q := Histogram(A, averageshifted = 4, style = polygon, color =
    "LightSlateGrey");
```

By default, frequencyscale is set to relative.

```
plots[display](P, Q);
```

An Ordinary histogram (Figure 3.24).

```
Histogram(A);
```

Varying bin width histogram. In this case, each bar has approximately the same area (Figure 3.25).

We can update our histogram to better account for the proportions of our data (Figure 3.26).

```
Histogram(A, binbounds = proportional);
```

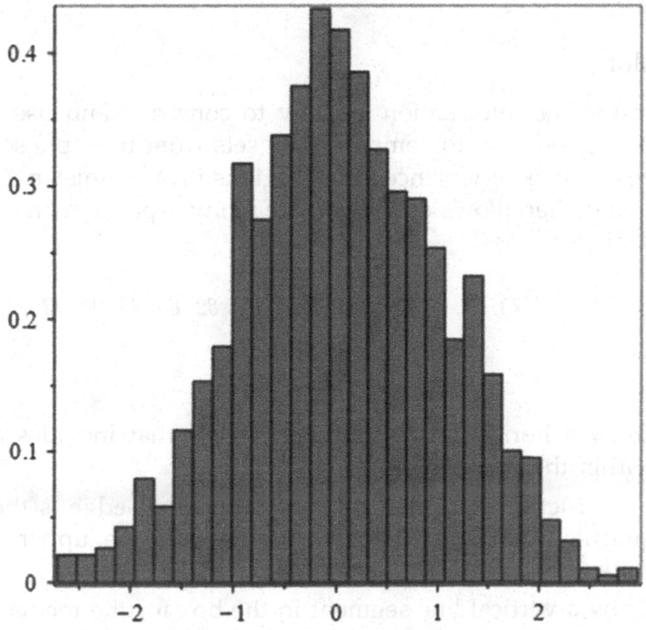


**FIGURE 3.25**
Ordinary histogram in Maple.

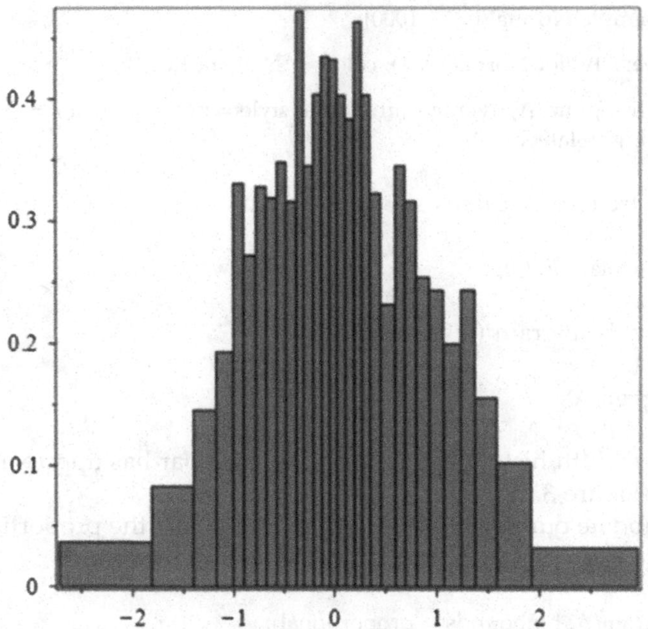


**FIGURE 3.26**
Example histogram in Maple with modified bin width.

### 3.3.8 Boxplot

We will present the information on how to construct and use a boxplot. Boxplots are a good way to compare data sets from multiple sources. For example, we can look at violence in ten regions in Afghanistan. Putting the ten boxplots together allows us to compare many aspects, such as medians, ranges, and dispersions.

*53, 55, 66, 69, 71, 73, 74, 75, 75, 76, 77, 78, 79, 82, 83, 85, 90, 92, 95, 99*

Boxplot

**Step 1**. Draw a horizontal measurement scale that includes all data within the range of data.

**Step 2**. Construct a rectangle (the box) whose left edge is the lower quartile value and whose right edge is the upper quartile value.

**Step 3**. Draw a vertical line segment in the box for the median value.

**Step 4**. Extend line segments from rectangle to the smallest and largest data values (these are called whiskers).

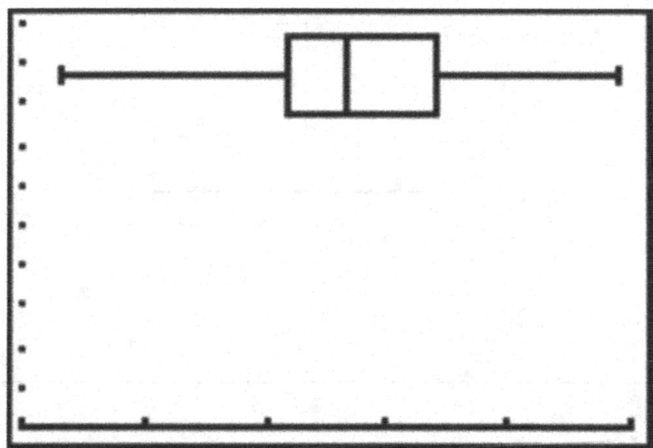

**FIGURE 3.27**
Example boxplot.

The values are in numerical order. What is needed are the range, the quartiles, and the median.

Range is the smallest and largest values from the data: 53 and 99.

The median is the middle value. It is the average of the 10th and 11th values, as we will see later: *(76 + 77)/ 2 = 76.5*

The quartiles values are the median of the lower and upper half of the data.

Lower quartile values: *53, 55, 66, 69, 71, 73, 74, 75, 75, 76*. Its median is 72.

Upper quartile values: *77, 78,79, 82, 83, 85, 90, 92, 95, 99*. Its median is 84.

You draw a rectangle from 72 to 84 with a vertical line at 76.5.

Then, draw a whisker to the left to 53 and to the right to 99.

It would look something like this (Figure 3.27):

We state that obtaining boxplots in R is much easier than in Excel.

#### *3.3.8.1 Boxplot in R*

The command to generate the boxplot (Figure 3.28) in R is

```
>boxplot(grades)
```

#### *3.3.8.2 Comparisons with Boxplot (Side by Side)*

Consider our data for casualties in Afghanistan through the years 2002–2009 (Figure 3.29). This is presented to you as a commander. What information is this telling you?

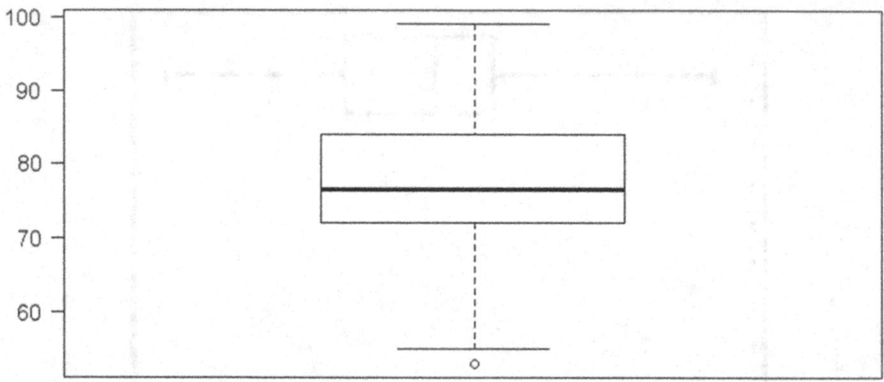


**FIGURE 3.28**
Example boxplot in R.

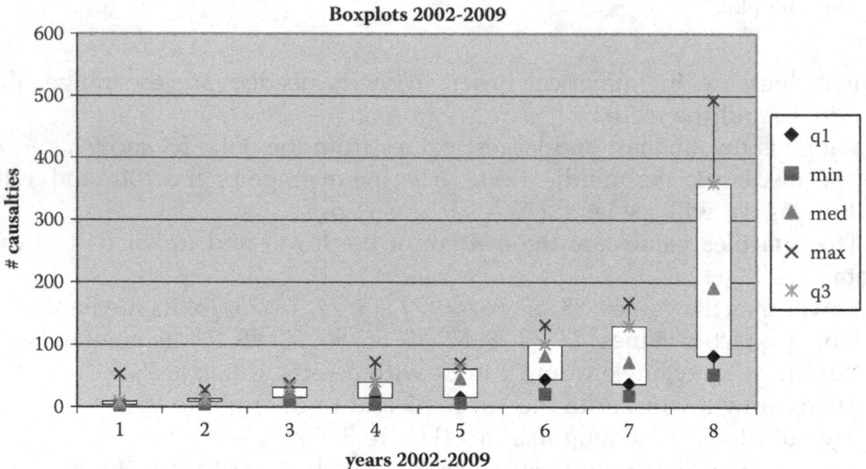


**FIGURE 3.29**
Example boxplot comparison of Afghanistan data.

### *3.3.8.3 Boxplots in Maple*

Plot options such as title are passed to the plots (Figure 3.30): -display command:

Enter the data in a matrix; call it M.

M := Matrix(A, scan = columns)%T;

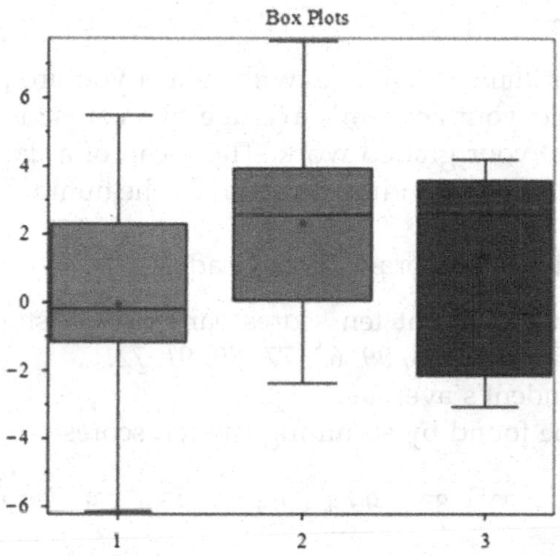


**FIGURE 3.30**
Example boxplot in Maple.

ADD MATRIX

```
BoxPlot(M, color = "Niagara Green", deciles = false, title = "Box Plots");
```

## 3.4 Statistical Measures

Statistical measures are simply a collection of techniques and processes dealing with the analysis, interpretations, and presentation of information or data that have numerical meaning. The most common statistical measures we use to describe our data include the mean, mode, median, variance, and standard deviation. We address several of these measures in this section.

### 3.4.1 Central Tendency or Location

#### *3.4.1.1 Describing the Data*

In addition to plots and tables, numerical descriptors are often used to summarize data. Three numerical descriptors, the *mean*, the *median*, and the *mode*, offer different ways to describe and compare data sets. These are generally known as the *measures of location*.

#### 3.4.1.2 The Mean

The mean is the arithmetic average, with which you are probably very familiar. For example, your academic average in a course is generally the arithmetic average of your graded work. The mean of a data set is found by summing all the data and dividing this sum by the number of data elements.

**Example 3.4:** Updated Algebra Course Grades

The following data represent ten scores earned by a student in a college algebra course: *55, 75, 92, 83, 99, 62, 77, 89, 91, 72.*

Compute the student's average.

The mean can be found by summing the ten scores.

$$55 + 75 + 92 + 83 + 99 + 62 + 77 + 89 + 91 + 72 = 795$$

and then dividing by the number of data elements (10), $795/10 = 79.5$

To describe this process in general, we can represent each data element by a letter with a numerical subscript. Thus, for a class of $n$ tests, the scores can be represented by $a_1, a_2, \ldots, a_n$. The mean of these $n$ values of $a_1, a_2, \ldots, a_n$ is found by adding these values and then dividing this sum by $n$, the number of values. The Greek letter Σ (called sigma) is used to represent the sum of all the terms in a certain group. Thus, we may see this written as

$$\sum_{i=1}^{n} a_i = a_1 + a_2 + \ldots + a_n \qquad \text{mean} = \bar{x} = \frac{\sum_{i=1}^{n} a_i}{n}$$

Think of the mean as the average. Notice that the mean does not have to equal any specific value of the original data set. The mean value of 79.5 was not a score ever earned by our student.

Batting average is the total number of hits divided by the total number of official at-bats. Is batting average a mean? Explain.

#### 3.4.1.3 The Median

The median locates the true middle of a numerically ordered list. The hint here is that you need to make sure that your data are in numerical order listed from smallest to largest along the $x$ number line. There are two ways to find the median (or middle value of an ordered list) depending on $n$ (the number of data elements):

1. If there is an odd number of data elements, then the middle (median) is the exact data element that is the middle value. For example, here are five ordered math grades earned by a student: *55, 63, 76, 84, 88.*

2. The middle value is 76 since there are exactly two scores on each side (lower and higher) of 76. Notice that with an odd number of values that the median is a real data element.
3. If there is an even number of data elements, then there is no true middle value within the data itself. In this case, we need to find the mean of the two middle numbers in the ordered list. This value, probably not a value of the data set, is reported as the median. We can illustrate this with several examples.
   a Here are six math scores for student a: *56, 62, 75, 77, 82, 85*
   The middle two scores are 75 and 77, because there are exactly two scores below 75 and exactly two scores above 77. We average 75 and 77. *(75+77)/2 = 152/2 = 76*, so
   76 is the median. Note that 76 is not one of the original data values.
   b Here are eight scores for student *b*: *72, 80, 81, 84, 84, 87, 88, 89*
   The middle two scores are 84 and 84, because there are exactly three scores lower than 84 and three scores higher than 84. The average of these two scores is 84. Note that this median is one of our data elements.
   It is also very possible for the mean to be equal to the median.

#### *3.4.1.4 The Mode*

The value that occurs the most often is called the mode. It is one of the numbers in our original data. The mode is found by collecting and organizing the data in order to count the frequency of each result. The result with the highest occurrences is the mode of the set. The mode does not have to be unique. It is possible for there to be more than one mode in a data set. As a matter of fact, if every data element is different from the other data elements, then every element is a mode.

For example, consider the data scores for a student in our mathematics class.

*75, 80, 80, 80, 80, 85, 85, 90, 90, 100*

Table 3.9 provides the number of occurrences for each value.

**TABLE 3.9**

Occurrence of Grades in the Course

| Value | Number of Occurrences |
|---|---|
| 75 | 1 |
| 80 | 4 |
| 85 | 2 |
| 90 | 2 |
| 100 | 1 |

Since 80 occurred four times and that is the largest value among the number of occurrences, then 80 is the mode.

### 3.4.2 Measures of Dispersion

#### *3.4.2.1 Variance and Standard Deviation*

*Measures of variation* or *measures of the spread* of the data include the variance and standard deviation. They measure the spread in the data, how far the data are from the mean. Variance is simply a measure of the dispersion of a set of data points around their mean value. Variance is a mathematical expectation of the average squared deviations from the mean.

The sample variance has notation $S^2$ and the sample deviation has notation $S$.

$$s^2 = \frac{\sum_{i=1}^{n} (x_i - \bar{x})^2}{n - 1}$$

where $n$ is the number of data elements.

$$s = \sqrt{s^2}$$

**Example 3.5:** Variance

Consider the following ten data elements:

50, 54, 59, 63, 65, 68, 69, 72, 90, 90.

The mean, $\bar{x}$, is 68. The variance is found by subtracting the mean, 68, from each point, squaring them, adding them up, and dividing by $n$–1.

$$\begin{aligned} S^2 &= [(50-68)^2 + (54-68)^2 + (59-68)^2 + (63-68)^2 + (65-68)^2 \\ &\quad + (68-68)^2 + (69-68)^2 + (72-68)^2 + (90-68)^2 + (90-68)^2]/9 \\ &= 180 \end{aligned}$$

$$s = \sqrt{s^2} = 13.42.$$

**Example 3.6:** Metabolic Rate Variance

Consider a person's metabolic rate at which the body consumes energy. Here are seven metabolic rates for men who took part in a study of dieting. The units are calories in a 24-hour period.

**TABLE 3.10**
Deviations of the Observations

| Observations | Deviations | Squared Deviations |
|---|---|---|
| $X_i$ | $x_i - \bar{x}$ | $(x_i - \bar{x})^2$ |
| 1,792 | 1,792 – 1,600 = 192 | 36,864 |
| 1,666 | 1,666 – 1,600 = 66 | 4,356 |
| 1,362 | 1,362 – 1,600 = –238 | 56,644 |
| 1,614 | 1,614 – 1,600 = 14 | 196 |
| 1,460 | 1,460 – 1,600 = –140 | 19,600 |
| 1,867 | 1,867 – 1,600 = 267 | 71,289 |
| 1,439 | 1,439 – 1,600 = –161 | 25,921 |
| | Sum = 0 | Sum = 214,870 |

1792 1666 1362 1614 1460 1867 1439

The researchers reported both $\bar{x}$ and S for these men.
The mean.

$$\bar{x} = \frac{1792 + 1666 + 1362 + 1614 + 1460 + 1867 + 1439}{7} = \frac{11{,}200}{7} = 1600$$

To clearly see the nature of the variance, start with a table of the deviations of the observations from the mean (Table 3.10).

The variance, $s^2 = 214{,}870/6 = 35{,}811.67$

The standard deviation, $s = \sqrt{35{,}811.67} = 189.24$

Some properties of the standard deviation are:

- S measures spread about the mean.
- S = *0* only when there is no spread.
- S is strongly influenced by extreme outliers.

### 3.4.3 Measures of Symmetry and Skewness

We define a measure, the coefficient of skewness, $S_k$. Mathematically, we determine this value from formula:

$$s_k = \frac{3(\bar{x} - \tilde{x})}{s}$$

We use the following rules for skewness and symmetry.

If Sk ≈ 0, the data are symmetric.

If Sk > 0 the data are positively skewed (skewed right).

If Sk < 0, the data are negatively skewed (skewed left).

We use the bell-shaped curve to denote symmetry. Figure 3.31 illustrates the concept of Skewness:

**Range** is a measure that takes the maximum and minimum values of the data. Often, this is provided a single number. Assume we have the following data (Table 3.11):

The maximum value is 1,867, and the minimum value is 1,362. If you take the difference, 1,867 – 1,362 = 505. What does 505 represent? I suggest you give the range as an interval [1,362, 1,867].

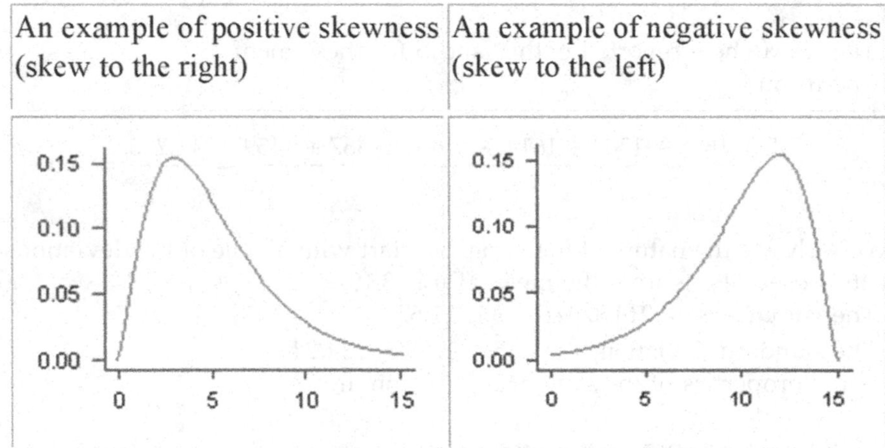


**FIGURE 3.31**
Positive and negative skewness.

**TABLE 3.11**
Range Data

| | | | | | |
|---|---|---|---|---|---|
| 1,792 | 80.0 | 103.0 | 116.1 | 112.3 | 120.0 |
| 1,666 | 72.5 | 95.5 | 100.6 | 96.7 | 104.4 |
| 1,362 | 65.0 | 88.0 | 85.0 | 81.1 | |
| 1,614 | 57.5 | 80.5 | 69.4 | 65.6 | |
| 1,460 | 50.0 | 73.0 | 53.9 | 50.0 | |
| 1,867 | | | | | |
| 1,439 | | | | | |

**TABLE 3.12**

Summary Statistics Data

| | | | | | |
|---|---|---|---|---|---|
| 91.5 | 80.0 | 103.0 | 116.1 | 112.3 | 120.0 |
| 84.0 | 72.5 | 95.5 | 100.6 | 96.7 | 104.4 |
| 76.5 | 65.0 | 88.0 | 85.0 | 81.1 | |
| 69.0 | 57.5 | 80.5 | 69.4 | 65.6 | |
| 61.5 | 50.0 | 73.0 | 53.9 | 50.0 | |

#### *3.4.3.1 Summary of Descriptive Measures with Excel*

We can obtain all this information quickly in Excel. Given a list of data, we can use the Data Analysis package, Descriptive Statistics, Summary Statistics option.

We first enter the "sample" data (Table 3.12) in a column in Excel (see Figure 3.32).

Then, go to **Data→Data Analysis** (Figure 3.33):

Then, highlight **Descriptive Statistics** (Figure 3.34):

Press OK; then fill in the **dialog box** and check the **summary statistics box** (Figure 3.35):

Press OK; then you will receive the descriptive statistics output (Figure 3.36):

The extracted (Figure 3.37) descriptive statistics are:

The histogram (Figure 3.38) appears somewhat symmetric. The values of the data's descriptive statsitcs support this. Now let's sumamrize the interpretations of the statsitics.

#### *3.4.3.2 Descriptive Statistics with R*

Descriptive statistics are usually: mean, median, mode, quartiles, variance, standard deviation, coefficient of skewness, and range.

We note that R does not easily provide the mode, so we suggest looking at the stem and leaf plot and seeing which elements are repeated the greatest number of times.

```
6 | 02
6 | 9
7 | 0011122234
7 | 778
```

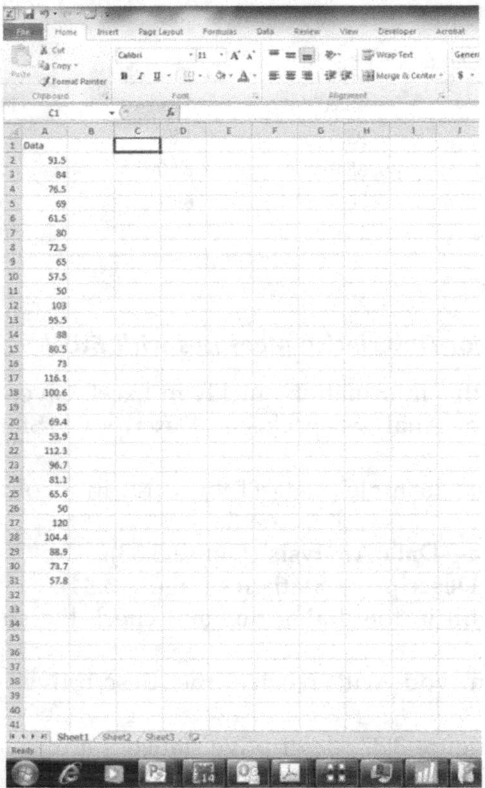


**FIGURE 3.32**
Excel screenshot of summary statistics data.

We see 71 and 72 each repeated three times. Therefore, we have two modes 71 and 72.

In R, some statistics can be obtained by the command >summary(name).

```
> summary(Height)
Min. 1st Qu. Median Mean 3rd Qu. Max.
60.00 70.00 71.50 71.16 73.25 78.00
```

Notice the information provided here.

Other commands are shown:

```
> mean(Height)
[1] 71.15625
```

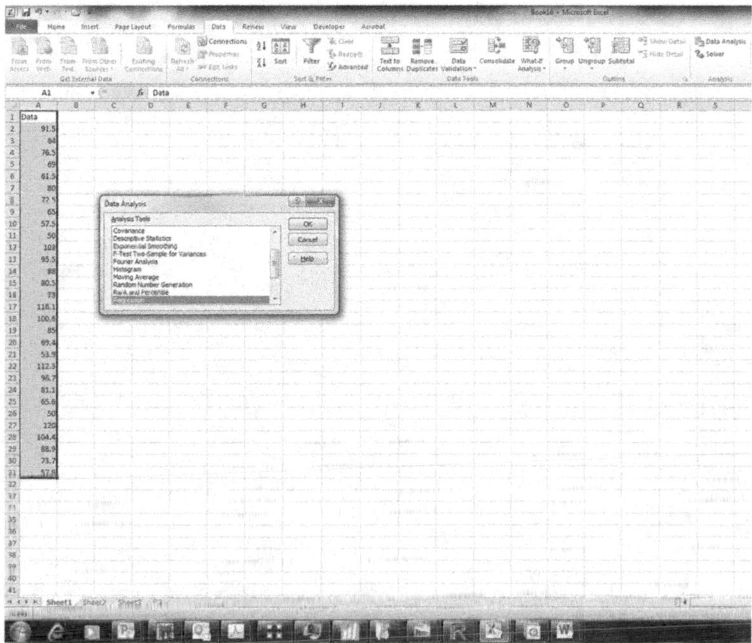

**FIGURE 3.33**
Excel screenshot of data analysis dialogue box.

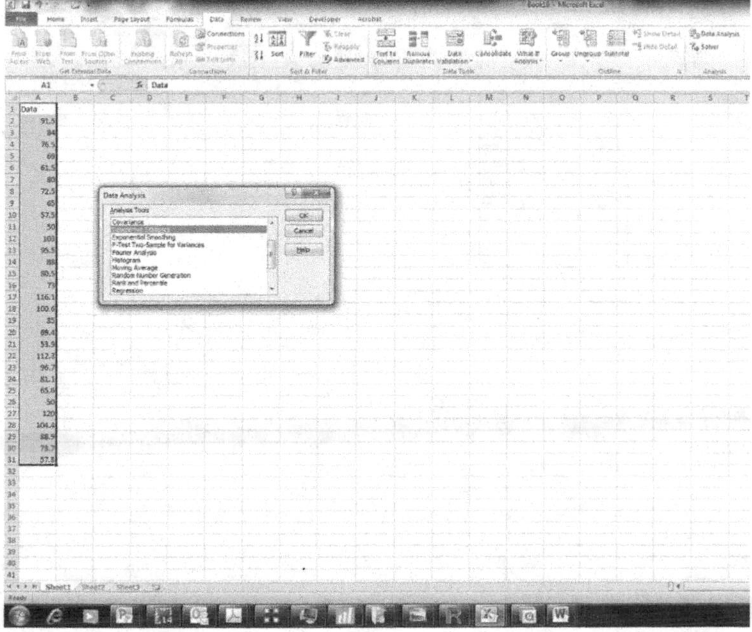

**FIGURE 3.34**
Descriptive statistics data analysis dialogue box.

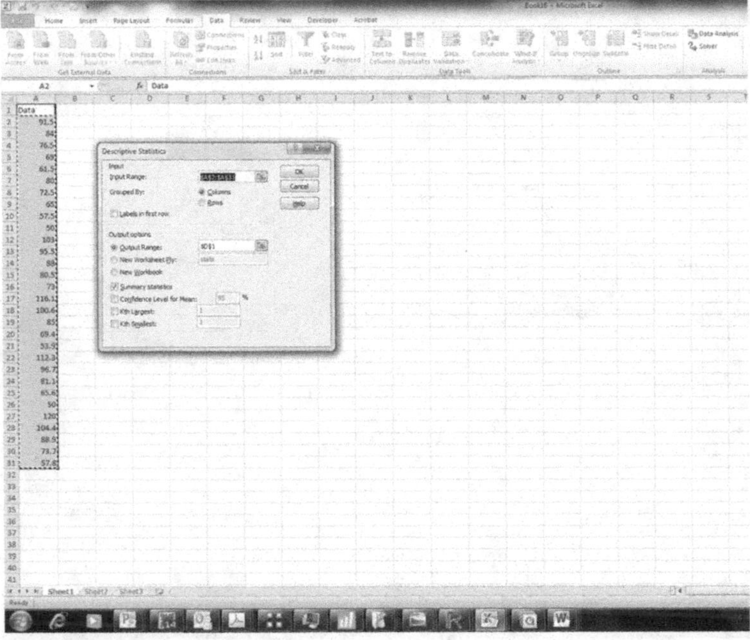

FIGURE 3.35
Descriptive dialog box.

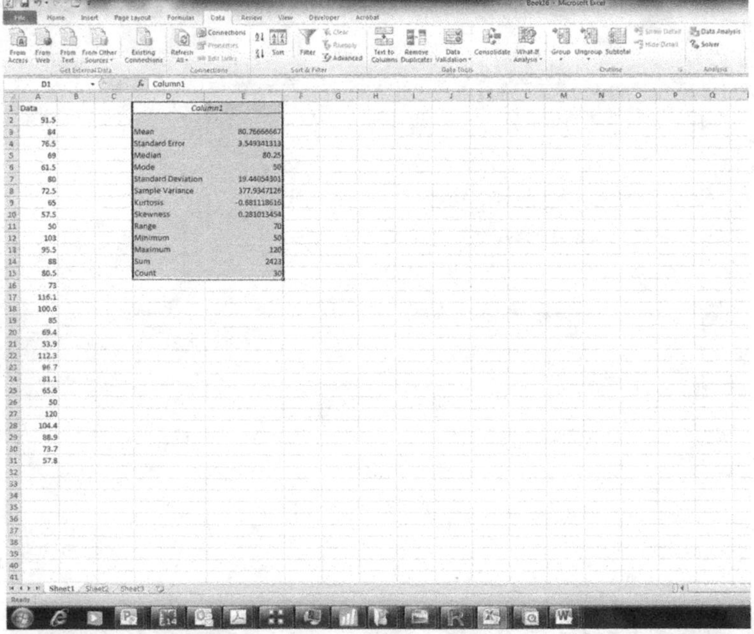

FIGURE 3.36
Solution output of statistics summary.

| *Sample Data* | |
|---|---|
| | |
| Mean | 81.57777778 |
| Standard Error | 3.830226661 |
| Median | 80.5 |
| Mode | 50 |
| Standard Deviation | 19.90244155 |
| Sample Variance | 396.1071795 |
| Kurtosis | -0.76095724 |
| Skewness | 0.225189193 |
| Range | 70 |
| Minimum | 50 |
| Maximum | 120 |
| Sum | 2202.6 |
| Count | 27 |

**FIGURE 3.37**
Descriptive statistics summary.

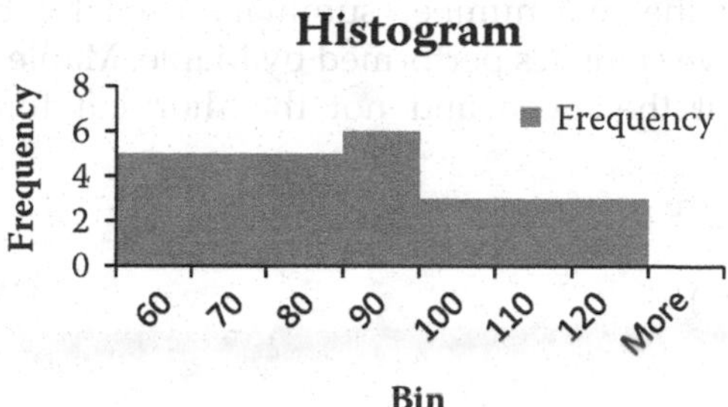


**FIGURE 3.38**
Histogram of statistics summary data.

```
> median(Height)
[1] 71.5
> var(Height)
[1] 22.99063
> sd(Height)
[1] 4.794854
```

Now, how about the range:

Go back to summary and write the Max-Min (don't subtract) as 78–60.

Finding and interpreting the coefficient of skewness.

This requires the formula:

```
s_k = (3*(mean-median))/(standard deviation)
> sk<-(3*(mean(Height)-median(Height))/sd(Height))
> sk
[1] -0.2150743
>
```

The negative value says that mathematically (with this shortcut formula) the data are skewed left.

#### *3.4.3.3 Descriptive Statistics in Maple*

Enter the data into Maple. Our example uses data_set as the data.

We obtain both the five number summary used for boxplots and the complete descriptive statistics performed by Maple. Maple's skewness is the 3rd moment about the mean and not the short-cut formula mentioned previously.

The commands are

```
>with(Statistics):
```

```
> data_set := [91.5, 84, 76.5, 69, 61.5, 80, 72.5, 65, 57.5, 50, 103, 95.5, 88, 80.5, 73, 116.1,
      100.6, 85, 69.4, 52.9, 112.3, 96.7, 81.1, 65.6, 50, 120, 104.4, 88.9, 73.7, 57.8];
data_set := [91.5, 84, 76.5, 69, 61.5, 80, 72.5, 65, 57.5, 50, 103, 95.5, 88, 80.5, 73, 116.1, 100.6,    (7)
    85, 69.4, 52.9, 112.3, 96.7, 81.1, 65.6, 50, 120, 104.4, 88.9, 73.7, 57.8]
> FivePointSummary(data_set)
                  minimum = 50.
            lowerhinge = 65.600000000000
               median = 80.250000000000                                                             (8)
            upperhinge = 95.500000000000
                  maximum = 120.
Compute the mean, standard deviation, skewness, kurtosis, etc.
> DataSummary(data_set)
                  mean = 80.7333333333333
     standarddeviation = 19.4889948314486
             skewness = 0.254991424516107
              kurtosis = 2.15915701871294                                                           (9)
                  minimum = 50.
                  maximum = 120.
            cumulativeweight = 30.
>
```

The command to find the mode is

>Mode(data_set)

## 3.5 Exercises

### 3.5.1 Basic Statistics

Determine whether the following variables would be quantitative or categorical. Provide an example of the value of such a variable and include the units, if any exist.

1. Flip a penny that lands as "heads" or "tails".
2. The color of M & M's.
3. The number of calories in the local fast-food selections.
4. The life expectancy for males in the United States.
5. The life expectancy for females in the United States.
6. The number of babies born on New Year's Eve.
7. The dollars spent each month out of the allocated supply budget.

**TABLE 3.13**

Sports Injuries in 2001

| Sport | Injuries | Participants | Sport | Injuries | Participants |
|---|---|---|---|---|---|
| Basketball | 646,678 | 26,200,000 | Fishing | 84,115 | 47,000,000 |
| Bicycling | 600,649 | 54,000,000 | Skateboard | 56,435 | 8,000,000 |
| Baseball | 459,542 | 36,100,000 | Hockey | 54,601 | 1,800,000 |
| Football | 453,684 | 13,300,000 | Golf | 38,626 | 24,700,000 |
| Soccer | 150,449 | 10,000,000 | Tennis | 29,936 | 16,700,000 |
| Swimming | 130,362 | 66,200,000 | Water skiing | 26,663 | 9,000,000 |
| Weightlifting | 86,398 | 39,200,000 | Bowling | 25,417 | 40,400,000 |

8. The number of hours that a soldier works per week.
9. The amount of car insurance paid per year.
10. Whether the bride is older, younger, or the same age as the groom.
11. The difference in ages of a couple at a wedding.
12. The following table represents the numbers of sports-related injuries treated in U.S. hospital emergency rooms in 2001, along with an estimate of the number of participants in that sport (Table 3.13).
    a. If we want to use the number of injuries as a measure of the hazardousness of a sport, which sport is more hazardous between bicycling and football? Between soccer and hockey?
    b. Use either a calculator or a computer to calculate the *rate* of injuries per thousand participants. *Rate* is defined as the average number of injuries out of the total participants.
    c. Rank-order this new measure for the sports.
    d. How do your answers in part (a) Compare if we do the hazardous analysis using the *rates* in (b)? If different, why are the results different?
13. Make a stem and leaf, histogram, and boxplot of the following data sets and comment about the shape.
    a. 100, 105, 111, 115, 121, 129, 131, 131, 133, 135, 137, 145, 146, 150, 160, 180
    b. 0.10, 0.15, 0.22, 0.23, 0.50, 0.62, 0.62, 0.65, 0.66, 0.69, 0.72
    c. 63, 65, 72, 81, 83, 85, 92, 93, 94, 105, 106, 121, 135
14. Make a pie chart for the following data (Table 3.14):
    What information is best displayed by a pie chart?
15. Make a bar chart for this data: female doctorates as a percent of graduates in that field that were females (Table 3.15):
    Can you make a pie chart? What do you have to do first?

**TABLE 3.14**

U.S. Material Status

| Marital Status | Count (in millions) |
|---|---|
| Never Married | 43.9 |
| Married | 116.7 |
| Widowed | 13.4 |
| Divorced | 17.6 |

**TABLE 3.15**

Percent of Female Doctorates

| | |
|---|---|
| Computer Science | 15.40% |
| Education | 60.80% |
| Engineering | 11.10% |
| Life Science | 40.70% |
| Physical Sciences | 21.70% |
| Psychology | 62.20% |
| Mathematics | 10.00% |

16. Display the following data: In 1995, there were 90,402 deaths from accidents in the United States. Among these, there were 43,363 from motor vehicles, 10,483 from falls, 9,072 from poisoning, 4,350 from drowning, and 4,235 from fires. How many deaths were due to other unknown causes?
    a. In Math I class last semester, the final averages were: 88, 63, 82, 98, 89, 72, 86, and 70. Display the data as a stem and leaf. Are they symmetric? Are they skewed?
    b. In Math II class last semester, the final averages were: 66, 61, 78, 54, 75, 40, 78, 91, 84, 82, 76, and 65. Display the data as a histogram. Are they symmetric? Are they skewed?
    c. Using the grades in Math I (#4) and Math II (#5), display the data as two boxplots side by side. Is each display symmetric? Is each display skewed? Can you compare the two data sets? Which class had the higher grades? Which class has grades that are more spread out? Does the symmetry and skewness of the data sets tell us anything about the grades?

### 3.5.2 Statistical Measures

1. The 1994 live birth rates per thousand population in the mountain states of Idaho, Montana, Wyoming, Colorado, New Mexico, Arizona, Utah, and Nevada were 12.9, 15.5, 13.5, 14.8, 16.7, 17.4, 20.1, and 16.4, respectively. What is the mean, variance, and standard deviation?
2. In five attempts, it took a person 11, 15, 12, 8, and 14 minutes to change a tire on a car. What is the mean, variance, and standard deviation?
3. A soldier is sent to the range to test a new bullet that the manufacturer says is very accurate. You send your best shooter with his weapon. He fires 10 shots with each using the standard ammunition and then the new ammunition. We measure the distance from the bull's eye to each shot location. Which appears to the better ammunition? Explain.
   Standard Ammunition: −3, −3, −1, 0, 0, 0, 1, 1, 1, 2
   New Ammunition −2, −1, 0, 0, 0, 0, 1, 1, 1, 2
4. AGCT Scores: AGCT-score
   AGCT stands for Army General Classification Test. These scores have a mean of 100, with a standard deviation of 20.0. Here are the AGCT scores for a unit:
   79, 100, 99, 83, 92, 110, 149, 109, 95, 126, 101, 101, 91, 71, 93, 103, 134, 141, 76, 108, 122, 111, 97, 94, 90, 112, 106, 113, 114, 117
   Find the mean, median, mode, standard deviation, variance, and coefficient of skewness for the data. Provide a brief summary to your S-1 about this data.
5. The following table shows automotive tire sales from 1966 to 1994. Set x to be the year (i.e. 1966) and y to be the tire sales in thousands of dollars (Table 3.16).

Make a scatterplot. What's the pattern here?

**TABLE 3.16**
Tire Sales in the United States

| Year | Tire Sales (in $1000) |
|---|---|
| 1966 | 33 |
| 1970 | 38.4 |
| 1974 | 64 |
| 1978 | 107 |
| 1982 | 179 |
| 1986 | 299 |
| 1990 | 499 |
| 1994 | 833 |

## References

Strunk, W., and E.B. White (1999). *The Elements of Style, 4th Edition.* Pearson, New York: New York for Strunk and White.

Tufte, Edward R. (1986). *The Visual Display of Quantitative Information,* Graphics Press, Cheshire, CT.

# 4

# *Modeling with Probability*

**OBJECTIVES**

1. Understand basic rules of classical probability.
2. Understand and use the normal distribution.
3. Understand the power of the central limit theorem.
4. Understand the construction of a hypothesis test.
5. Apply Bayesian probability.

A ***probability model*** is a mathematical representation of a random event or situation. We can define a probability model by its ***sample space***, the ***events*** within the sample space, and the ***probabilities*** associated with each event. In this chapter, we will continue our discussion of statistical and probability models, starting with understanding classical probability and moving through discrete and continuous distributions to finish with hypothesis testing.

## 4.1 Classical Probability

**Probability** is a measure of the likelihood of a random phenomenon or chance behavior. Probability describes the long-term proportion with which a certain **outcome** will occur in situations with short-term uncertainty. Probability deals with experiments that yield random short-term results or outcomes that reveal long-term predictability.

The long-term proportion with which a certain outcome is observed is the probability of that outcome.

### 4.1.1 The Law of Large Numbers

As the number of repetitions of a probability experiment increases, the proportion with which a certain outcome is observed gets closer to the probability of the outcome.

DOI: 10.1201/9781003298762-4

In probability, an **experiment** is any process that can be repeated in which the results are uncertain. A **simple event** is any single outcome from a probability experiment. Each simple event is denoted as $e_i$.

The **sample space,** $S$, of a probability experiment is the collection of all possible simple events. In other words, the sample space is a list of all possible outcomes of a probability experiment. An **event** is any collection of outcomes from a probability experiment. An event may consist of one or more simple events. Events are denoted using capital letters such as $E$ (Devore, 1995).

**Example 4.1:** Flipping a Coin

Consider the experiment of flipping a fair coin twice. We can then build a probability model to describe the experiment by

a. Identifying the simple events of the probability experiment.
b. Determining the sample space.
c. Defining the event $E$ = "having only one heads".

**Solution:**

a. Events for two flips: There are only two potential events or outcomes. H = heads and T = tails
b. Sample space {HH, HT, TH, TT}
c. Having one heads {HT, TH}

The **probability of an event**, denoted $P(E)$, is the likelihood of that event occurring.

**Properties of Probabilities**

1. The probability of any event $E$, $P(E)$, must be between 0 and 1 inclusive.
2. If an event is **impossible,** the probability of the event is 0.
3. If an event is a **certainty,** the probability of the event is 1. If $S = \{e_1, e_2, \ldots, e_n\}$, then

$$P(e_1) + P(e_2) + \ldots + P(e_n) = 1.$$

where $S$ is the sample space and $e_i$ are the events.

**P(only one heads in two flips)** = Number of outcomes with only one heads/ total number of outcomes = 2/4 = ½

The classical method of computing probabilities requires **equally likely outcomes.**

An experiment is said to have equally likely outcomes when each simple event has the same probability of occurring. An example of this is a flip of a fair coin where the chance of flipping heads is ½ and the chance of flipping tails is ½.

If an experiment has $n$ equally likely simple events and if the number of ways that an event $E$ can occur is $m$, then the probability of $E$, $P(E)$, is

$$P(E) = \frac{\text{Number of way that E can occur}}{\text{Number of Possible Outcomes}} = \frac{m}{n}$$

So, if $S$ is the sample space of this experiment, then

$$P(E) = \frac{N(E)}{N(S)}$$

**Example 4.2:** M & Ms and Probability

Suppose a "fun size" bag of M & Ms contains 9 brown candies, 6 yellow candies, 7 red candies, 4 orange candies, 2 blue candies, and 2 green candies. Suppose that a candy is randomly selected.

a. What is the probability that it is brown?
b. What is the probability that it is blue?
c. Comment on the likelihood of the candy being brown versus blue.

**Solution:**

a. *P(brown) = 9/30 = 0.3*
b. *P(blue) = 2/30 = 0.067*
c. Since there are more brown candies than blue candies, it is more likely to draw a brown candy than a blue candy.

### 4.1.2 Probability from Data

We can also develop probability models from observed data. The probability of an event $E$ is approximately the number of times event $E$ is observed divided by the number of repetitions of the experiment.

$$P(E) \approx \textit{relative frequency of } E = \frac{\textit{frequency of } E}{\textit{Number of trial in the experiment}}$$

**TABLE 4.1**

Cruise Ship Disaster

| | Men | Women | Boys | Girls | Total |
|---|---|---|---|---|---|
| Survived | 332 | 318 | 29 | 27 | 706 |
| Died | 1360 | 104 | 35 | 18 | 1517 |
| Total | 1692 | 422 | 64 | 45 | 2223 |

Now, we will return to our Mediterranean cruise ship disaster in Chapter 3 (Example 3.1) on the cruise ship (Table 4.1). We can use this method to compute the probabilities.

*P(Survived the attack) = 705/2,284 = 0.309*
*P(Died)= 1,579/2,284 = 0.691*

*P(Woman and children survived) = (310 + 30 + 25)/(411 + 68 + 45)*
*= 365/924 = 0.395*

*P(Men survived) = 340/1,760 = 0.193*

#### *4.1.2.1 Intersections and Unions*

Now, let *E* and *F* be two events.

***E* and *F*** is the event consisting of simple events that belong to *both* E and *F*. The notation is ∩ (intersection), **E ∩ F.**

***E* or *F*** is the event consisting of simple events that belong to *either* E *or* F or both.

The notation is ∪ (union), **E ∪ F.**

Suppose we have an experiment where a pair of dice are thrown (Figure 4.1). Let *E* = "the first die is a 2" and let *F* = "the sum of the dice is less than or equal to 5". Find *P(E ∩ F)* and P(E ∪ F) directly by counting the number of ways *E* or *F* could occur and dividing this result by the number of possible outcomes.

**Event E = {2–1, 2–2, 2–3, 2–4, 2–5, 2–6}**

Event F = {1–1, 1–2, 1–3, 1–4, 2–1, 2–2, 2–3, 3–1, 3–2, 4–1}

**There are 36 potential outcomes** (Figure 4.1).

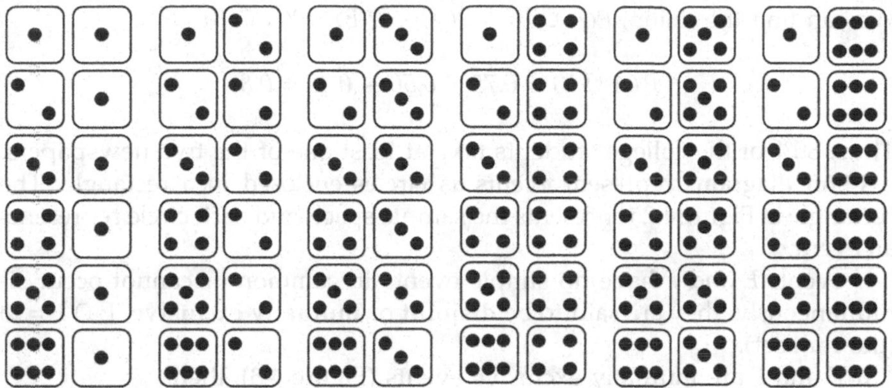

**FIGURE 4.1**
Potential outcomes of throwing two dice.

$P(E) = 6/36 = 1/6$
$P(F) = 10/36 = 5/18$
$(E \cap F) = \{2-1,\ 2-2,\ 2-3\}$
$(E \cup F) = \{1-1,\ 1-2,\ 1-3,\ 1-4,\ 2-1,\ 2-2,\ 2-3,\ 3-1,$
$3-2,\ 4-1,\ 2-4,\ 2-5,\ 2-6\}$
$P(E \cap F) = 3/36 = 1/12$
$\mathbf{P}(E \cup F) = 13/36$

### 4.1.2.2 The Addition Rule

For any two events *E* and *F*:

$$P(E \text{ or } F) = P(E) + P(F) - P(E \text{ and } F)$$
$$P(E \cup F) = P(E) + P(F) - P(E \cap F)$$

Consider the following example. Let event A be the event a college student takes the local newspaper and let event B be the event that a college student takes the *USA Today*. There are 1,000 college students living on post, and we know 750 take the local paper, and 500 take *USA Today*. We are told 450 take both papers.

$$P(A \cup B) = 450/1000 = 0.45$$
$$P(A) = 0.75$$
$$P(B) = 0.50$$

We can find the union, $P(A \cup B) = P(A) + P(B) - P(A \cap B)$

$$P(A \cup B) = 0.75 + 0.50 - 0.45 = 0.8$$

Thus, 80% of the college students take at least one of the two newspapers.

**Venn diagrams** represent events as circles enclosed in a rectangle. The rectangle in Figure 4.2 represents the sample space, and each circle represents an event.

If events $E$ and $F$ have no simple events in common or cannot occur simultaneously, they are said to be **disjoint** or **mutually exclusive**, $E \cap F = \varnothing$ (the null set).

If $E$ and $F$ are mutually exclusive events (Figure 4.3), then

$$P(E \text{ or } F) = P(E) + P(F).$$

In general, if $E, F, G, \ldots$ are mutually exclusive events, then

$$P(E \text{ or } F \text{ or } G \text{ or } \ldots) = P(E) + P(F) + P(G) + \ldots$$

Events $E$, $F$, and $G$ are mutually exclusive.

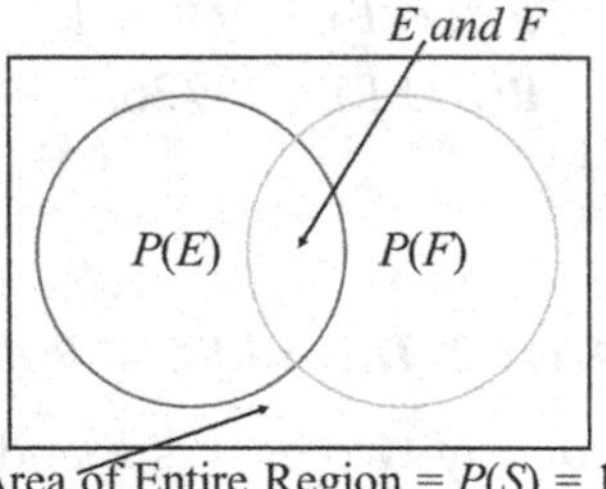


**FIGURE 4.2**
Sample space and events.

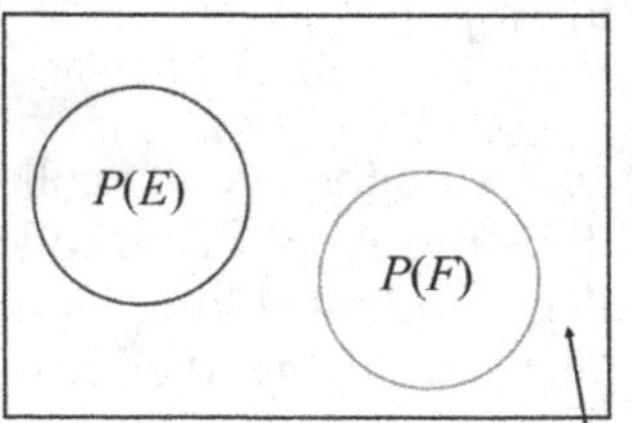


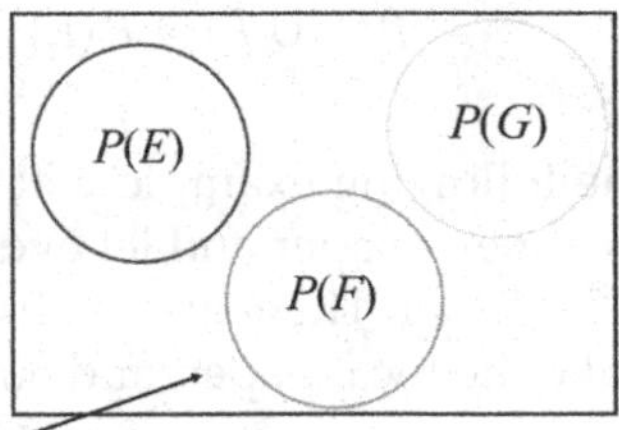


Area of Entire Region = $P(S) = 1$

**FIGURE 4.3**
Sample space and mutually exclusive events.

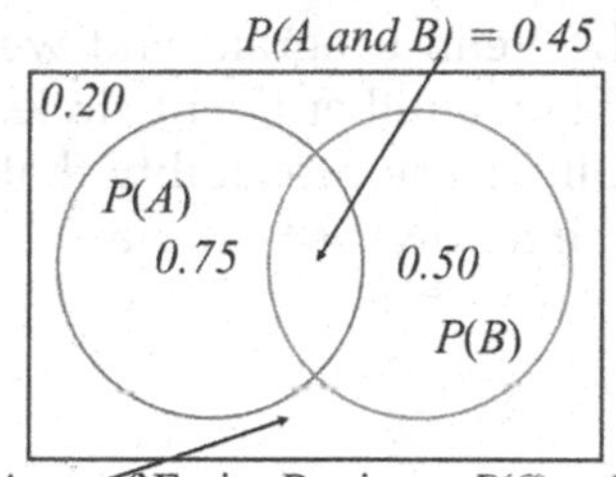


**FIGURE 4.4**
Venn diagram for Example 4.2.

### *4.1.2.3 Complement Rule*

If $E$ represents any event and $E'$ represent the complement of $E$, then

$$P(E') = 1 - P(E).$$

**Example 4.3: Single Die Roll**

Consider a roll of a single die, Let $S = \{1, 2, 3, 4, 5, 6\}$
Let event A be the roll in an even number. $A = \{2, 4, 6\}$
The event $A'$ would be $\{1, 3, 5\}$.

We can now return to our newspaper example (Example 4.2) for a closer look. The Venn Diagram would look like Figure 4.4.

The following probabilities can be used or found from the Venn Diagram (Figure 4.4). We always start filling in probabilities from inside the intersection of the events and move our way out. The sum of all probabilities within the Venn Diagram rectangle, $S$, the sample set is 1.0.

$$P(A) = 0.75$$
$$P(B) = 0.5$$
$$P(A \cap B) = 0.45$$
$$P(A \cup B) = P(A) + P(B) - P(A \cap B) = 0.8$$
$$P(\text{only } A) = 0.3$$
$$P(\text{only } B) = 0.05$$
$$P(\text{only take 1 paper}) = P(\text{only } A) + P(\text{only } B) = .3 + .05 = 0.35$$
$$P(\text{a college student does not get a paper}) = 0.2$$
$$P(\bar{A}) = 1 - 0.75 = 0.25$$

### *4.1.2.4 Conditional Probability*

The notation $P(F\,|\,E)$ is read as the "probability of event $F$ occurring given that event $E$ *has occurred*. It is the probability of an event $F$ given the occurrence of the event $E$. The idea in a Venn Diagram here is if an event has happened then we only consider that circle of the Venn Diagram and we look for the portion of that circle that is intersected by another Event circle.

The multiplication rule with conditional probability: The probability that two events A and B both occur is expressed by the following formula:

$$P(A \cap B) = P(A)P(B \mid A)$$

Think of this formula in a more useful way as either:

$$P(A \mid B) = \frac{P(A \cap B)}{P(B)}$$

$$P(B \mid A) = \frac{P(A \cap B)}{P(A)}$$

In most cases, these conditional probabilities led to different probabilities as answers.

Once again, we return to our newspaper example (Example 4.2). Find the P(A | B) and P(B | A).

$$P(A \cap B) = 0.45$$
$$P(A) = 0.75$$
$$P(B) = 0.5$$

$$P(A \mid B) = \frac{P(A \cap B)}{P(B)} = \frac{0.45}{0.50} = 0.90$$

$$P(B \mid A) = \frac{P(A \cap B)}{P(A)} = \frac{0.45}{0.75} = 0.60$$

Notice that the probabilities increased as we obtained more information about the events occurring. The probabilities do not always increase; they could decrease or remain the same. They do not have to be affected the same way.

#### *4.1.2.5 Independence*

Two events $E$ and $F$ are **independent** if the occurrence of event $E$ in a probability experiment does not affect the probability of event $F$. Two events are **dependent** if the occurrence of event $E$ in a probability experiment affects the probability of event $F$.

#### *4.1.2.6 Definition of Independent Events*

Two events $E$ and $F$ are independent if and only if

$$P(F|E) = P(F) \text{ or } P(E|F) = P(E)$$

Another way to see this is with the multiplication rule:

$P(A \cap B) = P(A) \cdot P(B)$ then the events A and B are independent.

If $P(A \cap B) \neq P(A) \cdot P(B)$ then the events are dependent.

**Example 4.4:** Test for Independence

Are the events from Example 4.2 of getting the local newspaper and *USA Today* independent events?

**Solution:**

$$P(A) = 0.75 \; P(B) = 0.5$$
$$P(A) * P(B) = (0.75) * (0.5) = 0.375$$
$$P(A \cap B) = 0.45$$

Since $P(A \cap B) \neq P(A) \cdot P(B)$ then these events are not independent.

**Example 4.5:** More Independence

Given the following information:

$$P(E) = 0.2 \; P(F) = 0.6 \; P(E \cup F) =$$

Are $E$ and $F$ independent events?

**Solution:**

$$P(E) * P(F) = 0.12$$

$P(E \cap F)$ is not given and must be found first. We do not assume independence and use the product rule. We use the addition rule where

$$P(E \cup F) = P(A) + P(B) - P(E \cap F) \textit{ and solve for } P(E \cap F).$$

$$0.68 = 0.2 + 0.6 - P(E \cap F)$$

$$P(E \cap F) = 0.12$$

Since $P(E \cap F)$ = 12 and $P(A)*P(B)$ = 12 then events E and F are independent.

**Example 4.6:** Golf Balls

Suppose we have a box full of 500 golf balls. In the box, there are 50 Titlist golf balls.

Suppose a golf ball is selected at random and then replaced. A second golf ball is then selected. What is the probability they are both Titlists?
*Note:* When sampling with replacement, the events are independent.

**Solution:**

When selecting two golf balls, the following can occur: both are Titlists, both are other, one of each.

We assume independence so $P(both\ Titlists) = P(T \cap T){=}P(T){*}P(T) = 0.1{*}0.1 = 0.01$

Note that mutually exclusive and independent are not synonymous.

## 4.2 Bayes' Theorem

We begin by providing the Theorem of Total Probability (Figure 4.5) (Sullivan, 2018)

Theorem of Total Probability (4.1)

Let $E$ be an event that is a subset of a sample space $S$. Let $A1, A2, \ldots, An$ be a partition of the sample space, $S$. Then,

$$P(E) = P(A_1) \cdot P(E \mid A_1) + P(A_2) \cdot P(E \mid A_2) + \ldots + P(A_n) \cdot P(E \mid A_n)$$

If we define $E$ to be any event in the sample space $S$, then we can write event $E$ as the union of the intersections of event $E$ with $A_1$ and event $E$ with $A_2$.

$$E = (E \cap A_1) + (E \cap A_2)$$

If we have more events, we just expand the union of the number of events that $E$ intersects with, as in the Figure 4.6.

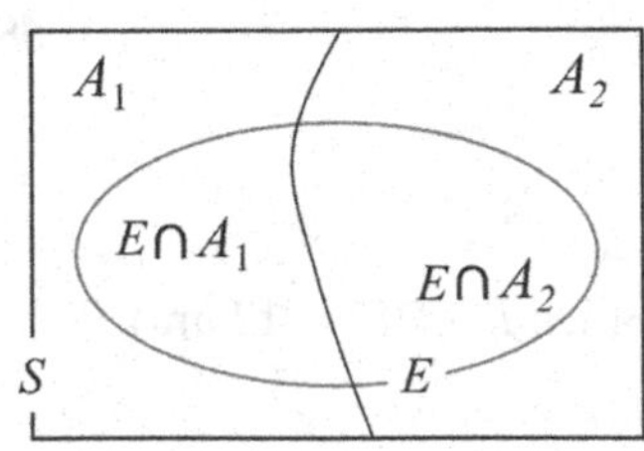


**FIGURE 4.5**
Graphical representation of Bayes Theorem.

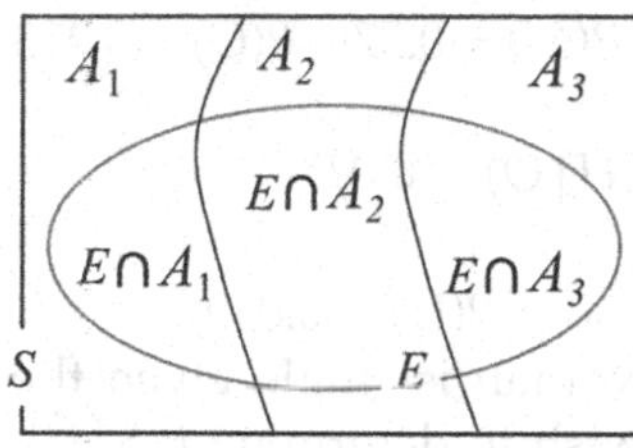


**FIGURE 4.6**
Increase events.

$$
\begin{aligned}
P(E) &= (E \cap A_1) + (E \cap A_2) + (E \cap A_3) \\
&= P(A_1) \cdot P(E \mid A_1) + P(A_2) \cdot P(E \mid A_2) + P(A_3) \cdot P(E \mid A_3)
\end{aligned}
$$

We now present Bayes' Theorem.

**Bayes' Theorem**
Let $A_1, A_2, \ldots, A_n$ be a partition of a sample space $S$. Then, for any event $E$ that is a subset of $S$ for which $P(E) > 0$, the probability of event $A_i$ for $i = 1, 2, \ldots, n$ given the event $E$, is

$$
\begin{aligned}
P(A_i \mid E) &= \frac{P(A_i) \cdot P(E \mid A_i)}{P(E)} \\
&= \frac{P(A_i) \cdot P(E \mid A_i)}{P(A_1) \cdot P(E \mid A_1) + P(A_2) \cdot P(E \mid A_2) + \ldots + P(A_n) \cdot P(E \mid A_n)}
\end{aligned}
$$

**Example 4.7:** Unemployed Women

**Problem:** According to the U.S. Census Bureau 21.1% of American adult women are single, 57.6% of American adult women are married, and 21.3% of American adult women are widowed or divorced (other). Of the single women, 7.1% are unemployed; of the married women, 2.7% are unemployed; of the "other" women, 4.2% are unemployed. Suppose that a randomly selected American adult woman is determined to be unemployed. What is the probability that she is single?

**Approach:** Define the following events:

$U$: unemployed
$S$: single
$M$: married
$O$: other

We have the following probabilities: $P(S) = 0.211$; $P(M) = 0.576$; $P(O) = 0.213$

$$P(U|S) = 0.071;\ P(U|M) = 0.027;\ P(U|O) = 0.042$$

and from the Theorem of Total probability, we know $P(U) = 0.039$

We wish to determine the probability that a woman is single given the knowledge that she is unemployed. That is, we wish to determine $P(S | U)$. We will use Bayes' Theorem as follows:

$$P(S|U) = \frac{P(S \cap U)}{P(U)} = \frac{P(S)\cdot P(U|S)}{P(U)}$$

**Solution:** $P(S|U) = \frac{0.211(0.071)}{= 0.039} = 0.384$

There is a 38.4% probability that a randomly selected unemployed woman is single.

We say that all the probabilities $P(A_i)$ are *a priori* probabilities. These are probabilities of events prior to any knowledge regarding the event. However, the probabilities $P(A_i | E)$ are *a posteriori* probabilities because they are probabilities computed after some knowledge regarding the event. In our example, the *a priori* probability of a randomly selected woman being single is 0.211. The *a posteriori* probability of a woman being single knowing that she is unemployed is 0.384. Notice the information that Bayes' Theorem gives us. Without any knowledge of the employment status of the woman, there is a 21.1% probability that she is single. But, with the knowledge that the woman is unemployed, the likelihood of her being single increases to 38.4%.

**Example 4.8:** Work Disability

**Problem:** A person is classified as work disabled if they have a health problem that prevents them from working in the type of work they can do. Table 4.2 contains the proportion of Americans that are 16 years of age or older that are work disabled by age.

If we let $M$ represent the event that a randomly selected American who is 16 years of age or older is male, then we can also obtain the following probabilities:

| | |
|---|---|
| $P(\text{male} \mid 16\text{–}24) = P(M \mid A_1) = 0.471$ | $P(\text{male} \mid 25\text{–}34) = P(M \mid A_2) = 0.496$ |
| $P(\text{male} \mid 35\text{–}44) = P(M \mid A_3) = 0.485$ | $P(\text{male} \mid 45\text{–}54) = P(M \mid A_4) = 0.497$ |
| $P(\text{male} \mid 55 \text{ and older}) = P(M \mid A_5) = 0.460$ | |

**TABLE 4.2**

Proportion of Disabled Workers

| Age | Event | Proportion Work Disabled |
|---|---|---|
| 16–24 | $A_1$ | 0.078 |
| 25–34 | $A_2$ | 0.123 |
| 35–44 | $A_3$ | 0.209 |
| 45–54 | $A_4$ | 0.284 |
| 55 and older | $A_5$ | 0.306 |

Source: U.S. Census Bureau.

a. If a work disabled American aged 16 years of age or older is randomly selected, what is the probability that the American is male?
b. If the work disabled American that is randomly selected is male, what is the probability that he is 25–34 years of age?

**Approach:**

$$P(A_2 \mid M) = \frac{P(A_2) \cdot P(M \mid A_2)}{P(M)}$$

where $P(M)$ is found from part (a).

a. We will use the Theorem of Total Probability to compute $P(M)$ as follows:

$$P(M) = P(A_1) \cdot P(M \mid A_1) + P(A_2) \cdot P(M \mid A_2) + P(A_3) \cdot P(M \mid A_3) + P(A_4) \cdot P(M \mid A_4) + P(A_5) \cdot P(M \mid A_5)$$

b. We use Bayes' Theorem to compute $P(25 - 34 \mid \text{male})$ as follows:

**Solution:**

a. Using

$$P(M) = P(A_1) \cdot P(M \mid A_1) + P(A_2) \cdot P(M \mid A_2) + P(A_3) \cdot P(M \mid A_3) + P(A_4) \cdot P(M \mid A_4) + P(A_5) \cdot P(M \mid A_5)$$

$$\begin{aligned} P(M) &= (0.078)(0.471) + (0.123)(0.496) + (0.209)(0.485) + (0.284)(0.497) \\ &\quad + (0.306)(0.460) \\ &= 0.481 \end{aligned}$$

There is a 48.1% probability that a randomly selected work disabled American is male.

$$P(A_2 \mid M) = \frac{P(A_2)\cdot P(M \mid A_2)}{P(M)} = \frac{0.123((0.496)}{0.481} = 0.127$$

There is a 12.7% probability that a randomly selected work disabled American who is male is 25–34 years of age.

Notice that the *a priori* probability (0.123) and the *a posteriori* probability (0.127) do not differ much. This means that the knowledge that the individual is male does not yield much information regarding the age of the work disabled individual.

## 4.3 Discrete Distributions in Modeling

We will also use several probability distributions for discrete random variables. A random variable is a rule that assigns a number to every outcome of a sample space. A discrete random variable takes on counting numbers 0,1,2,3, ... etc. These are either finite or countable. Then, a probability distribution gives the probability for each value of the random variable.

We can take another look at our coin flipping example (Example 4.1). Let the random variable $F$ be the number of heads of the two flips of the coin. The possible values of the random variable $F$ are 0, 1, and 2. We can count the number of outcomes that fall into each category of $F$ as shown in the *probability mass function* table (Table 4.3).

Note that the $\Sigma P(F) = 1/4 + 2/4 + 1/4 = 1$. This is a rule for any probability distribution.

We can summarize these rules:

1. *P(each event)* $\geq 0$
2. $\Sigma$*P(events)* $= 1$

Thus, the coin flip experiment is a probability distribution.

**TABLE 4.3**

Probability Mass Function Table of Coin Flipping Experiment

| Random Variable | 0 | 1 | 2 |
|---|---|---|---|
| Occurrences | 1 | 2 | 1 |
| Corresponding to Events | TT | TH, HT | HH |
| P(F) | ¼ | 2/4 | 1/4 |

All probability distributions have means, μ, and variances, $\sigma^2$. We can find the mean and the variance for a random variable X using the following formulas:

$$\mu = E[X] = \Sigma x\, P(X = x)$$
$$\sigma^2 = E[X^2] - (E[X])^2$$

For our example, we compute the mean and variance as follows:

$$\mu = E[X] = \Sigma x\, P(X = x) = 0(1/4) + 1(2/4) + 2(1/4) = 1$$
$$\sigma^2 = E[X^2] - (E[X])^2 = 0(1/4) + 1(2/4) + 4(1/4) - 1^2 = 0.5$$

We can also find the standard deviation, $\sigma$.

$$\sigma = \sqrt{\sigma^2}$$

Thus, we find the variance first and then take its square root.

$$\sigma = \sqrt{0.5}$$

There will be several discrete distributions that will arise in our modeling: Bernoulli, Binomial and Poisson.

Consider an experiment made up of a repeated number of independent and identical trials having only two outcomes, like tossing a fair coin {Head, Tail}, or a {red, green} stoplight. These experiments with only two possible outcomes are called *Bernoulli trials*. Often, they are found by assigning either a *S* (success) or *F* (failure) or a 0 or 1 to an outcome. Something either happened (1) or did not happen (0).

A binomial experiment is found counting the number of successes in *N* trials.

**Binomial** experiment:

a. Consists of *n* trials where *n* is fixed in advance.
b. Trials are identical and can result in either a success or a failure.

c. Trials are independent.
d. Probability of success is constant from trial to trail.

Formula: $b(x;\ n,\ p) = p(X = x) = \binom{n}{x} p^x (1 - p)^{n-x}$ for $x = 0, 1, 2, \ldots n$

Cumulative Binomial: $p(X \le x) = B(x;\ n,\ p) = \sum_{y=0}^{x} \binom{n}{y} p^y (1 - p)^{n-y}$
for $x = 0, 1, 2, \ldots n$

Mean: $\mu = n\ p$
Variance: $\sigma^2 = n\ p\ (1 - p)$

Example 4.1, our coin flip experiment follows these rules and is a binomial experiment. The probability that we got 1 heads in 2 flips is:

$$P(X = 1) = \binom{2}{1} (0.5)^1 (1 - 0.5)^{2-1} = 0.50$$

If we wanted 5 heads in 10 flips of a fair coin, then we can compute:

$$P(X = 5) = \binom{10}{5} (0.5)^5 (1 - 0.5)^{10-5} = 0.24610$$

Excel has built in commands to obtain these results, as we will see.

**Example 4.9:** Light Bulbs

Light bulbs are manufactured in a small local plant. In testing the light bulbs, prior to packaging and shipping, they either work, *S*, or fail to work, *F*. The company cannot test all the light bulbs but does test a random batch of 100 light bulbs per hour. In this batch, they found 2% that did not work, but all batches were shipped to distributors.

As a distributor, you are worried about past performance of these lights bulbs that you sell individually off the shelf. If a customer buys 20 lights bulbs, what is the probability that all work?

Problem ID: Predict the probability that *x* lights bulbs out of *N* work.

**Assumptions:** The light bulbs follow the binomial distribution rules stated earlier.
Model: Formula: $b(x;\ n,\ p) = p(X = x) = \binom{n}{x} p^x (1 - p)^{n-x}$ for $x = 0, 1, 2, \ldots n$

We can use Excel. The following is from the help page in Excel for the command to execute the Binomial distribution. The syntax:

*BINOMDIST (number_s, trials, probability_s, cumulative)*

Where

number_s are the value of $x$.

trials are the number of trials in the experiment known in advance, $n$

probability_s = the probability of success of our experiment (0 < p(s) < 1)

cumulative is a true or false response. False give the PFM value for $P(X = x)$ and true gives the CDF value for $P(X \leq x)$.

=BINOMDIST(*x,n,p cumulative*),

To obtain the probability that $P(X = 10)$ given that $n = 20$, $p(S) = 0.50$ we would use

=BINOMDIST(10,20,0.50, false)

To determine the CDF probability that $P(x \leq 10)$ given $n = 20$ and $P(s) = 0.50$ we would use the cumulative distribution, BINOMDIST(10, 20,…50, True).

0.588

To find $P(X > 16)$, we need to know $P(X \leq 15)$ first, so we use = 1-$P(X \leq 15)$.

=*1-BINOMDIST(15,20,0.5,true) = 0.00591*

If we have discrete data that follow a binomial distribution, then the histogram (Figure 4.7) might look as follows:

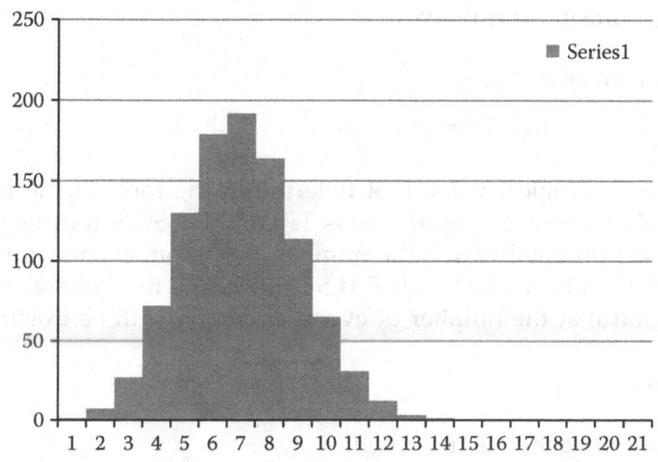


**FIGURE 4.7**
Histogram of statistics summary data.

It is symmetric. The keys to ensure a binomial distribution are that the binomial assumptions must hold and the data must be discrete.

### 4.3.1 Poisson Distribution

A random variable is said to have a Poisson distribution if the probability distribution function of X is:

$$p(x, \lambda) = \frac{e^{-\lambda}\lambda^{x}}{x!},$$

for x = 0, 1, 2, 3… for some $\lambda > 0$.

We consider $\lambda$ as a *rate per unit time or per unit area*. A key assumption is that with a Poisson distribution the mean and the variance are the same.

For example, let X represent the number of flaws on the surface of a randomly selected crystal glass. It has been found that on average, five flaws are found per glass surface. Find the probability that a randomly selected glass has exactly two flaws.

**Poisson**

Returns the Poisson distribution. A common application of the Poisson distribution is predicting the number of events over a specific time, such as the number of cars arriving at a toll plaza in 1 minute.

**Syntax**

POISSON(X, mean, cumulative)

X is the number of events.

Mean is the expected numeric value.

Cumulative is a logical value that determines the form of the probability distribution returned. If cumulative is TRUE, POISSON returns the cumulative Poisson probability that the number of random events occurring will be between 0 and *x* inclusive; if FALSE, it returns the Poisson probability mass function that the number of events occurring will be exactly *x*.

**Remarks**

If *x* is not an integer, it is truncated.

If *x* or mean is nonnumeric, POISSON returns the #VALUE! error value.

If $x < 0$, POISSON returns the #NUM! error value.

If mean < 0, POISSON returns the #NUM! error value.

POISSON is calculated as follows.
For cumulative = FALSE:

$$\text{Poisson} = p(x, \lambda) = \frac{e^{-\lambda}\lambda^x}{x!}$$

For cumulative = TRUE:

$$\text{Cumulative Poisson} = \sum_{k=0}^{x} \frac{e^{-\lambda}\lambda^k}{k!}$$

Consider the following data (Table 4.4):

=POISSON(A2, A3, TRUE) Cumulative Poisson probability with the terms above (0.124652)

=POISSON(A2, A3, FALSE) Poisson probability mass function with the terms above (0.084224)

POISSON(2, 5, false)
0.084224337

POISSON(2, 5, false)

0.084224337

$P(X = 2)$=

$\frac{e^{-5}5^2}{2!}$=

0.084=

A ***Poisson distribution*** has a mean, μ, of λ and variance $\sigma^2$ of λ.

A ***Poisson process*** is a Poisson distribution that varies over time (generally its time). There exists a rate, called α for a short time period. Over a longer period of time, λ becomes *αt*.

**TABLE 4.4**
Poisson Example Data

| | Data | Description |
|---|---|---|
| | A | B |
| 1. | 2 | Number of events |
| 2. | 5 | Expected mean value |

**Example 4.10:** Poisson Process Electronic Machine

Suppose your pulse is read by an electronic machine at a rate of 5 times per minute. Find the probability that your pulse is read 15 times in a 4 minute interval.

$\lambda = \alpha\ t$ = 5 times 4 minutes = 20 pulses in a 4-minute period

$$P(X = 15) = \frac{e^{-20}20^{15}}{15!} = 0.052$$

Poisson(15, 20, false) = 0.5165

Poisson data usually is slightly positively skewed.

Using R commands to find probabilities.

We can use R and the following commands.

P (X = x) → dbinom(x,n,p)

P(X ≤ x) → pbinom(x,n,p)

P(X > x) → 1 - pbinom(x,n,p)

P(X ≥ x) → 1 - pbinom(x-1,n,p)

P(a ≤ X ≤ b) = pbinom(b,n,p) - pbinom(a-1,n,p)

P(a ≤ X < b) = pbinom(b-1,n,p)- pbinom(a-1,n,p)

P(a < X ≤ b) = pbinom(b,n,p) - pionom(a,n,p)

P(a < X < b) = pbinom(b-1,n,p) - pbinom(a,n,p)

So, if $x = 5$, $n = 10$, $p(s) = 5$, then $P(X = 5)$ and $P(X \leq 5)$ are found with commands

```
dbinom(5, 10, 0.5)
[1] 0.2460938
> pbinom(5, 10, 0.5)
[1] 0.6230469
```

For Poisson we use

dpois(x, λ) for P(X = x) and ppois(x, λ) for P(X ≤ x)

For example, to obtain $P(X = 5)$ and $P(X \leq 5)$ in a Poisson distribution with mean, λ = 0.5, we use the following commands:

```
dpois(5, 0.5)
```

```
[1] 0.0001579507
> ppois(5, 0.5)
[1] 0.9999858
```

**Example 4.11:** Binomial and Poisson in MAPLE

We have 25 items and $p(s) = 0.435$. We would like to find the probability that $P(X = 10)$ and $P(X < 10)$.

```
> with(Statistics) :
> X := RandomVariable(Binomial(25, .435)) :
> ProbabilityFunction(X, u)
```

$$\begin{cases} 0 & u < 0 \\ \binom{25}{u} 0.435^u 0.565^{25-u} & \text{otherwise} \end{cases}$$

```
> ProbabilityFunction(X, 1)
                    0.00001217844989
> Mean(X)
                         10.875
> Variance(X)
                        6.144375
> ProbabilityFunction(X, 10);
                      0.1513569340
>
```

$P(X = 10) = 0.151369340$.

To find $P(X < 10)$, we need a different package in Maple (Figure 4.8). The commands are:

$P(X < 10) = 0.443519999349433$.

## 4.4 Continuous Probability Models

### 4.4.1 Introduction

Some random variables do not have a discrete range of values. In the previous sub-section, we saw examples of discrete random variables and discrete distributions. What if we were looking at time, as a random event? Time has a continuous range of values and thus, as a continuous random variable can be continuous probability distribution. We define a continuous random variable as any random variable measured on continuous scale. Other examples include altitude of a plane, the percent of alcohol in a person's blood, net weight of a package of frozen chicken wings, the

```
> with(Student[Statistics]) :
>
> Y := BinomialRandomVariable(25, 0.435) :
> ProbabilityFunction(Y, x, output = plot)
```

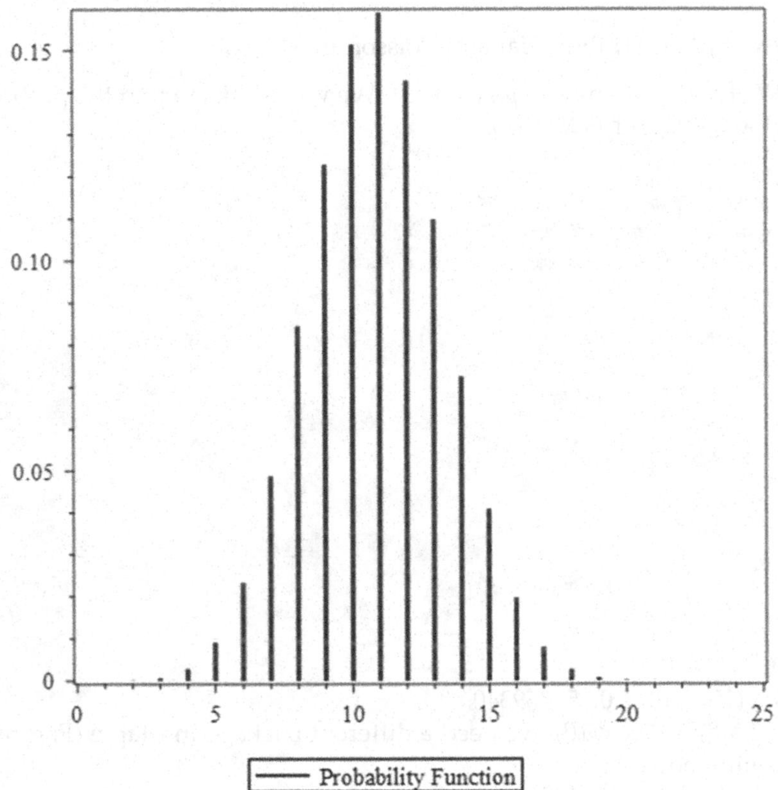


```
> CDF(Y, 10, numeric);
                    0.443519999349433
```

**FIGURE 4.8**
Probability graph in Maple.

distance a round misses a designated target, or the time to failure of an electric light bulb. We cannot list the sample space because the sample space is infinite. We need to be able to define a distribution as well as its domain and range.

For any continuous random variable, we can define the cumulative distribution function (CDF) as $F(b) = P(X \leq b)$.

For those that have seen calculus, the probability density function (PDF) of $f(x)$ is defined to be $P(a \leq x \leq b) = \int_a^b f(x)dx$.

To be a valid probability density function (PDF):

a. *f(x)* must be greater than or equal to 0 for all $x$ in its domain, and
b. the integral $\int_{-\infty}^{\infty} x{\cdot}f\,(x)dx = 1$= the area under the entire graph of *f(x)*.

Expected value or average value of a random variable x, with PDF defined as above, is defined as $E[X] = \int_{-\infty}^{\infty} x{\cdot}f\,(x)dx$.

In this section, we will see some modeling applications using many continuous distributions such as the exponential distribution and the normal distribution. For each of these two distributions, we will not have to use calculus to get our answers to probability questions with Excel.

### 4.4.2 The Exponential Distribution

Continuous distribution of a random variable X that has properties: $\mu = 1/\lambda$, variance $= \sigma^2 = 1/\lambda^2$

where $\lambda$ is the rate.

$$PDF = \lambda e^{-\lambda x} for\ x \geq 0$$
$$CDF = 1 - e^{-\lambda x} x \geq 0 (represents\ the\ area\ under\ the\ curve).$$

In probability theory and statistics, the **exponential distribution** (a.k.a. negative exponential distribution) is a family of continuous probability distributions. It describes the time between events in a Poisson process, i.e. a process in which events occur continuously and **independently at a constant average rate.**

The exponential distribution occurs naturally when describing the lengths of the inter-arrival times in a homogeneous Poisson process.

In real-world scenarios, the assumption of a constant rate (or probability per unit time) is rarely satisfied. For example, the rate of incoming phone calls differs according to the time of day. But if we focus on a time interval during which the rate is roughly constant, such as from 2 to 4 p.m. during workdays, the exponential distribution can be used as a good approximate model for the time until the next phone call arrives. Similar caveats apply to the following examples, which yield approximately exponentially distributed variables:

- The time until a radioactive particle decays, or the time between clicks of a Geiger counter.
- The time it takes before your next telephone call.
- The time until default (on payment to company debt holders) in reduced form credit risk modeling.

Exponential variables can also be used to model situations where certain events occur with a constant probability per unit length, such as the distance between mutations on a DNA strand, or between roadkills on a given road.

In queuing theory, the service times of agents in a system (e.g. how long it takes for a bank teller etc. to serve a customer) are often modeled as exponentially distributed variables. (The inter-arrival of customers for instance in a system is typically modeled by the Poisson distribution in most management science textbooks.)

Reliability theory and reliability engineering also make extensive use of the exponential distribution.

Reliability = 1 - Failure

Series -----(A)-----(B)----

P(A and B) must work. A and B are independent so

P(A and B) = P(A) * P(B)

Parallel events (Figure 4.9)

$$P(A\ or\ B) = P(A) + P(B) - P(A\ and\ B)$$
$$P(A\ or\ B) = P(A) + P(B) - P(A) * P(B)$$

**Example 4.12:** Postal Operations

*Illustrates the exponential distribution:* Let X = amount of time (in minutes) a postal clerk spends with their customer. The time is known to have an exponential distribution with the average amount of time equal to 4 minutes. The rate is 1 customer every 4 minutes or ¼ of a customer per minute.

X is a *continuous random variable* since time is measured. It is given that μ = 4 minutes. To do any calculations, you must know *m*, the decay parameter.

$$\lambda = 1/\mu \ Therefore,\ \lambda = ¼ = 0.25$$

The standard deviation, σ, is the same as the mean, μ = σ.

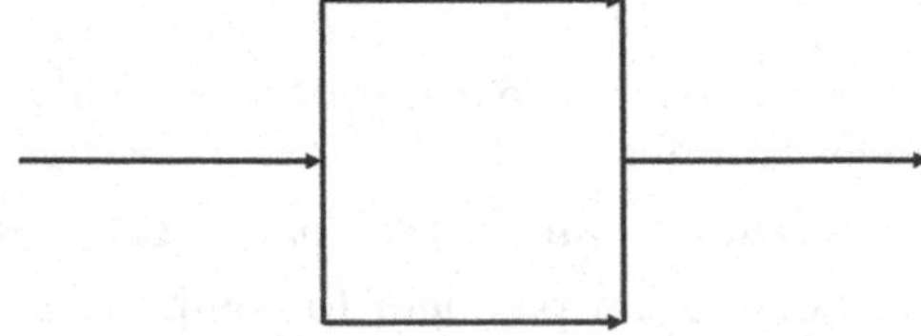

**FIGURE 4.9**
Parallel process.

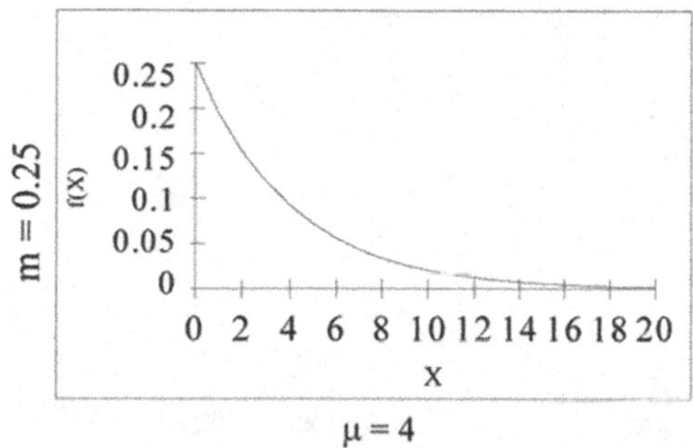


**FIGURE 4.10**
Exponential distribution for Example 4.8.

The distribution notation is X~Exp(λ). Therefore, X~Exp(0.25).

The probability density function is $f(X) = \lambda \cdot e^{-\lambda \cdot x}$

The number $e$ = 2.7182. It is a number that is used often in mathematics

$$f(X) = 0.25 \cdot e^{-0.25 \cdot X} \text{ where X s at least 0 and } \lambda = 0.25.$$
$$CDF = P(X < x) = 1 - e^{\square - x} = 1 - e^{-0.25x}$$

The graph is as follows (Figure 4.10):

Notice the graph is a declining curve. When X = 0,

Probabilities: Find the *P(X < 5), P(X > 5), P(2 < X < 6)*

$$P(X < 5) = 1 - e^{-.25*5} = 0.713495$$
$$P(X > 5) = 1 - P(X < 5) = 1 - 0.713495 = 0.2865$$
$$P(2 < X < 6) = P(X < 6) - P(X < 2) = 0.7768698 - 0.393469 = 0.3834008$$

In Excel, the command for exponential is simply = *exp(value)*. Thus, knowing how to find *P(X < a), P(X > a)*, and *P(a < X < b)* are key elements.

$$P(X < a) = 1 - exp(-\lambda a)$$
$$P(X > a) = 1 - (1 - exp(-\lambda a)) = exp(-\lambda a)$$
$$P(a < X < b) = 1 - exp - (\lambda b) - (1 - exp(-\lambda a))$$

| | | |
|---|---|---|
| λ | =0.25 | |
| a | 4 | |
| b | 5 | |
| p(X<5) | =1-EXP(-C3*C6) | |
| p(X<4) | =1-EXP(-C3*C5) | |
| P(4<X<5) | =C9-C10 | |

**FIGURE 4.11**
Excel solution for Example 4.8.

**Example 4.13:** Postal Clerk and Customers

**Problem 1:** Find the probability that a clerk spends four to five minutes with a randomly selected customer. We show this in parts in Excel. We found *l* as 0.25 before.

$$P(4 < X < 5)$$
$$\textit{Use the CDF } P(X < x) = 1 - e^{-0.25 \cdot x}$$
$$P(X < 5) = 1 - e^{-0.25 \cdot 5} = 0.\ 7135$$
$$P(X < 4) = 1 - e^{-0.25 \cdot 4} = 0.\ 6321$$
$$P(4 < X < 5) = P(X < 5) - P(X < 4) = 0.\ 7135 - 0.\ 6321 = 0.\ 0814$$

In Excel (Figure 4.11):

**Problem 2:** How long does it take for half of all customers to be finished? (Find the 50th percentile.)

| | | | |
|---|---|---|---|
| $P(X < k) = 0.50$ | $k\ P(X < k) = 0.50$ | $P(X < k) = 1 - e^{-0.25 \cdot k}$ | $0.50 = 1 - e^{-0.25 \cdot k}$ $e^{-0.25 \cdot k}$ $= 1 - 0.50 = 0.5$ |
| $\ln(e^{-0.25 \cdot k}) = \ln(0.50)$ | $-0.25 \cdot k = \ln(0.50)$ | $k = \ln(0.50)/-0.25$ $= 2.8$ minutes | |

The natural logarithm command in Excel is = *ln(a).*

**TABLE 4.5**
Failure Time

| Number of Units in Group | Time-to-Failure |
|---|---|
| 7 | 100 |
| 5 | 200 |
| 3 | 300 |
| 2 | 400 |
| 1 | 500 |
| 2 | 600 |

**Problem 3:** Which is larger, the mean or the median? Mean is $1/\lambda = 1/0.25 =$ 4 minutes (given), median is just found in Example 4.13 to be 2.8. The mean is larger.

**Example 4.14:** Exponential Distribution

Twenty units were reliability tested with the following results (Table 4.5).

We can also use our knowledge of Excel to develop a histogram (Figure 4.12) and the descriptive statistics (Table 4.6) for the problem.

OK, now what? Assume an exponential distribution with $\mu = 255$ hours or

$\lambda = 1/255 =$ 0□0039: or 0□0039 failures per hour.

So, the average lifetime is 255 hours.

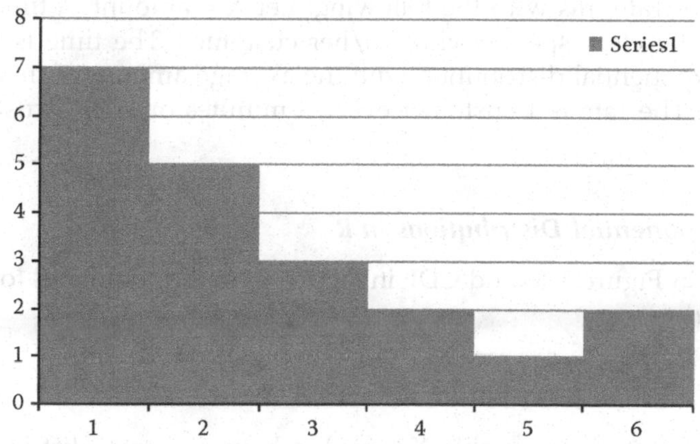


**FIGURE 4.12**
Histogram of failure times (Example 4.10)

**TABLE 4.6**

Time to Failure for Example 4.10

| | |
|---|---|
| Mean | 255 |
| Standard Error | 37.33 |
| Median | 200 |
| Mode | 100 |
| Standard Deviation | 166.94 |
| Sample Variance | 27,868.42 |
| Skewness | 0.959 |
| Min | 100 |
| Max | 600 |
| Range | 500 |
| Count | 20 |

$$P(X > 3) = 1 - P(X < 3) = 1 - exp(0.0039 * 3) = 0.01170$$

Or about a 1.2% chance of having more than 3 failures in a given hour.

So, what if we want the following:

*P*(more than 3 failures in a day)

$\lambda$ is now (0.0039*24) = 0.0941 per day

$$P(X > 3) = 1 - P(X < 3) = 1 - 0.754 = 0.24599$$

We can illustrate this with the following: Let X = amount of time (in minutes) a postal clerk spends with his/her customer. The time is known to have an exponential distribution with the average amount of time equal to 4 minutes. The rate is 1 customer every 4 minutes or ¼ of a customer per minute.

#### 4.4.2.1 Exponential Distributions in R

In R (PDF in Figure 4.13 and CDF in Figure 4.14) the command for $P(X < x) = pexp(x, \mu)$

### 4.4.3 The Normal Distribution

A continuous random variable ***X*** is said to have a normal distribution with parameters $\mu$ and $\sigma$ (or $\mu$ and $\sigma^2$), where $-\infty < \mu < \infty$ and $\sigma > 0$, if the pdf of ***X*** is

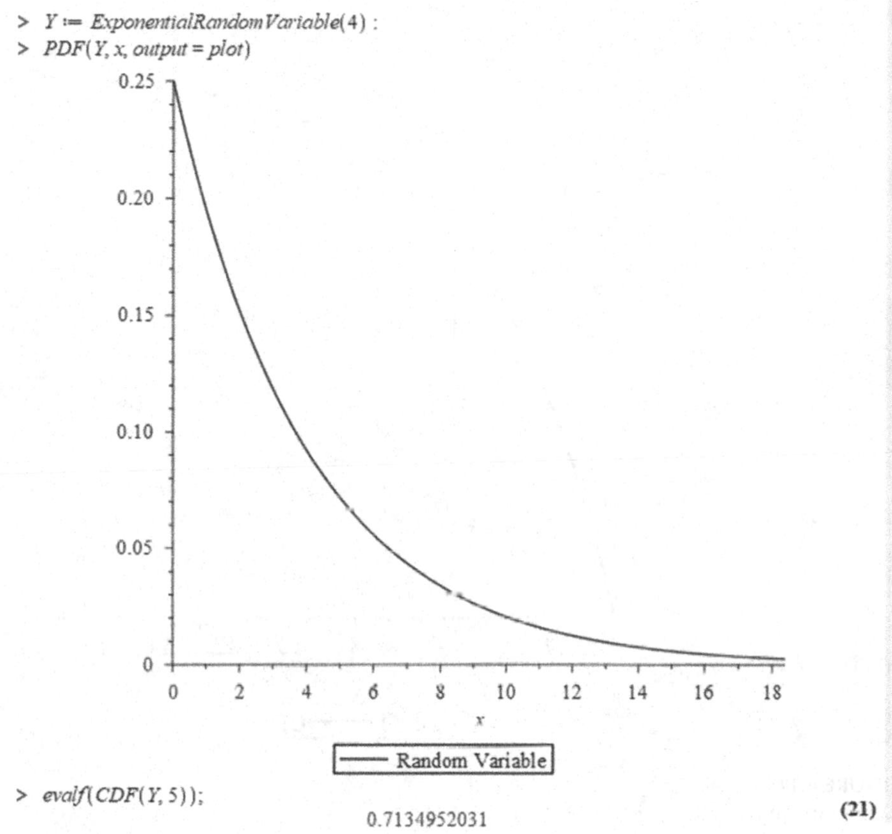


**FIGURE 4.13**
Exponential graph in R.

$$f(x; \mu, \sigma) = \frac{1}{\sqrt{2\pi\sigma}} e^{-\frac{(x-\mu)^2}{(2\sigma^2)}}, \quad -\infty \le x \le \infty$$

The plot of the normal distribution is our bell-shaped curve (Figure 4.15).

To compute *P(a < x < b)* when *X* is a normal random variable, with parameters μ and σ, we must evaluate $\int \frac{1}{\sqrt{2\pi\sigma}} e^{-\frac{(x-\mu)^2}{(2\sigma^2)}} dx$.

Since none of the standard integration techniques can be used to evaluate this integral, the standard normal random variable *Z* with parameters μ = 0 and σ = 1 has been numerically evaluated and tabulated for certain values. Since most applied problems do not have parameters of μ = 0 and σ = 1, "standardizing" transformation can be used

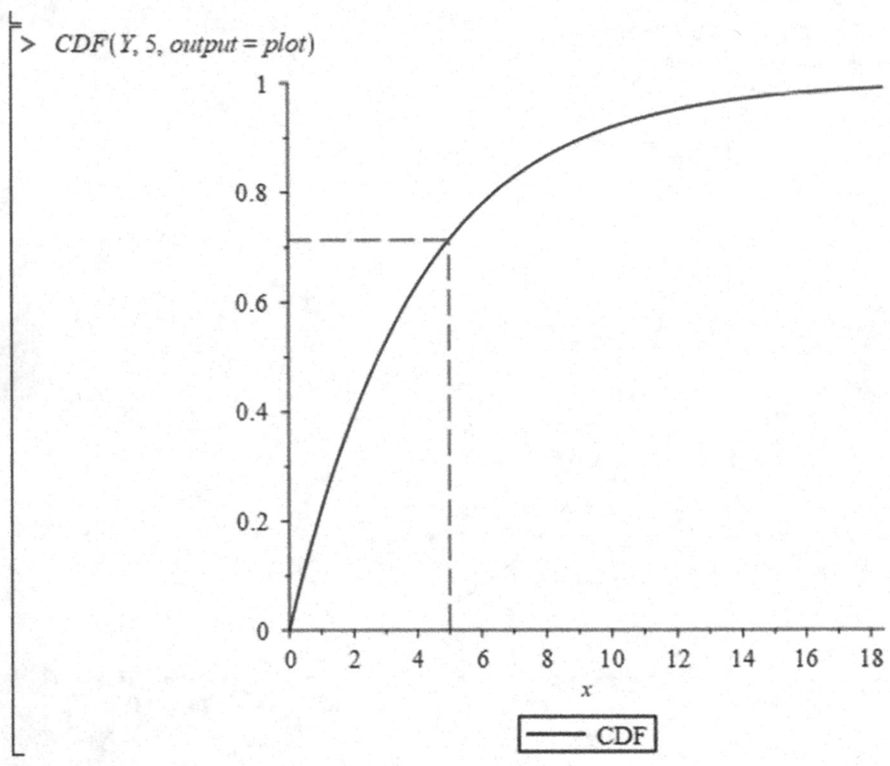


**FIGURE 4.14**
CDF graph in R.

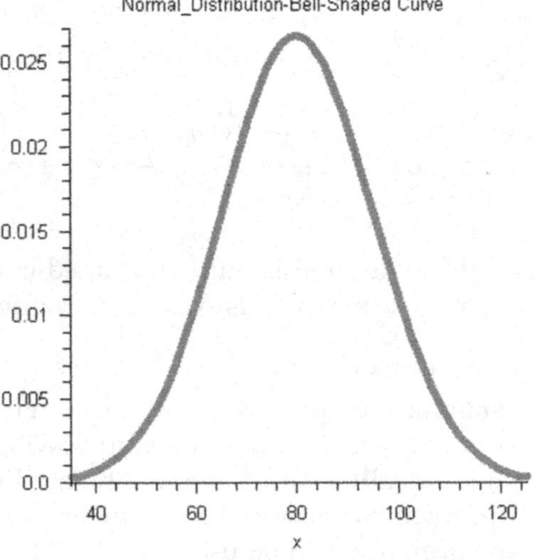


**FIGURE 4.15**
Bell-shaped curve of the normal distribution.

$$Z = \frac{(x - \mu)}{\sigma}.$$

**Example 4.15:** Diet Coke Probability

For example, the amount of fluid dispensed into a can of Diet Coke is approximately a normal random variable with mean 11.5 fluid ounces and a standard deviation of 0.5 fluid ounces. We want to determine the probability that between 11 and 12 fluid ounces, $P(11 < x < 12)$, are dispensed.

$$Z_1 = (11-11.5)/0.5 = -1$$
$$Z_2 = (12-11.5)/0.5 = 1$$

This probability statement $P(11 < x < 12)$ is equivalent to $P(-1 < Z < 1)$. If we used the tables, we can compute this to be 0.8413 − 0.1587 = 0.6826. However, we can use Excel to compute the area between 11 and 12. This is displayed in the Figure 4.16.

To find the $P(a < x < b)$, or in our case $P(11 < x < 12)$, we would using the following command in Excel:

a = 11, b = 12, mean = 11.5, standard deviation = 0.5, and "True" is yes.

**Normdist(value, mean, standard deviation, TRUE)**

= Normdist(12, 11.5, 0.5, True) - Normdist(11, 11.5, 0.5, True)

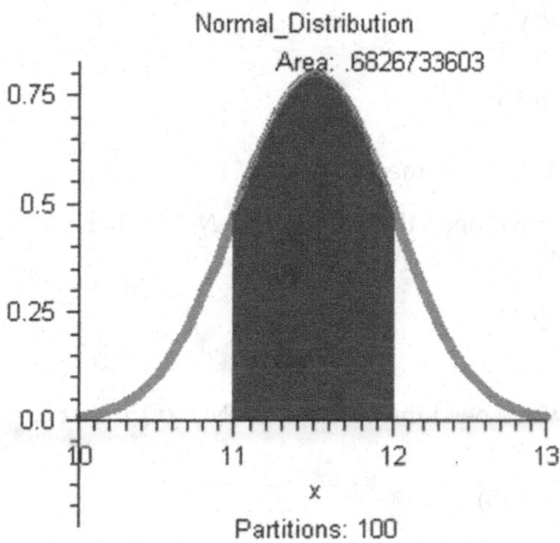


**FIGURE 4.16**
Normal distribution area from 11 to 12.

The probability is 0.682689.

Therefore, 68.26% of the time the cans are filled between 11 and 12 fluid ounces.

#### 4.4.3.1 Normal Distribution in R

Assume we have a RV X that is normally distributed with mean 5 and standard deviation 3.

We want to find the *P(X > 4.75)*. In R, the command is

*pnorm(4.75, 5, 3)*

The R command to find $P(X < x)$ and $P(X \leq x) \rightarrow pnorm(x,\mu,\sigma)$

And for $P(Z < z) \rightarrow$ pnorm(z)

$P( a \leq X \leq b) = P(A \leq X < b) = P( a < X \leq b) = P(A < X < b) \rightarrow pnorm(b, \mu, \sigma)- pnorm(a, \mu,\sigma)$

And in Z,

$P(z_1 < Z < z_2) \rightarrow nnorm(z1)\text{-}pnorm(z2)$

To find the *P (a < x < b)*, in our case *P (11 < x < 12)* = 0.682689

#### 4.4.3.2 Inverse of Normal Distribution

Often in either the RV *X* or *Z* we need the value of *x* or *z* that corresponds to a specific probability.

In R the command is

>qnorm, "probability", 'mean", "stdev")

For example, if we need the value of *Z~N(0,1)* that corresponds with a 0.95 probability.

>qnorm(0.95, 0, 1)

[1] 1.6448536

For example, if we need the value of *X~N(100,15)* that corresponds with a 0.95 probability.

>qnorm(0.95,100,15)

### *4.4.3.3* Normal Distribution in Maple

Assume we have a RV *X* that is normally distributed with mean 5 and standard deviation 3.

We want to find the *P(X > 4.75)*, and we want to find the inverse for a probability of 0.97.

The PDF and CDF are shown in Figures 4.17 and 4.18.

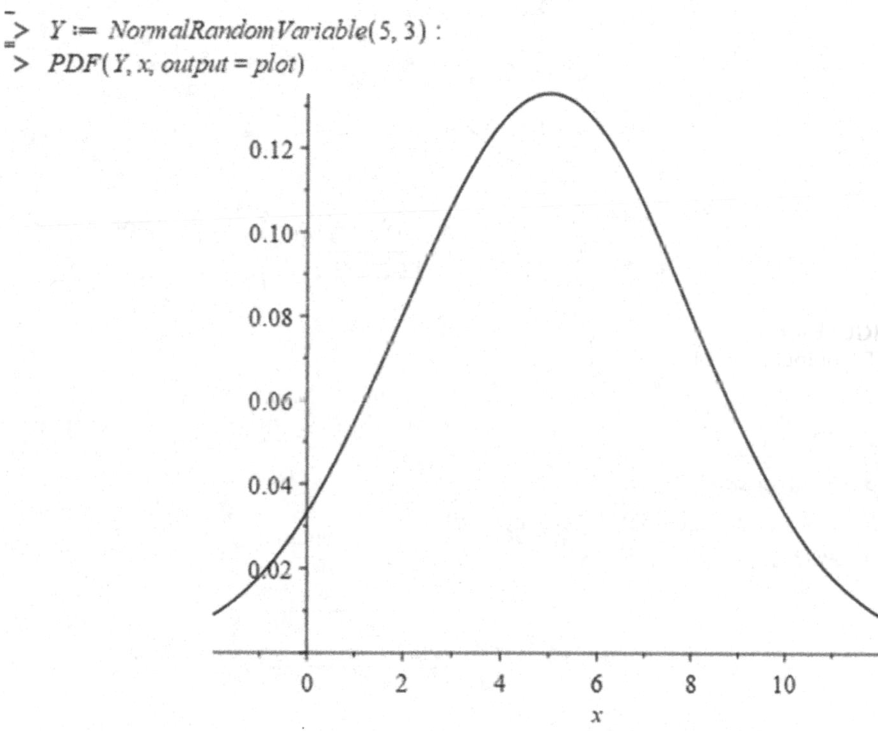


```
> evalf(CDF(Y, 4.75));
                    0.466793248147378
```

**FIGURE 4.17**
PDF output in Maple.

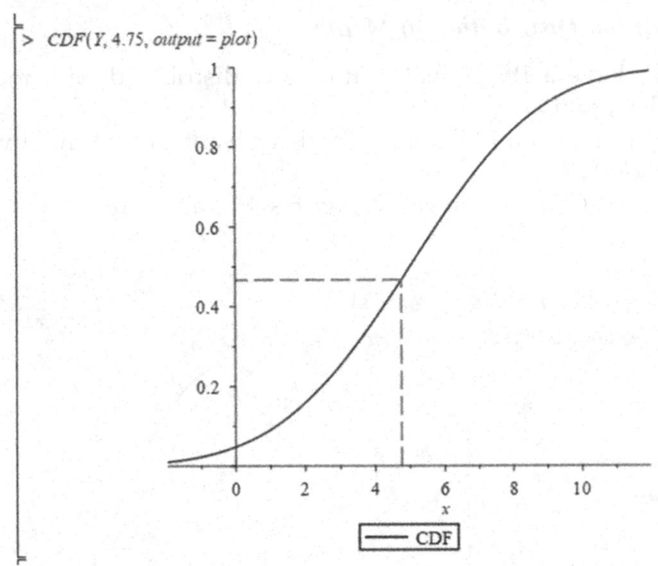


**FIGURE 4.18**
CDF output in Maple.

> *with*(*Student*[*Statistics*]) :

> *X* := *NormalRandomVariable*(μ, σ) :

> *PDF*(*X*, *u*)

$$\frac{\sqrt{2}\, e^{-\frac{(u-\mu)^2}{2\sigma^2}}}{2\sqrt{\pi}\,\sigma}$$

> *PDF*(*X*, 5)

$$\frac{e^{-\frac{5}{4}}}{4}$$

> *Mean*(*X*)

$$4$$

> *Variance*(*X*)

$$16$$

Inverse uses Quantile command as follows:

*Quantile*(*Normal*(5, 3), 0.97);

$$10.6423808244535$$

### 4.4.4 Central Limit Theorem

The Central Limit Theorem is one of the most important theorems in probability. It states that if $X_1, X_2, \ldots X_n$ are a random sample from a distribution with a mean μ and a standard deviation σ and $n$ is sufficiently large ($n > 30$), then the distribution of the average $\bar{X}$ or $TT$ (the total) has a normal distributions with parameters:

$$\mu_{\bar{x}} = \mu$$

$$\mu_T = n\mu$$

$$\sigma_{\bar{x}}^2 = \frac{\sigma^2}{n}$$

$$\sigma_T^2 = n\sigma^2$$

**Example 4.16:** Central Limit Theorem

When a batch of a certain pharmaceuticals is prepared, the amount of natural substance aloe is a random variable with mean value 4.0 g and standard deviation 0.35 g. If 50 batches are prepared, what is the probability that the sample average of the aloe is between 3.5 and 3.8 g?

Since $n = 50(n > 30)$, then the sample average aloe random variable, $\bar{X}$, follows a normal distribution with mean 4.0 and standard deviation, 0.4950.

$$P(3.5 < \bar{X} < 3.8) = P\left(\frac{(3.5 - 4.0)}{0.4950} < Z < \frac{(3.8 - 4)}{0.4950}\right) = 0.1869$$

Using our Excel commands, we must take the difference between the two calculated values in order to get the area between the two points and then graph it (Figure 4.19).

$a = 3.5$, $b = 3.8$, mean = 4, standard deviation = 0.495 and "True" is yes:

**= Normdist(3.8, 4, 0.4950, True) – Normdist(3.5, 4, 0.4950, True)**

The probability is 0.186868.

The normal distribution and the central limit (when applicable) are used in many applications of confidence intervals and hypothesis testing.

#### *4.4.4.1 The Central Limit Theorem in R*

The area is between 11 and 12, μ is the mean, and $s$ is the standard deviation.

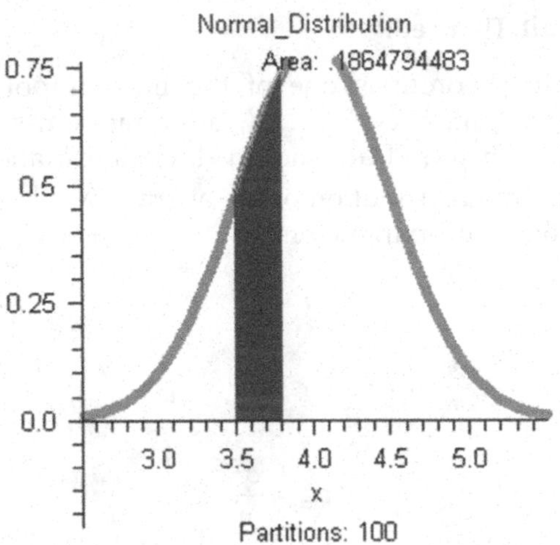


**FIGURE 4.19**
Normal curve with mean = 4, standard deviation = 0.4950 from 3.5 to 3.8.

The following provides the R command for our various probability requirements:

$P(X < x)$ and $P(X \le x)$ is *pnorm(x,μ,s)*

$P(Z < z)$ is *pnorm(z)*

$P(a \leq X \leq b) = P(A \leq X < b) = P(a < X \leq b) = P(A < X < b)$ is

*pnorm(b, μ, s) – pnorm(a,μ,s)*

and Z, which is $P(z1 < Z < z2)$ is *pnorm(z1) – pnorm(z2)*

To find the $P\,(a < x < b)$, in our case $P\,(11 < x < 12) = 0.6827$

## 4.5 Confidence Intervals and Hypothesis Testing

The basic concepts and properties of confidence intervals involve initially understanding and using two assumptions:

1. The population distribution is normal and
2. The standard deviation σ is known or can be easily estimated.

In its simplest form, we are trying to find a region for μ (and thus a confidence interval) that will contain the value of the true parameter of interest.

Theformula for finding the confidence interval for an unknown population mean from a sample is $\bar{x} \pm Z_{\alpha/2}\frac{\sigma}{\sqrt{n}}$

The value of $Z_{\alpha/2}$ is computed from the normality assumption and the level of confidence, $1 - \alpha$, desired.

We will consider a variation of the Diet Coke example (Example 4.15). For example, the amount of fluid dispensed into a can of Diet Coke is approximately a normal random variable with unknown mean fluid ounces and a standard deviation of 0.5 fluid ounces. We want to determine a 95% confidence interval for the true mean. A sample of 36 Diet Cokes was taken and a sample mean of $\bar{x} = 11.35$ was found.

Now, $1 - \alpha = 0.95$. Therefore, $\alpha = 0.5$, and since there are two regions, we need $\alpha/2 = 0.25$ and $Z_{\alpha/2} = 1.96$. This is seen in Figure 4.20.

Our confidence interval for the parameter, μ, is11.35 ± 1.96 (0.5/6).

We can interpret this or any confidence interval. If we took 100 experiments of 36 random samples each and calculated the 100 confidence intervals in the same manner, $\bar{x} \pm Z_{\alpha/2}\frac{\sigma}{\sqrt{n}}$.

Thus, 95 of the 100 confidence intervals would contain the true mean, μ. We do not know which of the 95 confidence intervals contain the true mean. Thus, to a modeler, each confidence interval built will either contain the true mean or it will not contain the true mean.

In Excel, the command is CONFIDENCE(alpha,st_dev,size), and it only proved the value of $Z_{\alpha/2}\frac{\sigma}{\sqrt{n}}$. We must still combine to get the interval $\bar{x} \pm Z_{\alpha/2}\frac{\sigma}{\sqrt{n}}$.

In Excel, using the Descriptive Statistics function, we find our confidence interval for our example:

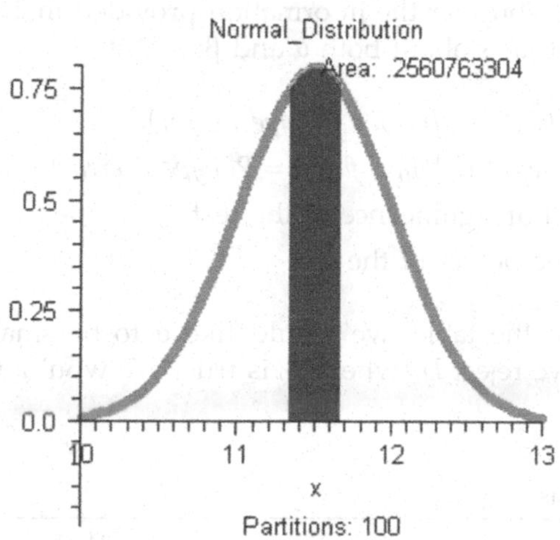


**FIGURE 4.20**
Confidence interval 11.35 ± 1.96 (0.5/6).

$$\bar{X} = 11.35$$
$$s = 0.50$$
$$n = 36$$
$$\alpha\,(alpha) = 0.05$$

### 4.5.1 Simple Hypothesis Testing

A more powerful technique for interring information about a parameter is a hypothesis test. A statistical hypothesis test is a claim about a single population characteristic or about values of several population characteristics. There is a null hypothesis (which is the claim initially favored or believe to be true) and is denoted by $H_0$. The other hypothesis, the alternate hypothesis, is denoted as $H_a$. We will always keep equality with the null hypothesis. The objective is to decide, based upon sample information, which of the two claims is correct. Typical hypothesis tests can be categorized by three cases:

| | | | | | |
|---|---|---|---|---|---|
| CASE 1 | $H_0$ | $\mu = \mu_0$ | versus | $H_a$ | $\mu \neq \mu_0$ |
| CASE 2 | $H_0$ | $\mu \leq \mu_0$ | versus | $H_a$ | $\mu > \mu_0$ |
| CASE 3 | $H_0$ | $\mu \geq \mu_0$ | versus | $H_a$ | $\mu < \mu_0$ |

There are two types of errors that can be made in hypothesis testing: Type 1 errors called α error and Type II errors called β errors. It is important to understand these. Consider the information provided in Table 4.7.

Some important facts about both α and β:

a. $\alpha = P(\text{reject } H_0 | H_0 \text{ is true}) = P(\text{Type I error})$
b. $\beta = P(\text{fail to reject } H_0 | H_0 \text{ is false}) = P(\text{Type II error})$
c. α is the level of significance of the test.
d. (4)1 − β is the power of the test

Thus, referring to the table, we would like α to be small, since it is the probability that we reject $H_0$ when $H_0$ is true. We would also want 1 − β to

**TABLE 4.7**
Status of Hypothesis

| | | **$H_o$ True** | **$H_a$ true** |
|---|---|---|---|
| Test Conclusion | Fail to Reject $H_o$ | 1−α | β |
| | Reject $H_o$ | α | 1−β |

be large since it represents the probability that we reject $H_0$ when $H_0$ is false. Part of the modeling process is to determine which of these errors is the costliest, and work to control that error as your primary error of interest.

The following template if provided for hypothesis testing:

STEP 1. Identify the parameter of interest.
STEP 2. Determine the null hypothesis, $H_0$.
STEP 3. State the alternative hypothesis, $H_a$.
STEP 4. Give the formula for the test statistic based upon the assumptions that are satisfied.
STEP 5. State the rejection criteria based upon the value of α.
STEP 6. Obtain your sample data and substitute into your test statistic.
STEP 7. Determine the region in which your test statistics lies (rejection region or fail to reject region).
STEP 8. Make your statistical conclusion. Your choices are to either reject the null hypothesis or fail to reject the null hypothesis. Ensure the conclusion is scenario oriented.

In this section, we present examples of hypothesis tests for means and proportions.

#### 4.5.1.1 Tests with One Sample Mean

$H_o$: $\mu = \mu_o$

$H_a$: This can be any of the following as required:

$\mu \neq \mu_o$ $\mu < \mu_o$ $\mu > \mu_o$

Test Statistic: $Z = \frac{xbar - \mu_0}{\sigma / \sqrt{n}}$

Decision: Reject the claim, *Ho* if and only if (iff) for

$\mu \neq \mu_o$ Either $Z \geq z\alpha/2$ or $Z \leq -z\alpha/2$

$\mu < \mu_o$ $Z \leq -z\alpha$

$\mu > \mu_o$ $Z \geq z\alpha$

#### 4.5.1.2 Tests with a Population Proportion (Large Sample)

Null Hypothesis: $H_o$: $p = p_o$

*Test Statistic:* $z = \frac{p_1 - p_0}{\sqrt{p_0(1-p_0)/n}}$

| *Alternative Hypothesis* | *Rejection Region* |
|---|---|
| *Ha: $p > p_0$* | $z \geq z_\alpha$ |
| *Ha: $p < p_0$* | $z \leq -z_\alpha$ |
| *Ha: $p \neq p^0$* | *either* $z \geq z_{\alpha/2}$ *or* $z \leq -z_{\alpha/2}$ |

These procedures are valid for $np_0 \geq 5$ and $n(1\text{-}p_o) \geq 5$.

### *4.5.1.3 Tests Comparing Two Sample Means*

$H_o$: $\mu_1 = \mu_2$ we write this as $\mu_1 - \mu_2 = 0$

$H_a$: This can be any of the following as required:

$\Delta\mu \neq 0$ , $\Delta\mu < 0$ $\Delta\mu > 0$

Test Statistic: $Z = \frac{xbar1 - xbar2}{\sqrt{\frac{s_1^2}{m} + \frac{s_2^2}{n}}}$

Decision: Reject the claim, *Ho* iff for

$\Delta\mu \neq 0$ Either $Z \geq z\alpha/2$ or $Z \leq -z\alpha/2$

$\Delta\mu < 0$ $Z \leq -z\alpha$

$\Delta\mu > 0$ $Z \geq z\alpha$

**Example 4.17:** Aviation

You run a small aviation transport company for a major corporation. You are tired of hearing management complain that your crews rest too much during the day. Aviation rules require a crew to get around 9 hours of rest each day. You collect a sample of 37 crew members and determine that their sample average, $\bar{x}$, is 8.94 hours with a sample deviation of 0.2 hours.

The parameter of interest is the true population mean, μ. We determine that the worst-case scenario is less than 9 hours, so our test is:

$$H_0:\ \mu > 9$$
$$H_a:\ \mu < 9$$

The test statistic is $Z = \frac{(\bar{x} - \mu)}{s / \sqrt{n}}$. This is a one-tailed test.

We select α to be 0.05.

We reject $H_0$ at $\alpha = 0.05$, if $Z < -1.645$.

From our sample of 36 aviators, we find $Z = \frac{(\bar{x} - \mu)}{s / \sqrt{n}} = \frac{(8.94 - 9)}{0.2 / \sqrt{36}} = -1.8$.

$$Z = -1.8.$$

Since, $-1.8 < -1.645$, then we reject null hypothesis that aviators rest 9 or more hours per day and conclude the alternate hypothesis is true, that your aviators rest less than 9 hours per day. Rejecting the null hypothesis is the better strategy because it is now concluded that we reject the null hypothesis that the aviator crews rest 9 or more hours a day.

***p*-value**. The *p*-value is the appropriate probability related to the test statistic. It is written so that the result is the smallest alpha ($\alpha$) level in which we may reject the null hypothesis. It is normal probability. From above our test statistic is $-1.8$. We are doing a lower tail test. *p*-value is

$P(Z < -1.8) = 0.0359$. Thus, we reject the null hypothesis for all values of alpha $> 0.0359$. Thus, we reject of alpha is 0.05 but fail to reject if alpha is 0.01.

One critical component of statistical significance testing is the *p*-value ($\alpha$), or the probability of calculating a test statistic that is at least as large as the actual observed value, assuming the null hypothesis is true. The common practice is to reject the null hypothesis when the *p*-value is less than the significance level $\alpha$. The most common tested significance levels are 0.05 or 0.01. When the null hypothesis is rejected, the result is said to be statistically significant.

The *p*-value is a probability that has a value ranging from zero to one. It supports answering the question: If the population really has the same mean as the observed data, what is the probability that the random sampling would lead to a difference between the sample mean as large as observed?

We usually use either a normal distribution directly or evoke the central limit theorem

(for $n > 30$) for testing means. We have the ability to conduct this test in Excel (Figure 4.21).

**Example 4.18:** Proportions

We believe that the mean of our distribution is ½. We want to test if our sample comes from this distribution.

$$H_0: \mu = 1/2$$
$$Ha: \mu \neq 1/2$$

The test statistic is key. From our data with sample size $n = 49$, we find that the mean is 0.41 and the standard deviation is 0.2.

The test statistic for a **one sample test of a proportion** is

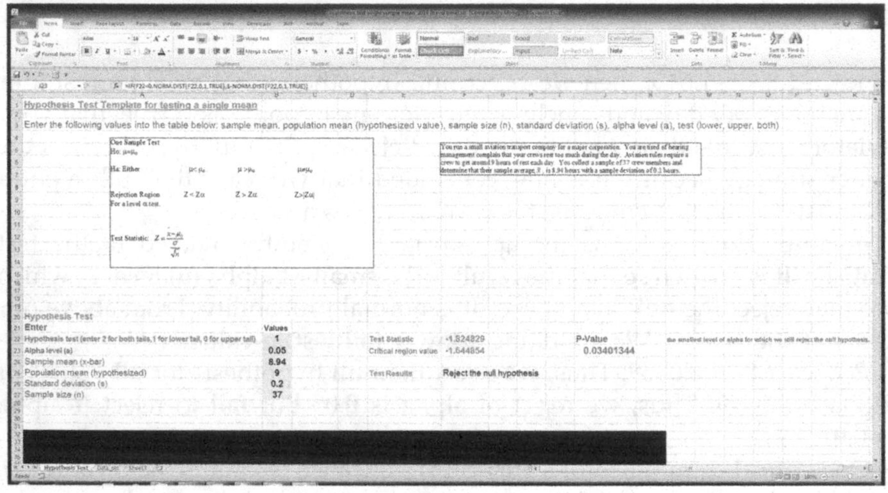


**FIGURE 4.21**
Excel solution of hypothesis test for Example 4.11.

$$z = \frac{p - p_0}{\sqrt{p_o(1 - p_o)/n}}$$

So we substitute $p = 1/2$, $p_o = 0.41$, $(1-p_o) = 0.59$, $n = 49$

$$z = \frac{0.5 - 0.41}{\sqrt{.41(1 - .41)/49}}$$

We find $z = 1.28$

We next need to find the probability that corresponds to the statement *P(Z ≥ 1.28)*.

If $z$ were negative, we would want *P(Z ≤ −1.28)*.

We need to find that probability. We find *P(Z ≤ 1.28)* = 0.8997. So *P(Z > 1.28)* = 1 − *p(Z < 1.28)* = 1 − 0.8997 = 0.1003.

We compare that value, *p*, to our level of significance (Figure 4.22).

If $p < a$, then we have significance ($a$ is usually either *0.05* or 0.01).

Statistical calculations can answer this question: If the populations really have the same mean, what is the probability of observing such a large difference (or larger) between sample means in an experiment of this size? The answer to this question is called the p-value.

The *p*-value is a probability, with a value ranging from zero to one. If the *p*-value is small, you may conclude that the difference between sample

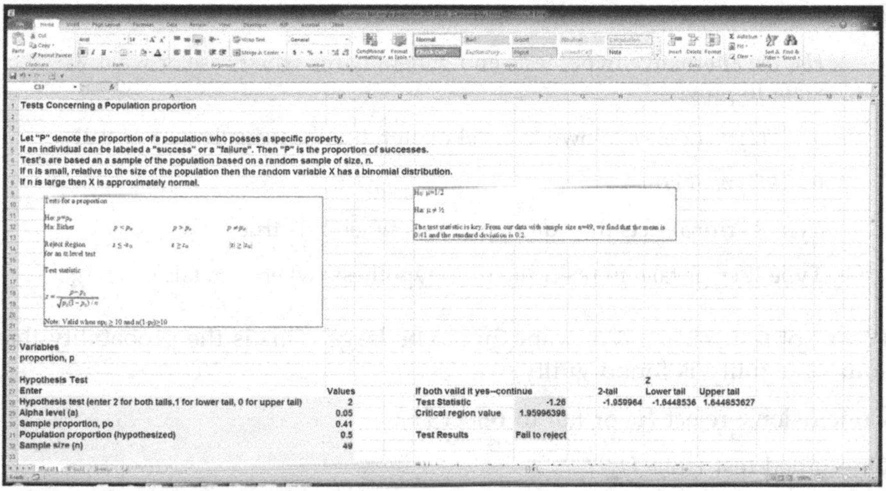

**FIGURE 4.22**
Excel solution of hypothesis test for Example 4.12.

means is unlikely to be a coincidence. Instead, you will conclude that the populations have different means.

We can build an Excel template (Figure 4.23) to help determine the values needed to conduct our hypothesis tests.

Given our hypothesis test above, the probability of a Type I error, α, is the area under the normal bell-shaped curve centered at $\mu_0$ corresponding to the rejection region. This value is 0.05.

### 4.5.2 Notation and Definitions

We will start with some basic definitions associated with hypothesis testing.

| 2 Tail Test | | Test Stat Value | | |
|---|---|---|---|---|
| Mean | 8.94 | $Z = \dfrac{\bar{x} - \mu}{s/\sqrt{n}}$ | -1.8 | |
| Population Mean, hypoth | 9 | | | |
| Standard Deviation, S | 0.2 | | | |
| N, sample size | 36 | | | |
| Alpha Level | 0.05 | -1.64485 | | Results |
| Enter tail information | 2 | | | Reject |
| Upper tail as 0 | | | | |
| Lower tail as 1 | | | | |
| Both tails as 2 | | | | |
| User inputs are in yellow | | | | |

**FIGURE 4.23**
Excel template for hypothesis data.

$H_0$: the null hypothesis and what we assume to be true.

*Ha:* the alternative hypothesis and generally the worst case or what we want to prove.

$\alpha$ = *P(Type I error),* known as level of significance (usually 0.05 or 0.01).

$\beta$ = *P(Type II error).*

Type I error rejects the null hypothesis when it is true.

Type II error fails to reject the null hypothesis when it is false.

Power of test = 1 − β. We want this to be large. This is the probability that someone guilty is found guilty.

Conclusions: reject $H_0$ or fail to reject $H_0$.

One-tailed test from $H_a$.

Two-tailed test from $H_a$.

Test statistic, $T_s$, comes from our data and is found by $z = \frac{\bar{x} - \mu}{\left(\frac{s}{\sqrt{n}}\right)}$

Rejected region: that area under the normal curve where we reject the null hypothesis.

*P*-value is the smallest level of significance at which $H_0$ would be rejected when a specified test procedure is used on a given data set. We compare *p*-value to our given α.

If *p*-value ≤ α, then we will reject $H_o$ at level α. If *p*-value > α, then we fail to reject $H_o$ at level α. It is usually thought of as the probability associated with the test statistic, $P(\bar{X} > T_S)$. Table 4.8 captures the relationship between our conclusions and type error.

**Example 4.19:** Type I and Type II Errors

The example of trial defendants is a classic case to demonstrate the ideas of Type I and Type II error.

Trial Defendant: $H_0$: The defendant is innocent

$H_A$: The defendant is guilty

**TABLE 4.8**

Hypothesis Testing Relationship

| | | $H_o$ True | $H_o$ is False |
|---|---|---|---|
| Test Conclusion | Reject $H_o$ | Type I Error, *P(Type I error)* = α | Correct Decision |
| | Fail to Reject $H_o$ | Correct Decision | Type II Error *P(Type II error)* = β |

What is a Type I error: Someone who is innocent is convicted – we want that to be small.

What is a Type II error: Someone who is guilty is cleared – we want that to be small also.

Safe Drug: $H_0$: The drug is not safe and effective.

$H_A$: The drug is safe and effective.

What is a Type I error: Unsafe/ineffective drug is approved.

What is a Type II error: Safe/effective drug is rejected.

The reason we do it this way is we want to prove that the drug is safe and effective.

Mathematically when we examine hypothesis tests we always put the = with $H_0$!!!!! (Figure 4.24)

**Example 4.20:** Two-Tailed Test. Being Too Big or Too Small Is Bad

A **machine that produces rifle barrels** is set so that the average diameter is 0.50 inch. In a sample of 100 rifle barrels, it was found that $\bar{x}$ = 0.51 inch. Assuming that the standard deviation is 0.05 inch, can we conclude at the 5% significance level that the mean diameter is not 0.50 inch?

$H_0$: $\mu = 0.50$

$H_A$: $\mu \neq 0.50$

*Rejection region*: $|z| > z_{\alpha/2} = z_{0.025} = \mathbf{1.96}$

Draw the picture and rejection region (Figure 4.25)

Test statistic: $z = (\bar{x} - \mu) / (\sigma / \sqrt{n}) = (0.51 - 0.50) / (0.05 / \sqrt{100}) = 0.01 / 0.005 = \mathbf{2.0}$

Conclusion: Reject $H_0$, Yes.

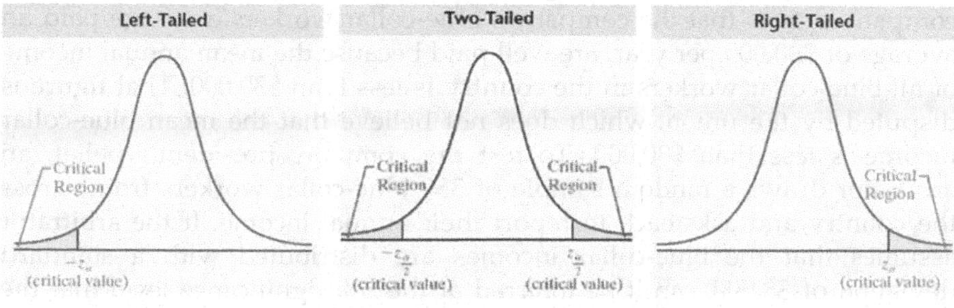


**FIGURE 4.24**
One versus two-tailed testing.

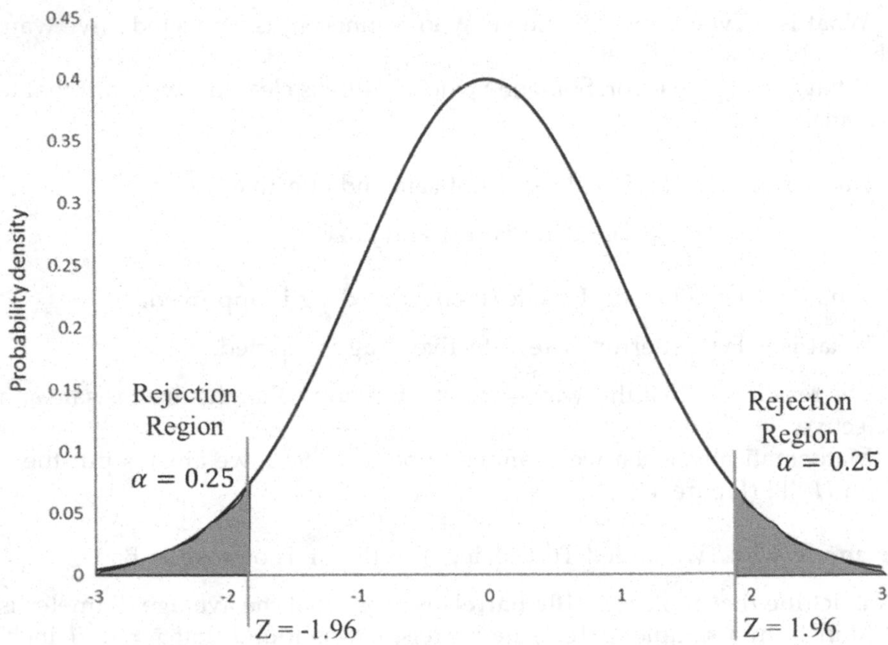


**FIGURE 4.25**
Rejection region for α = 0.25.

*P*-value: The probability of obtaining a sample result that is at least as unlikely as what is observed, or the observed level of significance. It is the probability in the tail associated with the value: $P(Z > 2)$ or $p(\bar{X} > X_{ts})$.

In the **rifle barrel case**: *p*-value = 0.5 – 0.4772 = **0.0228**

Using Excel: = *norm.dist(2,4,0,1)* = 0.022718

**Example 4.21:** Left-Tailed Test. Being Too Small Is the Worst Case

In the middle of **labor-management negotiations**, the president of a company argues that the company's blue-collar workers, who are paid an average of $30,000 per year, are well paid because the mean annual income of all blue-collar workers in the country is less than $30,000. That figure is disputed by the union, which does not believe that the mean blue-collar income is less than $30,000. To test the company president's belief, an arbitrator draws a random sample of 350 blue-collar workers from across the country and asks each to report their annual income. If the arbitrator assumes that the blue-collar incomes are distributed with a standard deviation of $8,000, can it be inferred at the 5% significance level that the company president is correct?

$$H_0\text{: } \mu \geq 30{,}000$$
$$H_A\text{: } \mu < 30{,}000$$

*Rejection region:* $z < z_\alpha = -z_{0.05} = \mathbf{-1.\,645}$

$\Uparrow \quad \Uparrow$

*One – tail test, draw picture again.*

$$\bar{x} = 29{,}120$$

*Test statistic:* $z = (\bar{x} - \mu)/(\sigma/\sqrt{n}) = (29{,}120 - 30{,}000)/(8{,}000/\sqrt{350})$
$= -880/427.618 = \mathbf{-2.058}$

Conclusion: Reject $H_0$. Yes.

*p*-value: the smallest value of that would lead to rejection of the null hypothesis.

$$p\text{-value} = P(z < -2.058) = 0.5 - 0.4803 = \mathbf{0.}0197$$

$\Uparrow$

$z \approx 2.06 \Rightarrow$

**Example 4.22:** Left-Tailed Test. Being Too Large Hurts Unit Performance and Is the Worst Case

We want to measure if the new regulations were effective, so we look to the left-tailed test to prove they were effective. In an attempt to reduce the number of person-hours lost as a result of non-combat related military accidents, the DOD has put in place new safety regulations. In a test of the effectiveness of the new regulations, a random sample of 50 units was chosen. The number of person-hours lost in the month prior to and the month after the installation of the safety regulations was recorded. Assume that the population standard deviation is = 5. What conclusion can you draw using a 0.05% significance level?

$\bar{x} = \mathbf{-1.2}$
$H_0\text{: } \mu \geq 0$
$H_A\text{: } \mu < 0$
*Rejection region:* $z < z_\alpha = -z_{0.05} = -1.644$

$\Uparrow \quad \Uparrow$

*One – tail test*

Draw the picture.

Test statistic: $z = (\bar{x} - \mu) / (\sigma / \sqrt{n}) = (-1.20 - 0) / (5 / \sqrt{50}) = -1.2 / 0.707 = \mathbf{-1.697}$

***p*-value** = 0.5 – 0.4554 = **0.0446**

Conclusion: Reject $H_0$, since $-1.697 < -1.644$

The new safety regulations are effective.

**Example 4.23:** Right-Tailed Test

Average, μ, of time spent reading newspapers: 8.6 minutes. Do people in military leadership positions spend more time than the national average time per day reading newspapers?

$$H_0: \mu \leq 8.6$$
$$H_A: \mu > 8.6$$

We sampled 100 officer and found that they spend 8.66 minutes reading the paper (or from the web) with a standard deviation of 0.1 minute.

$$Z = 8.66 - 8.6/(0.1/10) = 6$$
*Reject if* $Z > Z\alpha$, *assume* $\alpha = 0.01$. $Z\alpha = 2.32$
*Since* 6 > 2.32 *we reject Ho.*

a. **Type I Error:** Rejecting the null hypothesis, $H_0$, when it is true: Concluding that the mean newspaper-reading time for managers is greater than the national average of 8.6 minutes when in fact it is not.

**Possible consequences:** Wasted money on campaigns targeting managers who are *believed* to spend more time reading newspapers than the national average.

b. **Type II Error:** Failing to reject the null hypothesis, $H_0$, when it is false: Concluding that the mean newspaper-reading time for managers is less than or equal to the national average of 8.6 minutes when in fact it is greater than 8.6.

**Possible consequences:** Missed opportunity to potentially access managers who *may* spend more time reading newspapers than the national average.

**Example 4.24:** Small Sample from a Normal Ist

The population mean earnings per share for financial services corporations including American Express, E*Trade Group, Goldman Sachs, and Merrill

**TABLE 4.9**

Earning Per Share

| 1.92 | 2.16 | 3.63 | 3.16 | 4.02 | 3.14 | 2.20 | 2.34 | 3.05 | 2.38 |
|---|---|---|---|---|---|---|---|---|---|

Lynch was $3 (*Business Week*, August 14, 2000). In 2001, a sample of 10 financial services corporations provided the following earnings per share (Table 4.9):

Determine whether the population mean earnings per share in 2001 differ from $3 reported in 2000. α = 0.05

$$H_0: \mu = 3$$
$$H_A: \mu \neq 3$$

$t_{0.025,9}$ = **2.262** EXCEL: – (TINV(0.05,9) – **2.262159**

Reject if $t < -2.262$ or if $t > 2.262$

| | Mean Earnings | $(x - 2.8)$^2 | |
|---|---|---|---|
| | 1.92 | 0.77 | |
| | 2.16 | 0.41 | |
| | 3.63 | 0.69 | |
| | 3.16 | 0.13 | |
| | 4.02 | 1.49 | |
| | 3.14 | 0.12 | |
| | 2.20 | 0.36 | |
| | 2.34 | 0.21 | |
| | 3.05 | 0.06 | |
| | 2.38 | 0.18 | |
| **Sum** | **28.00** | **4.42** | |
| **Mean** | **2.8** | **0.4908** | **= Variance** |
| | | **0.7006** | **= St. Deviation** |

$t = (x^{bar} - \mu) / (s / \sqrt{n}) = (2.8 - 3) / (0.7006 / \sqrt{10}) = \mathbf{-0.9027}$

*p*-value: EXCEL: = (TDIST(0.9027, 9, 2)) = **0.390**

Do not reject $H_0$. We cannot conclude that the population mean earning per share has changed.

Again, utilizing a **confidence interval** to make a decision:

$$x^{bar} \pm t_{\alpha/2}(s/\sqrt{n}) = 2.\ 8 \pm 2.\ 262(0.\ 7006/\sqrt{10}) =$$
$$2.\ 8 \pm 0.\ 50 \Rightarrow \mathbf{2.\ 30.} \ldots 3.\ 30$$

As the claimed mean ($3) is within the range, we cannot reject the null hypotheses.

**Example 4.25:** Tests About a Population Proportion

In a television commercial, the manufacturer of a **toothpaste claims** that more than four out of five dentists recommend the ingredients in the product. To test that claim, a consumer-protection group randomly samples 400 dentists and asks each one whether they would recommend a toothpaste that contained certain ingredients. The responses are 0 = No and 1 = Yes. There were 71 No answers and 329 Yes answers. At the 5% significance level, can the consumer group infer that the claim is true or not true?

$\hat{p} = 329/400 - 0.8225$ p = 0.8

$H_0$: $p \leq 0.8$

$H_A$: $p > 0.8$ *Rejection region*: $z > z_\alpha = z_{0.05} = 1.645$

*Test statistic*: $z = (p - p)/\sqrt{(pq/n)} = (0.8225 - 0.8)/\sqrt{[(0.8 * 0.2.)/400]} =$
$0.0225/0.02 = \mathbf{1.125}$

Conclusion: Do not reject $H_0$. The claim is likely to be true.

If $\hat{p}$ (for some reason) remains 0.8225. How big would $n$ have to be for us to be able to support the claim?

$$1.645 = (0.8225 - 0.8)/\sqrt{(0.8 * 0.2)/n)}$$
$$n = 855.11 = 856$$

**Example 4.26:** A Windmill Example

The feasibility of constructing a profitable electricity-producing windmill depends on the average velocity of the wind. For a certain type of windmill, the average wind speed would have to exceed 20 mph in order for its construction to be feasible. To test whether or not a particular site is appropriate for this windmill, 50 readings of the wind velocity are taken, and the average is calculated. The test is designed to answer the question, is the site feasible? That is, is there sufficient evidence to conclude that the average wind velocity exceeds 20 mph? We want to test the following hypotheses.

$$H_0: A \leq 20$$
$$H_A: A > 20$$

If, when the test is conducted, a Type I error is committed (rejecting the null hypothesis when it is true), we would conclude mistakenly that the average wind velocity exceeds 20 mph. The consequence of this decision is that the windmill would be built on an inappropriate site. Because this error is quite costly, we specify a small value for *a*, α = 0.01.

If a Type II error is committed (not rejecting the null hypothesis when it is false), we would conclude mistakenly that the average wind velocity does not exceed 20 mph. As a result, we would not build the windmill on that site, even though the site is a good one. The cost of this error may not be very large, since, if the site under consideration is judged to be inappropriate, the search for a good site would simply continue.

But suppose that a site where the wind velocity is greater than or equal to 25 mph is extremely profitable. To judge the effectiveness of this test (to determine if our selection of α = 0.01 and $n = 50$ is appropriate), we compute the probability of committing this error.

Our task is to calculate β when μ = 25. (Assume that we know that (σ = 12 mph.)

Our first task is to set up the rejection region in terms of $\bar{x}$.

Rejection region: $z > z_{\alpha} = z_{0.01} = 2.33$

So we have $z = (\bar{x} - \mu) / (\sigma/\sqrt{n}) = (\bar{x} - 20) / (12/\sqrt{50}) > 2.33$

*Rejection region*: $\bar{x}^r > 23.95$

*Region where* $H_0$ *is not rejected*: $\bar{x} < 23.95$

*Thus*: $\beta = P(\bar{x} < 23.95$ (*given thatm* $\mu = 25$)

$= P\{[(\bar{x} - \mu)/(\sigma/\sqrt{n})] < [(23.95 - 25) / (12/\sqrt{50})] = P(z < -0.62)$

$= 0.5–0.2324 =$ **0.2672**

| | | | | | | |
|---|---|---|---|---|---|---|
| 23.95 | 22 | 1.95 | 1.697 | 1.15 | 0.3749 | **0.8749** |
| 23.95 | 22.5 | 1.45 | 1.697 | 0.85 | 0.3023 | **0.8023** |
| 23.95 | 23 | 0.95 | 1.697 | 0.56 | 0.2123 | **0.7123** |
| 23.95 | 23.5 | 0.45 | 1.697 | 0.27 | 0.1064 | **0.6064** |
| 23.95 | 24 | −0.05 | 1.697 | −0.03 | 0.0120 | **0.4880** |
| 23.95 | 24.5 | −0.55 | 1.697 | −0.32 | 0.1255 | **0.3745** |
| **23.95** | **25** | **−1.05** | **1.697** | **−0.62** | **0.2324** | **0.2676** |
| 23.95 | 25.5 | −1.55 | 1.697 | −0.91 | 0.3186 | **0.1814** |
| 23.95 | 26 | −2.05 | 1.697 | −1.21 | 0.3869 | **0.1131** |
| 23.95 | 26.5 | −2.55 | 1.697 | −1.50 | 0.4332 | **0.0668** |
| 23.95 | 27 | −3.05 | 1.697 | −1.80 | 0.4641 | **0.0359** |

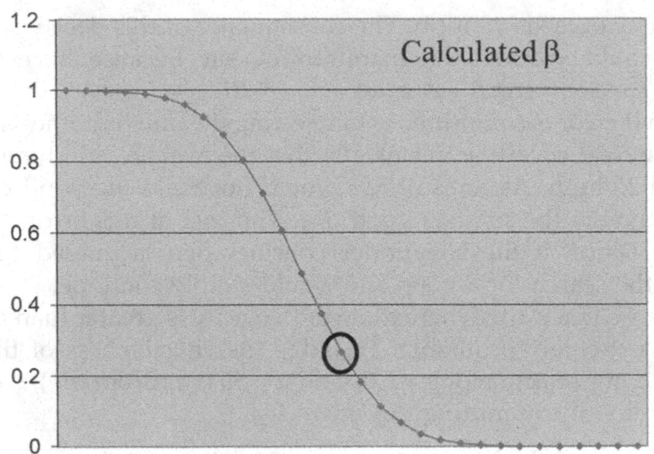


**FIGURE 4.26**
Graph β associated with numbers from 18 to 32.5.

This is the graph for β associated with numbers from 18 to 32.5 (Figure 4.26):

The probability of not rejecting the null hypothesis when $\mu = 25$ is 0.2676 (Figure 4.26).

This means that, when the mean wind velocity is 25 mph, there is a 26.76% probability of erroneously concluding that the site is not profitable. If this probability is considered too large, we can reduce it by either increasing $\alpha$ or increasing $n$.

For example, if we increase $\alpha$ to 0.10 and leave $n = 50$, then $\beta = 0.0475$.

Rejection region: $(\bar{x}\text{-}20) \;/\; (12/\sqrt{50}) > 1.28 \Rightarrow \bar{x} > 22.17$

| | | | | | | |
|---|---|---|---|---|---|---|
| 22.17 | 24.5 | −2.33 | 1.697 | −1.37 | 0.4147 | **0.0853** |
| ***22.17*** | ***25*** | ***−2.83*** | ***1.697*** | ***−1.67*** | ***0.4525*** | ***0.0475*** |
| 22.17 | 25.5 | −3.33 | 1.697 | −1.96 | 0.4750 | **0.0250** |

With $\alpha = 0.10$, however, the probability of building on a site that is not profitable is too large (Figure 4.27).

If we let $\alpha = 0.01$ but increase $n$ to 100, then $\beta = 0.0329$.

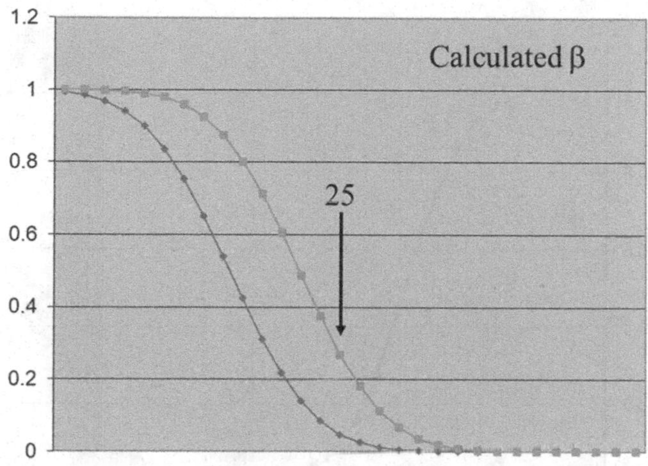


**FIGURE 4.27**
Update graph β.

| | | | | | | |
|---|---|---|---|---|---|---|
| 22.796 | 23 | −0.204 | 1.20 | −0.17 | 0.0675 | **0.4325** |
| 22.796 | 23.5 | −0.704 | 1.20 | −0.59 | 0.2224 | **0.2776** |
| 22.796 | 24 | −1.204 | 1.20 | −1.00 | 0.3413 | **0.1587** |
| 22.796 | 24.5 | −1.704 | 1.20 | −1.42 | 0.4222 | **0.0778** |
| ***22.796*** | ***25*** | ***−2.204*** | ***1.20*** | ***−1.84*** | ***0.4671*** | ***0.0329*** |
| 22.796 | 25.5 | −2.704 | 1.20 | −2.25 | 0.4878 | **0.0122** |
| 22.796 | 26 | −3.204 | 1.20 | −2.67 | 0.4962 | **0.0038** |
| 22.796 | 26.5 | −3.704 | 1.20 | −3.09 | 0.4990 | **0.0010** |
| 22.796 | 27 | −4.204 | 1.20 | −3.50 | 0.5000 | **0.0000** |

Now, both α and β are quite small, but the cost of sampling has increased (Figure 4.28).

Nonetheless, the cost of sampling is small in comparison to the costs of making Type I and Type II errors in this situation.

Another way of judging a test is to measure its power – the probability of its leading us to reject the null hypothesis when it is false – rather than measuring the probability of a Type II error.

Thus, the power of the test is equal to 1 – β. In the present example, the power of the test with $n$ = 50 and α = 0.01 is 1 – 0.2676 = 0.7324.

| | | |
|---|---|---|
| 2.34 | 3.05 | 2.38 |

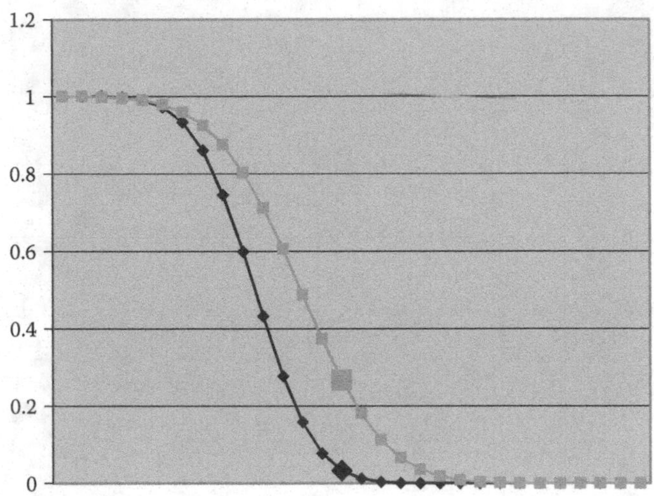


**FIGURE 4.28**
Updated graph β.

## 4.6 Exercises

### 4.6.1 Classical Probability

- **Basic Skills**

In Problems 1–14, find the indicated probabilities by referring to the tree (Figure 4.29) diagram given below and by using Bayes' Theorem.

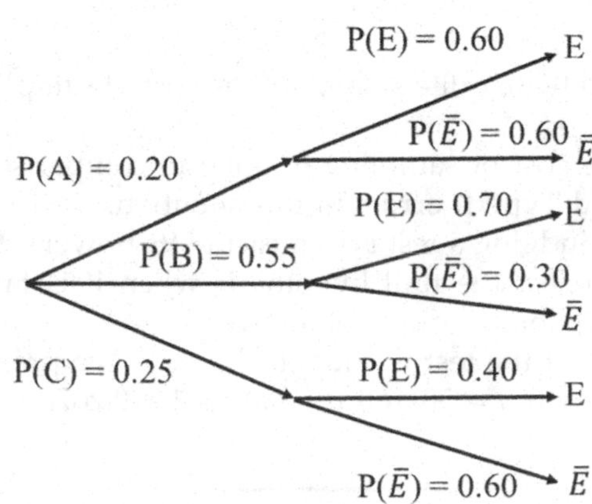


**FIGURE 4.29**
Tree diagram for Questions 1–14.

1. $P(E \mid A)$
2. $P(E \mid B)$
3. $P(\bar{E} \mid A)$
4. $P(\bar{E} \mid B)$
5. $P(E \mid C)$
6. $P(\bar{E} \mid C)$
7. $P(E)$
8. $P(\bar{E})$
9. $P(A \mid E)$
10. $P(A \mid \bar{E})$
11. $P(C \mid E)$
12. $P(B \mid \bar{E})$
13. $P(B \mid E)$
14. $P(C \mid \bar{E})$
15. Suppose that events $A_1$ and $A_2$ form a partition of the sample space $S$ with $P(A_1) = 0.55$ and $P(A_2) = 0.45$. If $E$ is an event that is a subset of $S$ and $P(E \mid A_1) = 0.06$ and $P(E \mid A_2) = 0.08$, find $P(E)$.
16. Suppose that events $A_1$ and $A_2$ form a partition of the sample space $S$ with $P(A_1) = 0.35$ and $P(A_2) = 0.65$. If $E$ is an event that is a subset of $S$ and $P(E \mid A_1) = 0.12$ and $P(E \mid A_2) = 0.09$, find $P(E)$.
17. Suppose that events $A_1$, $A_2$, and $A_3$ form a partition of the sample space $S$ with $P(A_1) = 0.35$, $P(A_2) = 0.45$, and $P(A_3) = 0.2$. If $E$ is an event that is a subset of $S$ and $P(E \mid A_1) = 0.25$, $P(E \mid A_2) = 0.18$, and $P(E \mid A_3) = 0.14$, find $P(E)$.
18. Suppose that events $A_1$, $A_2$, and $A_3$ form a partition of the sample space $S$ with $P(A_1) = 0.3$, $P(A_2) = 0.65$, and $P(A_3) = 0.05$. If $E$ is an event that is a subset of $S$ and $P(E \mid A_1) = 0.05$, $P(E \mid A_2) = 0.25$, and $P(E \mid A_3) = 0.5$, find $P(E)$.
19. **Colorblindness** The most common form of colorblindness is so-called "red-green" colorblindness. People with this type of colorblindness cannot distinguish between green and red. Approximately 8% of all males have red-green colorblindness, whereas only about 0.64% of women have red-green colorblindness. In 2000, 49.1% of all Americans were male and 50.9% were female according to the U.S. Census Bureau.

    a. What is the probability that a randomly selected American is colorblind?

    b. What is the probability that a randomly selected American who is colorblind is female?

**TABLE 4.10**

Murder Victims in 2000

| Level | Event | Proportion |
|---|---|---|
| Less than 17 years | $A_1$ | 0.082 |
| 17–29 | $A_2$ | 0.424 |
| 30–44 | $A_3$ | 0.305 |
| 45–59 | $A_4$ | 0.125 |
| At least 60 years | $A_5$ | 0.064 |

Source: Federal Bureau of Investigation.

20. **The Elias Test** The standard test for the HIV virus is the Elias test that tests for the presence of HIV antibodies. If an individual does not have the HIV virus, the test will come back negative for the presence of HIV antibodies 99.8% of the time and will come back positive for the presence of HIV antibodies 0.2% of the time (a false positive). If an individual has the HIV virus, the test will come back positive 99.8% of the time and will come back negative 0.2% of the time (a false negative). The latest reports available indicate that approximately 0.7% of the world population has the HIV virus.
    a. What is the probability that a randomly selected individual has a test that comes back positive?
    b. What is the probability that a randomly selected individual has the HIV virus if the test comes back positive?
21. **Murder Victims** The following data (Table 4.10) represent the proportion of murder victims at the various age levels in 2000.

If we let $M$ represent the event that a randomly selected murder victim was male, then we can also obtain the following probabilities:

| | | |
|---|---|---|
| $P(M \mid A_1) = 0.622$ | $P(M \mid A_2) = 0.843$ | $P(M \mid A_3) = 0.733$ |
| $P(M \mid A_4) = 0.730$ | $P(M \mid A_5) = 0.577$ | |

a. What is the probability that a randomly selected murder victim was male?
b. What is the probability that a randomly selected male murder victim was 17–29 years of age?
c. What is the probability that a randomly selected male murder victim was less than 17 years of age?

22. **Espionage** Suppose that the CIA suspects that one of its operatives is a double agent. Past experience indicates that 95% of all operatives suspected of espionage are, in fact, guilty. The CIA decides to administer a polygraph to the suspected spy. It is known that the polygraph returns results that indicate a person is guilty 90% of the time if they are guilty. The polygraph returns results that indicate a person is innocent 99% of the time if they are innocent. What is the probability that this particular suspect is innocent given that the polygraph indicates that he is guilty?

### 4.6.2 Discrete Distributions

1. If 75% of all purchases at Wal-Mart are made with cash andX is the number among ten randomly selected purchases made with cash, then find the following:
   a. $p(X = 5)$
   b. $p(X \le 5)$
   c. $\mu$, and $\sigma^2$
2. Russell Stover's produces fine chocolates and it's known from experience that 10% of its chocolate boxes have flaws and must be classified as "seconds."
   a. Among six randomly selected chocolate boxes, how likely is it that one is a second?
   b. Among the six randomly selected boxes, what is the probability that at least two are seconds?
   c. What is the mean and variance for "seconds"?
3. Consider the following TV ad for an exercise program: 17% of the participants lose 3 pounds, 34% lose 5 pounds, 28% lose 6 pounds, 12% lose 8 pounds, and 9% lose 10 pounds. Let X = the number of pounds lost on the program.
   a. Give the probability mass function of X in a table.
   b. What is the probability that the number of pounds lost is at most 6? At least 6?
   c. What is the probability that the number of pounds lost is between 6 and 10?
   d. What are the values of $\mu$ and $\sigma^2$?
4. A machine fails on average 0.4 times a month (30 consecutive days). Determine the probability that there are 10 failures in the next year.

### 4.6.3 Continuous Probability Models

Find the following probabilities:

1. $X \sim N(\mu = 10, \sigma = 2)$, $P(X > 6)$
2. $X \sim N(\mu = 10, \sigma = 2)$, $P(6 < x < 14)$
3. Determine the probability that lies within one standard deviation of the mean, two standard deviations of the mean, and three standard deviations of the mean. Draw a sketch of each region.
4. A tire manufacturer thinks that the amount of wear per normal driving year of the rubber used in their tire follows a normal distribution with mean = 0.05 inches and standard deviation 0.05 inches. If 0.10 inches is considered dangerous, then determine the probability that $P(X > 0.10)$.

### 4.6.4 Hypothesis Testing

Discuss how to set up each of the following as a hypothesis test.

1. Does drinking coffee increase the risk of getting cancer?
2. Does taking aspirin every day reduce the chance of a heart attack?
3. Which of two gauges is more accurate?
4. Why is a person "innocent until proven guilty"?
5. Is the drinking water safe to drink?
6. Which error, Type I or Type II, is the worst error? Numerous complaints have been made that a certain hot coffee machine is not dispensing enough hot coffee into the cup. The vendor claims that on average the machine dispenses at least 8 ounces of coffee per cup. You take a random sample of 36 hot drinks and calculate the mean to be 7.65 ounces with a standard deviation of 1.05 ounces. Find a 95% confidence interval for the true mean.
7. Numerous complaints have been made that a certain hot coffee machine is not dispensing enough hot coffee into the cup. The vendor claims that on average the machine dispenses at least 8 ounces of coffee per cup. You take a random sample of 36 hot drinks and calculate the mean to be 7.65 ounces with a standard deviation of 1.05 ounces. Set up and conduct a hypothesis test to determine if the vendors claim is correct. Use an $\alpha = 0.05$ level of significance. Determine the Type II error if the true mean were 7.65 ounces.
8. An intelligence agency claims that the proportion of the population who have access to computers in Afghanistan is at least 30%. A sample of 500 people is selected, and 125 of these said they had access to a computer. Test the claim at a 5% level of significance.

9. A manufacturer of AA batteries claims that the mean lifetime of their batteries is 800 hours. We randomly select 40 batteries and find their mean is 790 hours with a standard deviation of 22 hours. Test the claim at both a 5% and a 1% level of significance.
10. As a commander you are asked to test a new weapon in the field. This weapon is claimed to 95% reliable. You issue 250 of these weapons to your soldiers, and of these 15 did not work properly, i.e. failed to meet military specifications. Perform a hypothesis test at a 5% level of significance.
11. You need steel cables for an upcoming mission that are at least 2.2 cm in diameter. You procure 35 of the cables and find through measurement that the mean diameter is only 2.05 cm. The standard deviation is 0.3 cm. Perform a hypothesis test of the cables at a 5% level.
12. For safety reason, it is important that then mean concentration of a chemical used to make, a volatile substance not exceed 8mg/L. A random sample of 34 containers have a sample mean of 8.25 mg/L with a standard deviation of 0.9 mg/L. Do you conform to the safety requirements?
13. In a survey in 2003, adult Americans were asked which invention they hated the most but could not do without; 30% chose the cell phone. In a more recent survey, 363 of 1,000 adult Americans surveyed stated that the cell phone was the invention they hated the most and could not do without. Test at the 5% level of significance if the proportion of adult Americans who hate and have a cell phone is the same as it was in 2003.

## References

Devore, J. (1995). Probability and Statistics for Engineeringand Sciences, 4th Edition. Wadsworth Publishing, Boston, CA.

Sullivan, M. (2018). Fundamental of Statistics: Informeddecision using data, 5th Edition. Pearson Publishers, Boston, MA.

# 5

# *Differential Equations*

**OBJECTIVES**

1. Understand the typical models of ordinary differential equations.
2. Understand the analytical solution techniques for separable and linear ODEs.
3. Know the numerical methods.
4. Understand the importance of slope fields.
5. Understand the concept of correlation and linearity.
6. Build and interpret nonlinear regression models.
7. Build and interpret logistics regression models.
8. Build and interpret Poisson regression models.

## 5.1 Introduction

In this section, we introduce many mathematical models from a variety of disciplines. Our emphasis in this section is building the mathematical model, or expression, that will be solved later in the chapter. Recall previously that we discussed the modeling process. In this section, we will confine ourselves to the first three steps of the modeling process: (1) Identifying the problem, (2) Assumptions and Variables, and (3) Building the model.

**Example 5.1:** Competition between Species

Imagine a small fishpond supporting both trout and bass. Let *T(t)* denote the population of trout at time *t* and *B(t)* denote the population of bass at time *t*. We want to know if both can coexist in the pond. Although population growth depends on many factors, we will limit ourselves to basic isolated growth and the interaction with the other competing species for the scarce life-support resources.

DOI: 10.1201/9781003298762-5

We assume that the species grow in isolation. The level of the population of the trout or the bass, *B(t)* and *T(t)*, depend on many variables such as their initial numbers, the amount of competition, the existence of predators, their individual species birth and death rates, and so forth. In isolation we assume the following proportionality models (following the same arguments as the basic populations models that we have discussed before) to be true where the environment can support an unlimited number of trout and/or bass. Later, we might refine this model to incorporate the limited growth assumptions of the logistics model:

$$\frac{dB}{dt} = mb$$
$$\frac{dT}{dt} = aT$$

Next, we modify the proceeding differential equations to take into account the competition of the trout and bass for living space, oxygen, and food supply. The effect is that the interaction decreases the growth of the species. The interaction terms for competition led to decay rate that we call $n$ for bass and $b$ for trout. This leads to following simplified model:

$$\frac{dB}{dt} = mb - nBT$$
$$\frac{dT}{dt} = aT - bBT$$

If we have the initial stocking level, $B_0$ and $T_0$, we determine how the species coexist over time.

If the model is not reasonable, we might try logistic growth instead of isolated growth. Logistic growth in isolation were discussed in first order ODE models as a refinement.

**Example 5.2: Economics:** Basic Supply and Demand Models

Suppose we are interested in the variation of the price of a specific product. It is observed that a high price for the product attracts more suppliers. However, if we flood the market with the product the price is driven down. Over time there is an interaction between price and supply. Recall the "Tickle Me Elmo" from Christmas a few years ago.

Problem Identification: Build a model for price and supply for a specific product.

Assumptions and variables:

Assume the price is proportional to the quantity supplied. Also, assume the change in the quantity supplied is proportional to the price. We define the following variables.

$$P(t) = \text{the price of the product at time, } t$$
$$Q(t) = \text{the quantity supplied at time, } t$$

We define two proportionalities constants as $a$ and $b$. The constant $a$ is negative and represents a decrease in price as quantity increases.

With our limited assumptions, the model could be

$$\frac{dP}{dt} = -aQ$$
$$\frac{dQ}{dt} = bP$$

**Example 5.3:** Predator-Prey Relationships

We now consider a model of population growth for two species in which one animal is hunted by another animal. An example of this might be wolves and rabbits where the rabbits are the primary food source for the wolves.

Let $R(t)$ = the population of the rabbits at time $t$ and $W(t)$ = the population of the wolves at time $t$.

We assume that rabbits grow in isolation but are killed by the interaction with the wolves. We further assume that the constants are proportionality constants.

$$\frac{dR}{dt} = aR - bRW$$

We assume that the wolves will die out without food and grow through their interaction with the rabbits. We further assume that these constants are also proportionality constants.

$$\frac{dW}{dt} = -mW + nRW$$

**Example 5.4:** An Electrical Network

Electrical networks with more than one loop give rise to systems of differential equations. Consider the electrical network displayed in Figure 5.1 where there are two resisters and two inductors. We apply Kirchhoff's law (the sum of the voltage drops in a closed circuit is equal to the impressed voltage) to each loop. We assume that no other factors interact with the flow of electricity in this circuit.

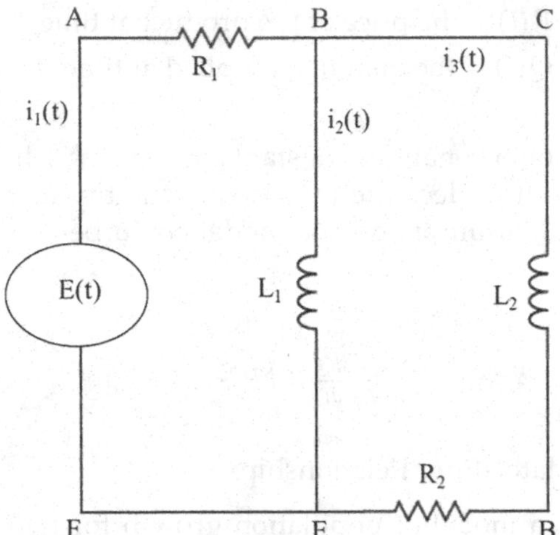


**FIGURE 5.1**
Electrical circuit.

Loop ABEF

$$E(t) = i_1 R_1 + L_1 \frac{di_2}{dt}$$

Loop ABCDEF

$$E(t) = i_1 R_1 + L_2 \frac{di_3}{dt} + i_3 R_2$$

We know that $i_1(t) = i_2(t) + i_3(t)$. We substitute this expression for $i_1$ into the laws to obtain the model:

$$\frac{di_2}{dt} = -\frac{R_1}{L_1} i_2 - \frac{R_1}{L_1} i_3 + \frac{E(t)}{L_1}$$

$$\frac{di_3}{dt} = -\frac{R_1}{L_2} i_2 - \frac{R_1 + R_2}{L_2} i_3 + \frac{E(t)}{L_2}$$

**Example 5.5:** Diffusion Models

Diffusion through a membrane leads to a first-order system of ordinary linear differential equations. For example, consider the situation in which two solutions of substance are separated by a membrane of permeability, $P$. Assume the amount of substance that passes through the membrane at any time is proportional to the difference between the concentrations of the two solutions. Therefore, if we let $x_1$ and $x_2$ represent the two concentrations, and $V_1$ and $V_2$ represent their corresponding volumes, then the system of differential equations is given by:

$$\frac{dx_1}{dt} = \frac{P}{V_1}(x_2 - x_1)$$
$$\frac{dx_2}{dt} = \frac{P}{V_2}(x_2 - x_1)$$

where the initial amounts of $x_1$ and $x_2$ are given.

If this model does not yield satisfactory results in terms of realism we might try a refinement of the diffusion model as follows:

Diffusion through a double-walled membrane, where the inner wall has permeability $P_1$ and the outer wall has permeability $P_2$ with $0<P_1<P_2$. Suppose the volume of the solution within the inner wall is $V_1$ and between the two walls is $V_2$. We let $x$ represent the concentration of the solution within the inner wall and $y$, the concentration between the two walls. This leads to the following system:

$$\frac{dx}{dt} = \frac{P_1}{V_1}(y - x)$$
$$\frac{dy}{dt} = \frac{1}{V_2}(P_2(C - y) + P_1(x - y))$$
$$x(0) = 2,\ y(0) = 1,\ C = 10$$

**Example 5.6:** S-I-R Models in a Pandemic

Consider a disease that is spreading throughout the Unites States such as the new flu or COVID-19. The CDC is interested in knowing and experimenting with a model for this new disease prior to it actually becoming a "real" epidemic or part of the pandemic. Let us consider the population being divided into three categories: susceptible, infected, and removed. We make the following assumptions for our model:

- No one enters or leaves the community, and there is no contact outside the community.
- Each person is either susceptible, $S$ (able to catch this new flu); infected, $I$ (currently has the flu and can spread the flu); or removed, $R$ (already had the flu and will not get it again; that includes death).
- Initially every person is either $S$ or $I$.
- Once someone gets the flu this year they cannot get again.
- The average length of the disease is 2 weeks, over which the person is deemed infected and can spread the disease.
- Our time period for the model will be per week.

The model we will consider is the SIR model (Allman & Rhodes, 2004).

Let's assume the following definition for our variables.

$S(n)$ = number in the population susceptible after period $n$.

$I(n)$ = number infected after period $n$.

$R(n)$ = number removed after period $n$.

Let's start our modeling process with $R(n)$. Our assumption for the length of time someone has the flu is 2 weeks. Thus, half the infected people will be removed each week,

$$dR/dt = 0.5I(t)$$

The value, 0.5, is called the removal rate per week. It represents the proportion of the infected persons who are removed from infection each week. If real data is available, then we could do "data analysis" in order to obtain the removal rate.

$I(t)$ will have terms that both increase and decrease its amount over time. It is decreased by the number that are removed each week, $0.5 * Itn)$. It is increased by the numbers of susceptible people that come into contact with an infected person and catch the disease, $aS(t)I(t)$. We define the rate, $a$, as the rate in which the disease is spread or the transmission coefficient. We realize this is a probabilistic coefficient. We will assume, initially, that this rate is a constant value that can be found from initial conditions.

Let's illustrate as follows. Assume we have a population of 1,000 students in the dorms. Our nurse found three students reporting to the infirmary initially. The next week, five students came into the infirmary with flu-like symptoms. $I(0) = 3$, $S(0) = 997$. In week 1, the number of newly infected is 30.

$$\begin{aligned} 5 &= a\,I(n)S(n) = a(3) * (995) \\ a &= 0.00167 \end{aligned}$$

Let's consider $S(t)$. This number is decreased only by the number that becomes infected. We may use the same rate, $a$, as before to obtain the model:

$$dS/dt = -0.00167\,S(t)I(t)$$

Our coupled SIR model is shown in the systems of differential equations below:

$$\begin{aligned} dR/dt &= 0.5\,I(t) \\ dI/dt &= -0.5\,i(t) + 0.00167\,I(t)S(t) \\ dS/dt &= -0.00167\,S(t)I(t) \\ I(0) &= 3,\ S(0) = 997,\ R(0) = 0 \end{aligned}$$

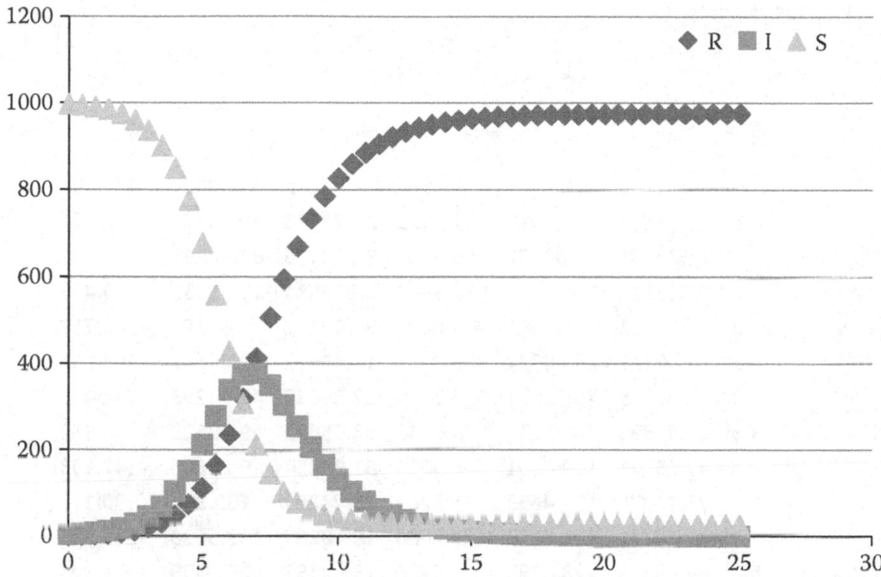


FIGURE 5.2
SIR model of the spread of a disease.

The SIR model above can be solved iteratively and viewed graphically. Let's iterate the solution and obtain the graph (Figures 5.2 and 5.3) to observe the behavior to obtain some insights.

In this example, we see that the maximum number of inflected persons occurs at about Day 7.

Everyone survives, and everyone will get the flu, which might not be the case in COVID-19, where we have a S-I-R-D model. In a S-I-R-D model a percent of the infected die. Under the current COVID-19 scenarios about 1%–2% die from the disease.

## 5.2 Qualitative Assessment of Autonomous Systems of First-Order Differential Equations

Consider the system of differential equations.

$$\frac{dx}{dt} = f(x, y)$$
$$\frac{dy}{dt} = g(x, y)$$

These are autonomous systems because they do not include $t$.

| | A | B | C | D | E | F | G |
|---|---|---|---|---|---|---|---|
| 1 | SIR Model | | Step Size: | 0.5 | | | |
| 2 | | | | | | | |
| 3 | t | R | I | S | R' | I' | S' |
| 4 | 0 | 0 | 3 | 997 | 1.5 | 3.49497 | -4.99497 |
| 5 | 0.5 | 0.75 | 4.747485 | 994.5025 | 2.373743 | 5.510972 | -7.88471 |
| 6 | 1 | 1.936871 | 7.502971 | 990.5602 | 3.751485 | 8.660195 | -12.4117 |
| 7 | 1.5 | 3.812614 | 11.83307 | 984.3543 | 5.916534 | 13.53551 | -19.452 |
| 8 | 2 | 6.770881 | 18.60082 | 974.6283 | 9.300412 | 20.97483 | -30.2752 |
| 9 | 2.5 | 11.42109 | 29.08824 | 959.4907 | 14.54412 | 32.06541 | -46.6095 |
| 10 | 3 | 18.69315 | 45.12094 | 936.1859 | 22.56047 | 47.98299 | -70.5435 |
| 11 | 3.5 | 29.97338 | 69.11244 | 900.9142 | 34.55622 | 69.42529 | -103.982 |
| 12 | 4 | 47.25149 | 103.8251 | 848.9234 | 51.91254 | 95.2805 | -147.193 |
| 13 | 4.5 | 73.20776 | 151.4653 | 775.3269 | 75.73267 | 120.384 | -196.117 |
| 14 | 5 | 111.0741 | 211.6573 | 677.2686 | 105.8287 | 133.5639 | -239.393 |
| 15 | 5.5 | 163.9884 | 278.4393 | 557.5723 | 139.2197 | 120.0479 | -259.268 |
| 16 | 6 | 233.5983 | 338.4633 | 427.9385 | 169.2316 | 72.6536 | -241.885 |
| 17 | 6.5 | 318.2141 | 374.7901 | 306.9959 | 187.395 | 4.753497 | -192.149 |
| 18 | 7 | 411.9116 | 377.1668 | 210.9216 | 188.5834 | -55.7305 | -132.853 |
| 19 | 7.5 | 506.2033 | 349.3016 | 144.4952 | 174.6508 | -90.3619 | -84.2889 |
| 20 | 8 | 593.5287 | 304.1206 | 102.3507 | 152.0603 | -100.078 | -51.982 |
| 21 | 8.5 | 669.5588 | 254.0815 | 76.3597 | 127.0407 | -94.6401 | -32.4006 |
| 22 | 9 | 733.0792 | 206.7614 | 60.15938 | 103.3807 | -82.6082 | -20.7725 |
| 23 | 9.5 | 784.7696 | 165.4573 | 49.77312 | 82.72867 | -68.9757 | -13.753 |
| 24 | 10 | 826.1339 | 130.9695 | 42.89662 | 65.48475 | -56.1024 | -9.38231 |
| 25 | 10.5 | 858.8763 | 102.9183 | 38.20546 | 51.45914 | -44.8926 | -6.56651 |
| 26 | 11 | 884.6058 | 80.47196 | 34.92221 | 40.23598 | -35.5428 | -4.69313 |
| 27 | 11.5 | 904.7238 | 62.70054 | 32.57564 | 31.35027 | -27.9393 | -3.41099 |
| 28 | 12 | 920.399 | 48.7309 | 30.87015 | 24.36545 | -21.8532 | -2.51223 |
| 29 | 12.5 | 932.5817 | 37.80429 | 29.61403 | 18.90214 | -17.0325 | -1.86963 |
| 30 | 13 | 942.0327 | 29.28803 | 28.67922 | 14.64402 | -13.2413 | -1.40273 |
| 31 | 13.5 | 949.3548 | 22.66739 | 27.97785 | 11.33369 | -10.2746 | -1.05909 |
| 32 | 14 | 955.0216 | 17.53009 | 27.44831 | 8.765043 | -7.96149 | -0.80356 |
| 33 | 14.5 | 959.4041 | 13.54934 | 27.04653 | 6.774671 | -6.16268 | -0.61199 |
| 34 | 15 | 962.7915 | 10.468 | 26.74054 | 5.234001 | -4.76654 | -0.46747 |

**FIGURE 5.3**
Screenshot of SIR model of the spread of a disease in Excel.

The solution is a pair of parametric equations, $x = x(t)$, $y = y(t)$. The solution is also a curve that varies over time. We call the solution curve a *trajectory, path,* or *orbit*. The $x$-$y$ plane is called the phase-plane. We can also obtain plots of $x$ versus $t$ and $y$ versus $t$. *Rest points* or *equilibrium points* are points that satisfy both $f(x, y) = 0$ and $g(x, y) = 0$, *simultaneously*. Once we have the equilibrium values, we desire information about their stability.

Rules of Stability: We classify equilibrium values as stable, asymptotically stable, or unstable. We define these as follows:

- **Stable:** If a trajectory starts close to a rest point it remains close for all future time.
- **Asymptotically Stable:** If a trajectory starts close then it tends toward the rest point as $t \to \infty$.
- **Unstable:** Does not follow either stable or asymptotically stable rules.

The following results are useful in investigating solutions to autonomous systems. We offer these without proof:

1. There is at most one trajectory through any point in the phase-plane.
2. A trajectory that starts at a point other than a rest point cannot reach a rest point in a finite amount of time.
3. No trajectory can cross itself unless it is a closed curve. If it is a closed curve, then it is a periodic solution.
4. Implications and properties of motion from a starting point (not a rest point):
   a. will move along the same path regardless of starting time.
   b. cannot return to a starting point unless motion is periodic.
   c. can NEVER cross another trajectory.
   d. can only approach and never can reach a rest point.

Consider the following autonomous competitive hunter system of differential equations:

$$dy/dt = 0.24\, x(t) - 0.08\, y(t)x(t)$$
$$dx/dt = y(t)(4.5 - 0.9\, x(t))$$

### 5.2.1 Qualitative Graphical Assessment

First, we plot (Figure 5.4) *dx/dt* and *dy/dt*, respectively. We want to see where $dx/dt = 0$ and $dy/dt = 0$, simultaneously. In the equation $dx/dt = 0$, we

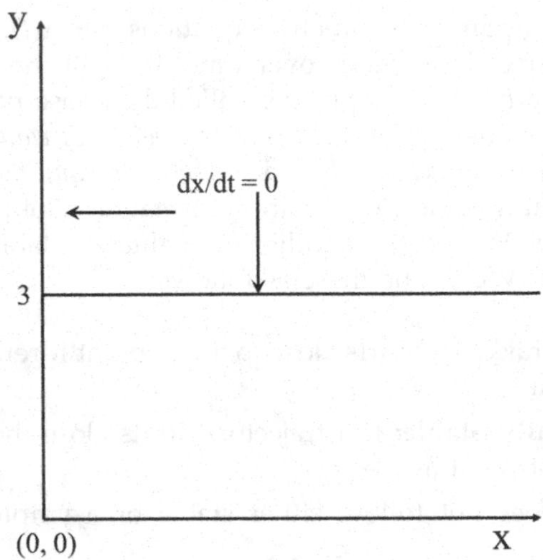


**FIGURE 5.4**
Plot of dx/dt and dy/dt.

find that either *y(t)* = 0 (which is the *x*-axis) or *x* = 5 (the horizontal line). In the equation *y(t)* = 0, we find either *x(t)* = 0 (the *y* axis) or the line *y(t)* = 3. There are two equilibrium points. They are the points (0,0) where the *x* axis and *y* axis intersect and the point (5,3) where the vertical line *x* = *5* intersect the horizontal line *y* = *3* (Figure 5.5).

Analysis shows that both (0,0) and (5,3) are unstable equilibrium values.

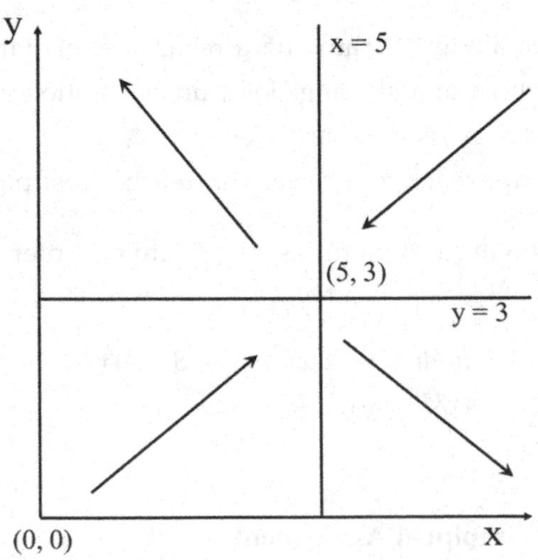


**FIGURE 5.5**
Plot of *dx/dt* and *dy/dt* and direction of movement.

## 5.3 Solving Homogeneous and Non-homogeneous Systems

We can solve systems of differential equations of the form:

$$dx/dt = ax + by + g(t)$$
$$dy/dt = mx + ny + h(t)$$

where (1) a, $b$, $m$, $n$ are constants and (2) the functions $g(t)$ and $h(t)$ can either be 0 or functions of $t$ with real coefficients.

When $g(t)$ and $h(t)$ are both 0, then the system of differential equations is called a homogeneous system; otherwise it is non-homogeneous. We will begin with homogeneous systems. The method we will use involves eigenvalues and eigenvectors.

**Example 5.7:** Homogeneous System

Consider the following homogeneous system: with initial conditions:

$$x' = 2x - y + 0$$
$$y' = 3x - 2y + 0$$
$$x(0) = 1, \; y(0) = 2$$

Basically, if we rewrite the system of differential equation in matrix form: $\mathbf{X' = A}\ \boldsymbol{x}$, where

$$A = \begin{bmatrix} 2 & 1 \\ 3 & -2 \end{bmatrix}$$

$$X' = \begin{bmatrix} dx/dt \\ dy/dt \end{bmatrix}, \quad x = \begin{bmatrix} x \\ y \end{bmatrix}$$

then we can solve $X' = Ax$. This form is highly suggestive of the first-order separable equation that we saw in the previous chapter. We can assume the solution to have a similar form: $\mathbf{X} = \mathbf{K}e^{\lambda t}$, where $\lambda$ is a constant and $\mathbf{X}$ and $K$ are vectors. The values of $\mathbf{X}$ are called **eigenvalues,** and the components of $K$ are the corresponding **eigenvectors**. We note that a full discussion of the theory and applications of eigenvalues and eigenvectors can be found in linear algebra textbooks as well as many differential equations textbooks.

Since we have a 2 × 2 system there are two linearly independent solutions that we call $X_1$ and $X_2$. The *complementary solution* or *general solution* $X = c_1X_1 + c_2X_2$ where $c_1$ and $c_2$ are arbitrary constants. We use the initial conditions to find specific values for $c_1$ and $c_2$.

The following steps can be used when we have real distinct eigenvalues:

Step 1. Set up the system as a matrix, $X' = AX, X(0) = X_0.$
Step 2. Find the eigenvalues, $K_1$ and $K_2$.
Step 3. Find the corresponding eigenvectors, $K_1$ and $K_2$.
Step 4. Set up the complementary solution $X_c = c_1X_1 + c_2X_2$ where

$$X_1 = K_1e^{\lambda_1 t}$$
$$X_2 = K_2e^{\lambda_2 t}$$

Step 5. Solve for $c_1$ and $c_2$. and rewrite the solution for $X_c$.

**Step 2.** Finding the eigenvalues. We set up the characteristic polynomial by finding the determinant of $A - \lambda I = 0$.

$$\det\left(\begin{bmatrix} 2-\lambda & -1 \\ 3 & -2-\lambda \end{bmatrix}\right) = 0$$

$$(2 - \lambda)(-2 - \lambda) + 3 = 0$$
$$\lambda^2 - 1 = 0$$
$$\lambda = 1, -1$$

**Step 3.** Finding the eigenvectors.

We substitute each solution for $l$ back in $A * k = 0$ and solve the system of equations for $k$, the eigenvectors.

Let $\lambda = 1$.

Let $k_1$ and $k_2$ be the components of eigenvector $\mathbf{K_1}$.

$$k_1 - k_2 = 0$$
$$3k_1 - 3k_2 = 0$$

We arbitrarily make $k_1 = 1$, thus $k_2 = 1$.

$$\mathbf{K_1} = [1, 1]$$

Let $\lambda = -1$.

Let $k_1$ and $k_2$ be the components of eigenvector $K_1$.

$$3k_1 - k_2 = 0$$
$$3k_1 - k_2 = 0$$

We arbitrarily make $k_2 = 3$, thus $k_1 = 1$.

$$\mathbf{K}_2 = [1, 3]$$

**Step 4.** We set up the complementary solution.

$\boldsymbol{X_c = c_1X_1 + c_2X_2}$ where

$$X_1 = K_1e^{\lambda_1 t}$$
$$X_2 = K_2e^{\lambda_2 t}$$
$$X_C = c_1\begin{bmatrix}1\\1\end{bmatrix}e^t + c_2\begin{bmatrix}1\\3\end{bmatrix}e^{-t}$$

We find the complimentary solution by setting $Xc$ = initial condition:

Since we only had a homogeneous system, we will solve for $c_1$ and $c_2$ now using the initial conditions, $x(0) = 1$, $y(0) = 2$.

We solve the system

$$c_1 + c_2 = 1$$
$$c_1 + 3c_2 = 2$$

We find $c_1$ and $c_2$ both equal 0.5.

The particular solution is

$$X_C = 0.5\begin{bmatrix}1\\1\end{bmatrix}e^t + 0.5\begin{bmatrix}1\\3\end{bmatrix}e^{-t}$$

We might plot the solutions to the components $\boldsymbol{X_1}$ and $\boldsymbol{X_2}$, each a function of $t$. We note that both solutions grow without bound as $t\to\infty$, Figure 5.6.

**Example 5.8:** Complex Eigenvalues (eigenvalues of the form $\lambda = a \pm bi$)

We note here that we do not use the form $e^{a\pm bi}$ and that complex eigenvalues always appear in conjugate pairs. The key to finding two real linearly independent solutions from complex solutions is Euler's identity:

$$e^{i\theta} = \cos\theta + \sin\theta)$$

We can rewrite the solutions for $\boldsymbol{X_1}$ and $\boldsymbol{X_2}$ using Euler's identity:

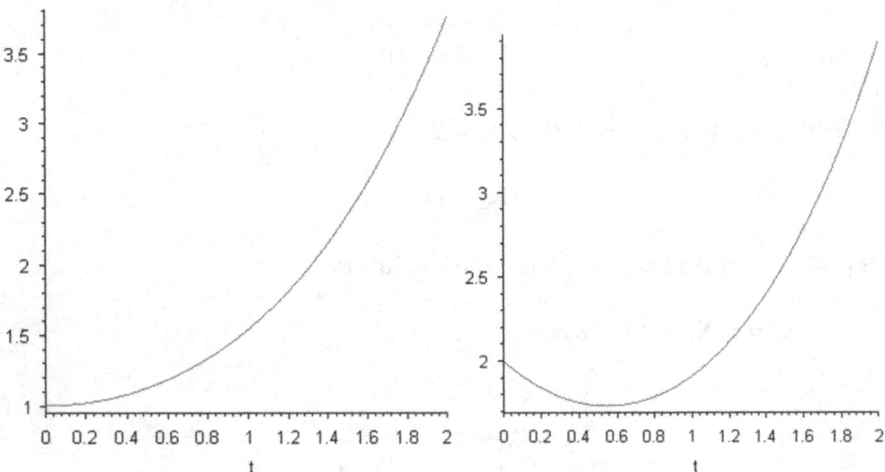


**FIGURE 5.6**
Plots of x(t) and y(t).

$$Ke^{\lambda t} = Ke^{(a+bi)} = Ke^{at}(\cos bt + i \sin bt)$$
$$Ke^{\lambda t} = Ke^{(a-bi)} = Ke^{at}(\cos bt - i \sin bt)$$

Consider the following steps as a summary when we get complex eigenvalues.

Step 1. Find the complex eigenvalues, $\lambda = a \pm bi$)

Step 2. Find the complex eigenvector, $K$

$$K = \begin{bmatrix} u_1 + i\,v_1 \\ u_2 + i\,v_2 \end{bmatrix}$$

Step 3. Form the real vectors

$$B_1 = \begin{bmatrix} u_1 \\ u_2 \end{bmatrix}$$

$$B_2 = -\begin{bmatrix} v_1 \\ v_2 \end{bmatrix}$$

Step 4. Form the linearly independent set of real solutions:

$$X_1 = e^{at}(B_1 \cos(bt) + B_2 \sin(bt))$$
$$X_2 = e^{at}(B_2 \cos(bt - B_1 \sin(bt))$$

Step 1. $X' = \begin{bmatrix} 6 & -1 \\ 5 & 4 \end{bmatrix} X$

Step 2. We set up and solve for the eigenvalue. We solve the characteristic polynomial

$$(6 - \lambda)(4 - \lambda) + 5 = 0$$
$$29 - 10\lambda + \lambda^2 = 0$$
$$\lambda = 5 \pm 2I$$

We find the eigenvalues are 5 + 2I and 5 – 2I.

Step 3. We find the eigenvectors we are substituting λ as we did before. We then create the two vectors B1 and B2.

$$\text{Let } \lambda = -5 + 2\text{I}.$$

Let $k_1$ and $k_2$ be the components of eigenvector $K_1$.

$$(1 - 2I)k_1 - k_2 = 0$$
$$5k_1 + (1 - 2I)k_2 = 0$$

We arbitrarily make $k_2 = (1 - 2I)$, thus $k_1 = 1$.

$$K1 = [1,\ 1 - 2I]$$
$$B1 = real(K1) = [1, 1],\ \ B2 = Imaginary(K1) = [0 - 2].$$
$$B1 = [1, 1]$$
$$B2 = [0, -2]$$

By substitution, we find the complementary solution:

$$X_c = c_1 e^{5t}\left(\begin{bmatrix} 1 \\ 1 \end{bmatrix}\cos(2t) - \begin{bmatrix} 0 \\ -2 \end{bmatrix}\sin(2t)\right) + c_2 e^{5t}\left(\begin{bmatrix} 0 \\ -2 \end{bmatrix}\cos(2t) + \begin{bmatrix} 1 \\ 1 \end{bmatrix}\sin(2t)\right)$$

Since we only had a homogeneous system, we will solve for $c_1$ and $c_2$ now using the initial conditions, $x(0) = 1$, $y(0) = 2$.

We get two equations

$$c_1 = 1 \text{ and}$$
$$c_1 - 2c_2 = 2$$

whose solutions are $c_1 = 1$, $c_2 = -0.5$.

$$X_p = e^{5t}\left(\begin{bmatrix}1\\1\end{bmatrix}\cos(2t) - \begin{bmatrix}0\\-2\end{bmatrix}\sin(2t)\right) - 0.5e^{5t}\left(\begin{bmatrix}0\\-2\end{bmatrix}\cos(2t) + \begin{bmatrix}1\\1\end{bmatrix}\sin(2t)\right)$$

Again, we obtain plots of $X_1$ and $X_2$ as functions of $t$, Figure 5.7.

**Example 5.9:** Repeated Eigenvalues Solution

When eigenvalues are repeated, we must find a method to obtain independent solutions. The following is a summary for repeated real eigenvalues.

$$X_1 = Ke^{\lambda t}$$

and the second linearly independent solution is given by

$$X_2 = Kt\,e^{\lambda t} + Pe^{\lambda t}$$

where the components of $P$ must satisfy the system

$$k_1 = (a - \lambda)p_1 + bp_2$$
$$k_2 = cp_1 + (d - \lambda)p_2$$

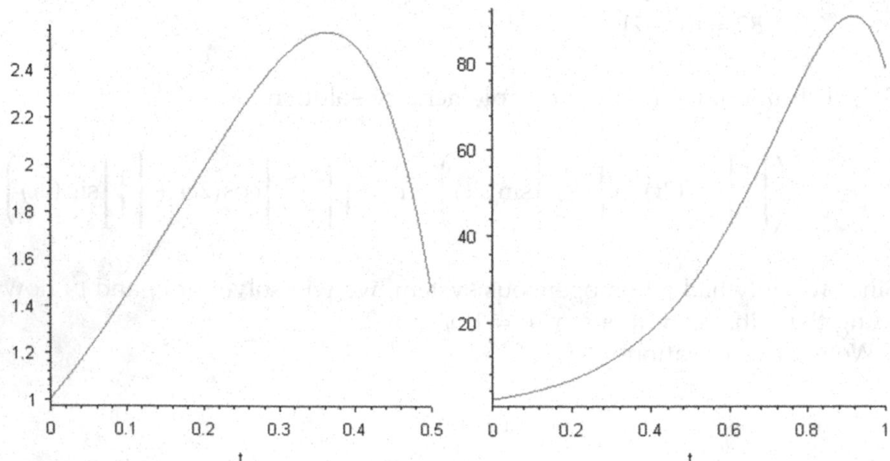


**FIGURE 5.7**
Plots of $X_1$ and $X_2$ as functions of $t$.

and

$$A = \begin{bmatrix} a & b \\ c & d \end{bmatrix}$$

Step 1. $X' = \begin{bmatrix} 3 & -18 \\ 2 & -9 \end{bmatrix} X$

Step 2. Solve the characteristic equation $(3 - \lambda)(-9 - \lambda) + 36 = 0$.
We find we have repeated roots and $\lambda = -3, -3$.

Step 3. We find **K** easily as the vector. Then, we solve for the vector **P**.

$$\begin{bmatrix} 6 & -18 \\ 2 & -6 \end{bmatrix}\begin{bmatrix} p_1 \\ p_2 \end{bmatrix} = \begin{bmatrix} 3 \\ 1 \end{bmatrix}$$

Step 1. Find the repeated eigenvalues, $\lambda_1 = \lambda_2, = \lambda$

Step 2. One solution is

We find $p_1$ and $p_2$ must solve $p_1 - 3p_2 = 1/2$ or $2p_1 - 6p_2 = 1$. We select $p_1 = 1$ then $p_2 = 1/6$.

Our complementary solution is

$$X_c = c_1 \begin{bmatrix} 3 \\ 1 \end{bmatrix} e^{-3t} + c_2 \left( \begin{bmatrix} 3 \\ 1 \end{bmatrix} te^{-3t} + \begin{bmatrix} 1 \\ \frac{1}{6} \end{bmatrix} e^{-3t} \right)$$

Since we only had a homogeneous system, we will solve for $c_1$ and $c_2$ now using the initial conditions, $x(0) = 1$, $y(0) = 2$.

We obtain two equations:

$$3c_1 + c_2 = 1$$
$$c_1 + (1/6)c_2 = 2$$
$$c_1 = 5/3c_2 = -4$$

$$X_c = \frac{5}{3} \begin{bmatrix} 3 \\ 1 \end{bmatrix} e^{-3t} - 4 \left( \begin{bmatrix} 3 \\ 1 \end{bmatrix} te^{-3t} + \begin{bmatrix} 1 \\ \frac{1}{6} \end{bmatrix} e^{-3t} \right)$$

We plot the solutions shown in Figure 5.8.

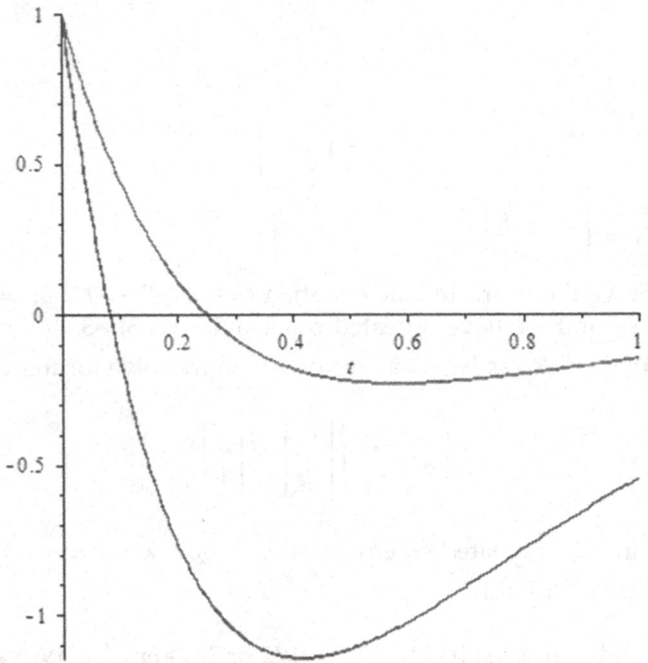


**FIGURE 5.8**

Plot of $X_c = \frac{5}{3}\begin{bmatrix}3\\1\end{bmatrix}e^{-3t} - 4\left(\begin{bmatrix}3\\1\end{bmatrix}te^{-3t} + \begin{bmatrix}1\\\frac{1}{6}\end{bmatrix}e^{-3t}\right)$.

## 5.4 Technology Examples for Systems of Ordinary Differential Equations

### 5.4.1 Excel for System of Ordinary Differential Equations

We can use the techniques described in Chapter 2 to input a predator-prey system and initial conditions in Excel and then iterate to determine the numerical estimates using Euler's method for a predator prey system of ODEs (Figure 5.9).

We can plot (Figure 5.10) this data to gain a visual qualification of the results:

The power of Euler's method is two-fold. First, it is easy to use and second as a numerical method it can be used to estimate a solution to a system of differential equations that does not have a closed form solution.

Assume we have the following predator-prey system that does not have a closed from analytical solution:

| | A | B | C | D | E | F | G | H |
|---|---|---|---|---|---|---|---|---|
| 1 | t | x | y | x' | y' | | step size | 0.1 |
| 2 | 0 | 3 | 6 | -3 | -9 | | | |
| 3 | 0.1 | 2.7 | 5.1 | -2.1 | -6.9 | | | |
| 4 | 0.2 | 2.49 | 4.41 | -1.35 | -5.19 | | | |
| 5 | 0.3 | 2.355 | 3.891 | -0.717 | -3.789 | | | |
| 6 | 0.4 | 2.2833 | 3.5121 | -0.1743 | -2.6319 | | | |
| 7 | 0.5 | 2.26587 | 3.24891 | 0.29979 | -1.66629 | | | |
| 8 | 0.6 | 2.295849 | 3.082281 | 0.722985 | -0.84988 | | | |
| 9 | 0.7 | 2.368148 | 2.997293 | 1.109856 | -0.14843 | | | |
| 10 | 0.8 | 2.479133 | 2.98245 | 1.4725 | 0.465867 | | | |
| 11 | 0.9 | 2.626383 | 3.029036 | 1.821077 | 1.01577 | | | |
| 12 | 1 | 2.808491 | 3.130613 | 2.164246 | 1.520001 | | | |
| 13 | 1.1 | 3.024915 | 3.282613 | 2.509519 | 1.994123 | | | |
| 14 | 1.2 | 3.275867 | 3.482026 | 2.86355 | 2.451234 | | | |
| 15 | 1.3 | 3.562222 | 3.727149 | 3.232369 | 2.902515 | | | |
| 16 | 1.4 | 3.885459 | 4.017401 | 3.621576 | 3.357694 | | | |
| 17 | 1.5 | 4.247617 | 4.35317 | 4.036511 | 3.825404 | | | |
| 18 | 1.6 | 4.651268 | 4.73571 | 4.482383 | 4.313498 | | | |
| 19 | 1.7 | 5.099506 | 5.16706 | 4.964398 | 4.82929 | | | |
| 20 | 1.8 | 5.595946 | 5.649989 | 5.48786 | 5.379773 | | | |
| 21 | 1.9 | 6.144732 | 6.187967 | 6.058263 | 5.971794 | | | |
| 22 | 2 | 6.750558 | 6.785146 | 6.681383 | 6.612208 | | | |
| 23 | 2.1 | 7.418697 | 7.446367 | 7.363356 | 7.308016 | | | |
| 24 | 2.2 | 8.155032 | 8.177168 | 8.11076 | 8.066488 | | | |
| 25 | 2.3 | 8.966108 | 8.983817 | 8.930691 | 8.895273 | | | |
| 26 | 2.4 | 9.859177 | 9.873345 | 9.830843 | 9.802509 | | | |
| 27 | 2.5 | 10.84226 | 10.8536 | 10.81959 | 10.79693 | | | |
| 28 | 2.6 | 11.92422 | 11.93329 | 11.90609 | 11.88795 | | | |
| 29 | 2.7 | 13.11483 | 13.12208 | 13.10032 | 13.08582 | | | |
| 30 | 2.8 | 14.42486 | 14.43067 | 14.41326 | 14.40165 | | | |
| 31 | 2.9 | 15.86619 | 15.87083 | 15.8569 | 15.84762 | | | |
| 32 | 3 | 17.45188 | 17.45559 | 17.44445 | 17.43702 | | | |

**FIGURE 5.9**
Screenshot of Euler's method for a system of equations in Excel.

$$dx/dt = 3x - xy$$
$$dy/dt = xy - 2y$$
$$x(0) = 1, \; y(0) = 2$$
$$t_o = 1, \; \Delta t = 0.1$$

We will obtain an estimated the solution using Euler's method (Figure 5.11). (Figures 5.12 and 5.13).

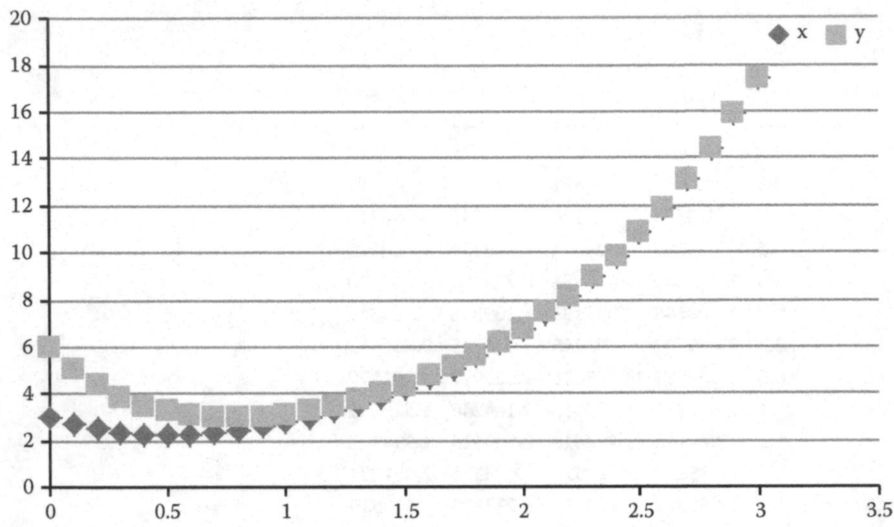


**FIGURE 5.10**
Visualization of Euler's method for a system of equations.

We experiment and find that when we plot $x(t)$ versus$y(t)$ we have an approximately a closed loop.

We can use the improved Euler method and Runge-Kutta 4 methods to iterate solutions to systems of differential equations as well. The vector version of the iterative formula for Runge-Kutta 4 is

$$X_{n+1} = X_n + (h/6)(K_1 + 2K_2 + 2K_3 + K_4)$$

*where*

$$K_1 = f(t_n, X_n)$$
$$K_2 = f(t_n + (h/2), X_n + (h/2)K_1)$$
$$K_3 = f(t_n + (h/2), X_n + (h/2)K_2)$$
$$K_4 = f(t_n + (h/2), X_n + (h/2)K_3)$$

We repeat our example with Runge–Kutta 4 (Figure 5.14).

$$dx/dt = 3x - xy$$
$$dy/dt = xy - 2y$$
$$x(0) = 1, \ y(0) = 2$$
$$t_o = 0, \ \Delta t = 0.25$$

| | A | B | C | D | E | F |
|---|---|---|---|---|---|---|
| 1 | Predator-Prey | | | | step size | 0.1 |
| 2 | | | | | | |
| 3 | | | | | | |
| 4 | t | x | y | x' | y' | |
| 5 | 0 | 1 | 2 | 1 | -2 | |
| 6 | 0.1 | 1.1 | 1.8 | 1.32 | -1.62 | |
| 7 | 0.2 | 1.232 | 1.638 | 1.677984 | -1.25798 | |
| 8 | 0.3 | 1.399798 | 1.512202 | 2.082618 | -0.90763 | |
| 9 | 0.4 | 1.60806 | 1.421439 | 2.538421 | -0.55712 | |
| 10 | 0.5 | 1.861902 | 1.365727 | 3.042856 | -0.1886 | |
| 11 | 0.6 | 2.166188 | 1.346867 | 3.580997 | 0.223833 | |
| 12 | 0.7 | 2.524288 | 1.36925 | 4.116482 | 0.717881 | |
| 13 | 0.8 | 2.935936 | 1.441038 | 4.577012 | 1.348719 | |
| 14 | 0.9 | 3.393637 | 1.57591 | 4.832044 | 2.196247 | |
| 15 | 1 | 3.876921 | 1.795535 | 4.669617 | 3.370078 | |
| 16 | 1.1 | 4.343883 | 2.132543 | 3.768134 | 4.998431 | |
| 17 | 1.2 | 4.720697 | 2.632386 | 1.735396 | 7.161922 | |
| 18 | 1.3 | 4.894236 | 3.348578 | -1.70602 | 9.691575 | |
| 19 | 1.4 | 4.723634 | 4.317735 | -6.2245 | 11.75993 | |
| 20 | 1.5 | 4.101184 | 5.493728 | -10.2272 | 11.54333 | |
| 21 | 1.6 | 3.07846 | 6.648062 | -11.2304 | 7.16967 | |
| 22 | 1.7 | 1.955419 | 7.365029 | -8.53546 | -0.32834 | |
| 23 | 1.8 | 1.101873 | 7.332195 | -4.77353 | -6.58524 | |
| 24 | 1.9 | 0.62452 | 6.67367 | -2.29428 | -9.1795 | |
| 25 | 2 | 0.395092 | 5.75572 | -1.08876 | -9.2374 | |
| 26 | 2.1 | 0.286216 | 4.83198 | -0.52434 | -8.28097 | |
| 27 | 2.2 | 0.233782 | 4.003883 | -0.23469 | -7.07173 | |
| 28 | 2.3 | 0.210313 | 3.29671 | -0.0624 | -5.90008 | |
| 29 | 2.4 | 0.204072 | 2.706702 | 0.059854 | -4.86104 | |
| 30 | 2.5 | 0.210058 | 2.220598 | 0.16372 | -3.97474 | |
| 31 | 2.6 | 0.22643 | 1.823124 | 0.26648 | -3.23344 | |
| 32 | 2.7 | 0.253078 | 1.49978 | 0.379672 | -2.62 | |
| 33 | 2.8 | 0.291045 | 1.23778 | 0.512885 | -2.11531 | |
| 34 | 2.9 | 0.342334 | 1.026249 | 0.675681 | -1.70118 | |

**FIGURE 5.11**
Screenshot of predator-prey system without a closed solution.

### 5.4.2 Maple for System of Ordinary Differential Equations

Again, we start by accessing the ***with(DEtools):*** command.

MAPLE allows us to find the homogeneous solution in matrix form. The following commands illustrate this:

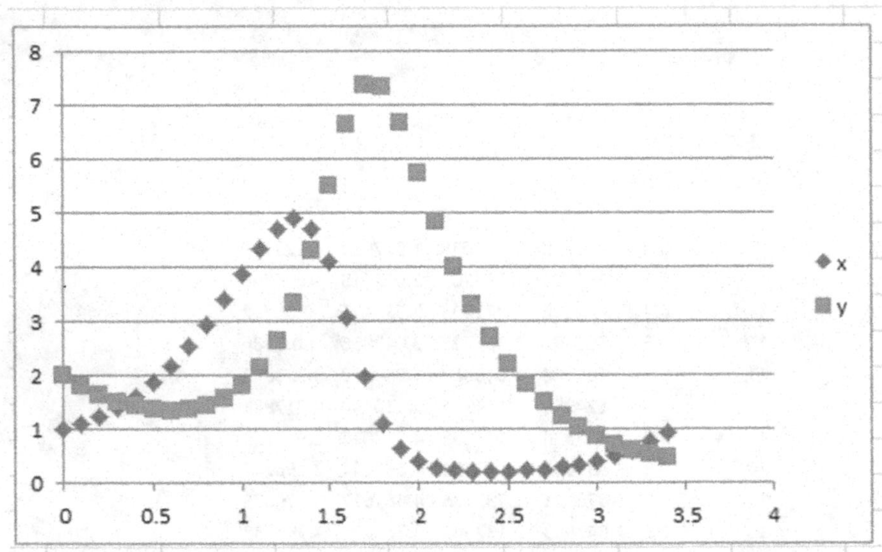


**FIGURE 5.12**
Status of predator and prey over time.

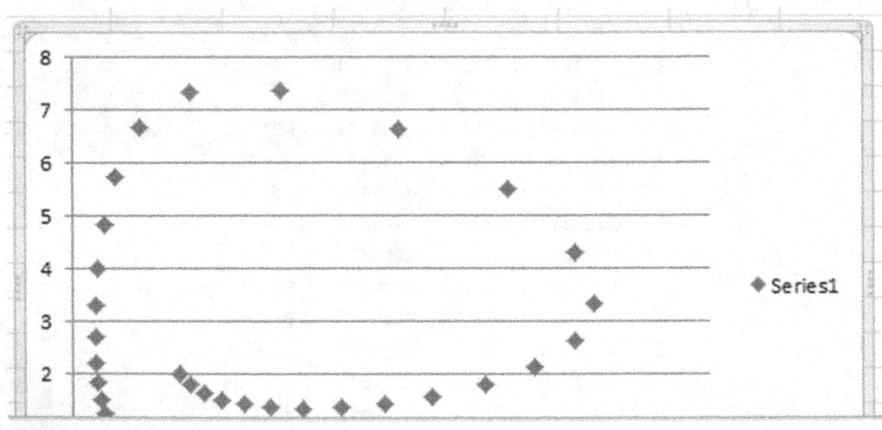


**FIGURE 5.13**
Visualization of predator-prey system without a closed solution.

```
> with(DEtools):with(plots):with(linalg):
> M:=array([[2,-1],[3,-2]]);
> lambda:=eigenvects(M);
> homsol:=matrixDE(M,t);
```

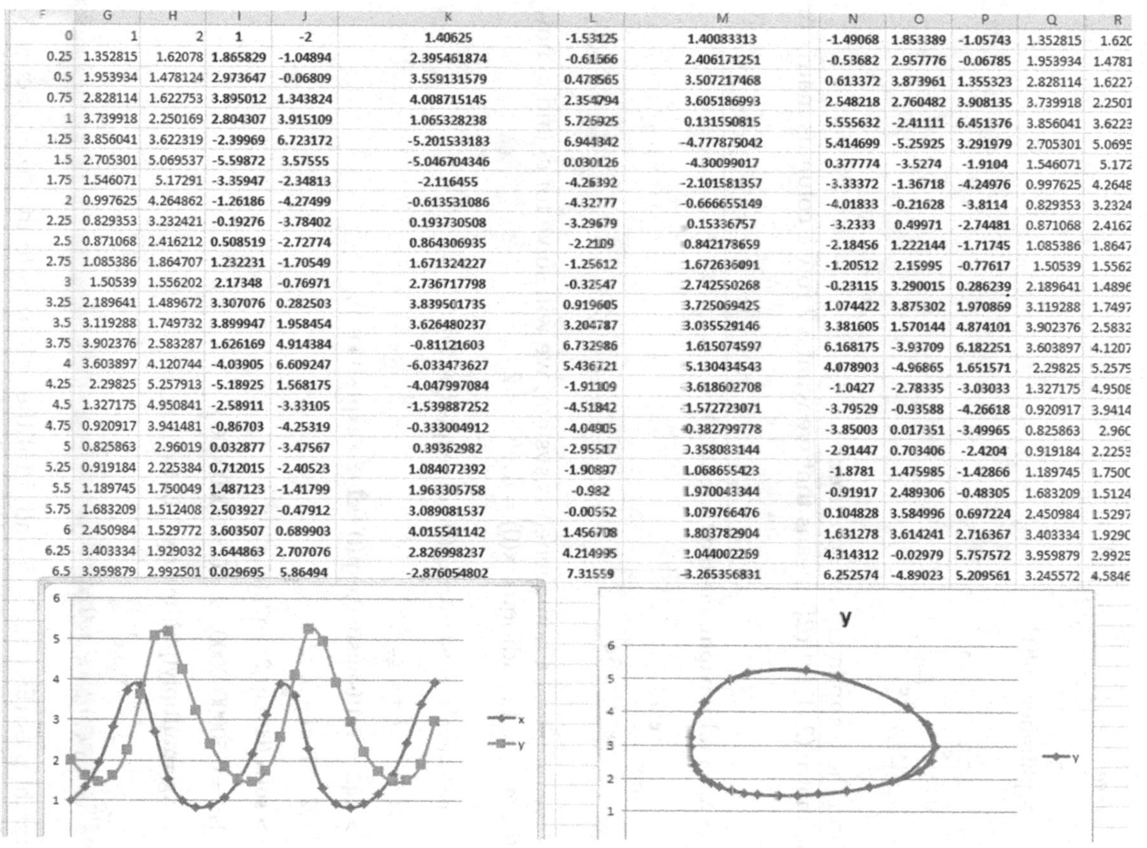

| F | G | H | I | J | K | L | M | N | O | P | Q | R |
|---|---|---|---|---|---|---|---|---|---|---|---|---|
| 0 | 1 | 2 | 1 | -2 | 1.40625 | -1.53125 | 1.40083313 | -1.49068 | 1.853389 | -1.05743 | 1.352815 | 1.620 |
| 0.25 | 1.352815 | 1.62078 | 1.865829 | -1.04894 | 2.395461874 | -0.61566 | 2.406171251 | -0.53682 | 2.957776 | -0.06785 | 1.953934 | 1.4781 |
| 0.5 | 1.953934 | 1.478124 | 2.973647 | -0.06809 | 3.559131579 | 0.478565 | 3.507217468 | 0.613372 | 3.873961 | 1.355323 | 2.828114 | 1.6227 |
| 0.75 | 2.828114 | 1.622753 | 3.895012 | 1.343824 | 4.008715145 | 2.354794 | 3.605186993 | 2.548218 | 2.760482 | 3.908135 | 3.739918 | 2.2501 |
| 1 | 3.739918 | 2.250169 | 2.804307 | 3.915109 | 1.065328238 | 5.726925 | 0.131550815 | 5.555632 | -2.41111 | 6.451376 | 3.856041 | 3.6223 |
| 1.25 | 3.856041 | 3.622319 | -2.39969 | 6.723172 | -5.201533183 | 6.944342 | -4.777875042 | 5.414699 | -5.25925 | 3.291979 | 2.705301 | 5.0695 |
| 1.5 | 2.705301 | 5.069537 | -5.59872 | 3.57555 | -5.046704346 | 0.030126 | -4.30099017 | 0.377774 | -3.5274 | -1.9104 | 1.546071 | 5.172 |
| 1.75 | 1.546071 | 5.17291 | -3.35947 | -2.34813 | -2.116455 | -4.26392 | -2.101581357 | -3.33372 | -1.36718 | -4.24976 | 0.997625 | 4.2648 |
| 2 | 0.997625 | 4.264862 | -1.26186 | -4.27499 | -0.613531086 | -4.32777 | -0.666655149 | -4.01833 | -0.21628 | -3.8114 | 0.829353 | 3.2324 |
| 2.25 | 0.829353 | 3.232421 | -0.19276 | -3.78402 | 0.193730508 | -3.29679 | 0.15336757 | -3.2333 | 0.49971 | -2.74481 | 0.871068 | 2.4162 |
| 2.5 | 0.871068 | 2.416212 | 0.508519 | -2.72774 | 0.864306935 | -2.2109 | 0.842178659 | -2.18456 | 1.222144 | -1.71745 | 1.085386 | 1.8647 |
| 2.75 | 1.085386 | 1.864707 | 1.232231 | -1.70549 | 1.671324227 | -1.25612 | 1.672636091 | -1.20512 | 2.15995 | -0.77617 | 1.50539 | 1.5562 |
| 3 | 1.50539 | 1.556202 | 2.17348 | -0.76971 | 2.736717798 | -0.32547 | 2.742550268 | -0.23115 | 3.290015 | 0.286239 | 2.189641 | 1.4896 |
| 3.25 | 2.189641 | 1.489672 | 3.307076 | 0.282503 | 3.839501735 | 0.919605 | 3.725069425 | 1.074422 | 3.875302 | 1.970869 | 3.119288 | 1.7497 |
| 3.5 | 3.119288 | 1.749732 | 3.899947 | 1.958454 | 3.626480237 | 3.204787 | 3.035529146 | 3.381605 | 1.570144 | 4.874101 | 3.902376 | 2.5832 |
| 3.75 | 3.902376 | 2.583287 | 1.626169 | 4.914384 | -0.81121603 | 6.732986 | 1.615074597 | 6.168175 | -3.93709 | 6.182251 | 3.603897 | 4.1207 |
| 4 | 3.603897 | 4.120744 | -4.03905 | 6.609247 | -6.033473627 | 5.436721 | 5.130434543 | 4.078903 | -4.96865 | 1.651571 | 2.29825 | 5.2579 |
| 4.25 | 2.29825 | 5.257913 | -5.18925 | 1.568175 | -4.047997084 | -1.91109 | -3.618602708 | -1.0427 | -2.78335 | -3.03033 | 1.327175 | 4.9508 |
| 4.5 | 1.327175 | 4.950841 | -2.58911 | -3.33105 | -1.539887252 | -4.51842 | -1.572723071 | -3.79529 | -0.93588 | -4.26618 | 0.920917 | 3.9414 |
| 4.75 | 0.920917 | 3.941481 | -0.86703 | -4.25319 | -0.333004912 | -4.04905 | -0.382799778 | -3.85003 | 0.017351 | -3.49965 | 0.825863 | 2.960 |
| 5 | 0.825863 | 2.96019 | 0.032877 | -3.47567 | 0.39362982 | -2.95517 | 0.358083144 | -2.91447 | 0.703406 | -2.4204 | 0.919184 | 2.2253 |
| 5.25 | 0.919184 | 2.225384 | 0.712015 | -2.40523 | 1.084072392 | -1.90897 | 1.068655423 | -1.8781 | 1.475985 | -1.42866 | 1.189745 | 1.7500 |
| 5.5 | 1.189745 | 1.750049 | 1.487123 | -1.41799 | 1.963305758 | -0.982 | 1.970043344 | -0.91917 | 2.489306 | -0.48305 | 1.683209 | 1.5124 |
| 5.75 | 1.683209 | 1.512408 | 2.503927 | -0.47912 | 3.089081537 | -0.00552 | 3.079766476 | 0.104828 | 3.584996 | 0.697224 | 2.450984 | 1.5297 |
| 6 | 2.450984 | 1.529772 | 3.603507 | 0.689903 | 4.015541142 | 1.456708 | 3.803782904 | 1.631278 | 3.614241 | 2.716367 | 3.403334 | 1.9290 |
| 6.25 | 3.403334 | 1.929032 | 3.644863 | 2.707076 | 2.826998237 | 4.214995 | 2.044002269 | 4.314312 | -0.02979 | 5.757572 | 3.959879 | 2.9925 |
| 6.5 | 3.959879 | 2.992501 | 0.029695 | 5.86494 | -2.876054802 | 7.31559 | -3.265356831 | 6.252574 | -4.89023 | 5.209561 | 3.245572 | 4.5846 |



**FIGURE 5.14**
Screenshot of predator-prey system in Excel.

$$M := \begin{bmatrix} 2 & -1 \\ 3 & -2 \end{bmatrix}$$

$$\lambda := [1, 1, \{[1, 1]\}], [-1, 1, \{[1, 3]\}]$$

$$homsol := \left[\begin{bmatrix} e^t & e^{(-t)} \\ e^t & 3\,e^{(-t)} \end{bmatrix}, [0, 0]\right]$$

**> phi:=homsol[1];**

$$\phi := \begin{bmatrix} e^t & e^{(-t)} \\ e^t & 3\,e^{(-t)} \end{bmatrix}$$

We find the complimentary solution:

The matrix(2, 1, [c1,c2]) means that we want a 2 row 1 columns matrix of $c_1$, $c_2$.

**> Xc:=multiply(phi, matrix(2,1,[c1,c2]));**

$$Xc := \begin{bmatrix} e^t\,c1 + e^{(-t)}\,c2 \\ e^t\,c1 + 3e^{(-t)}\,c2 \end{bmatrix}$$

Since we only had a homogeneous system, we will solve for $c_1$ and $c_2$ now using the initial conditions, x(0) = 1,y(0) = 2.

**> eq1:=evalf(subs(t=0, exp(t) * c1+exp(−t) * c2));**

$eq1 := 1.\,c1 + 1.\,c2$

**> eq2:=evalf(subs(t=0,exp(t) * c1+3 * exp(−t) * c2));**

$eq2 := 1.\,c1 + 3.\,c2$

**> solve({1=eq1,2=eq2},{c1,c2});**

$\{c2 = 0.5000000000,\ c1 = 0.5000000000\}$

**> Xg:=multiply(phi, matrix(2,1,[.5,.5]));**

$$Xg := \begin{bmatrix} 0.5\,e^t + 0.5\,e^{(-t)} \\ 0.5\,e^t + 1.5\,e^{(-t)} \end{bmatrix}$$

We might plot (Figures 5.15 and 5.16) the solutions to the components $X_1$ and $X_2$, each a function of $t$. We note that both solutions grow without bound as $t \to \infty$.

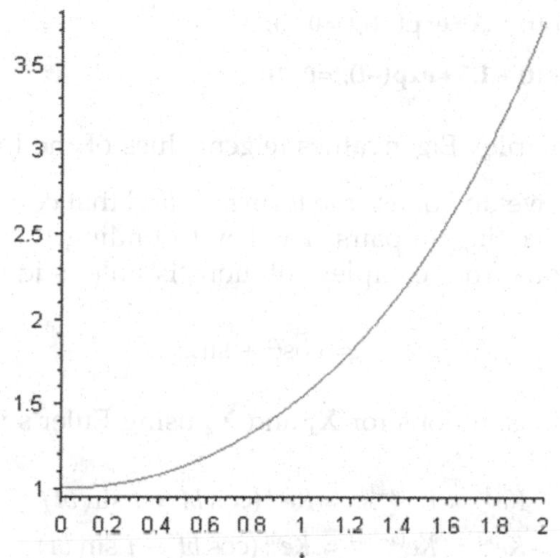


**FIGURE 5.15**
Plot of (.5*exp(t) + .5*exp(−t), t = 0..2) in Maple.

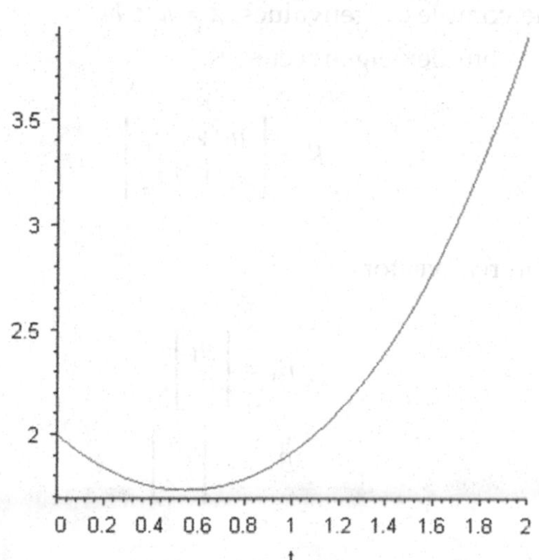


**FIGURE 5.16**
Plot of (.5*exp(t) + 1.5*exp(−t),t = 0..2) in Maple.

```
> plot(.5 * exp(t) + .5 * exp(-t),t = 0..2);
> plot(.5 * exp(t) + 1.5 * exp(-t),t=0..2);
```

**Example 5.10:** Complex Eigenvalues (eigenvalues of the form $\lambda = a \pm bi$)

We note here that we do not use the form $e^{a \pm bi}$ and that complex eigenvalues always appear in conjugate pairs. The key to finding two real linearly independent solutions from complex solutions is Euler's identity:

$$e^{i\theta} = \cos\theta + \sin\theta$$

We can rewrite the solutions for $X_1$ and $X_2$ using Euler's identity:

$$Ke^{\lambda t} = Ke^{(a+bi)} = Ke^{at}(\cos bt + i \sin bt)$$
$$Ke^{\lambda t} = Ke^{(a-bi)} = Ke^{at}(\cos bt - i \sin bt)$$

Consider the following steps as a summary when we get complex eigenvalues.

$$X_1 = e^{at}(B_1 \cos(bt) + B_2 \sin(bt))$$
$$X_2 = e^{at}(B_2 \cos(bt - B_1 \sin(bt))$$

Step 1. Find the complex eigenvalues, $\lambda = a \pm bi$

Step 2. Find the complex eigenvector, K

$$K = \begin{bmatrix} u_1 + i\, v_1 \\ u_2 + i\, v_2 \end{bmatrix}$$

Step 3. Form the real vectors

$$B_1 = \begin{bmatrix} u_1 \\ u_2 \end{bmatrix}$$

$$B_2 = -\begin{bmatrix} v_1 \\ v_2 \end{bmatrix}$$

Step 4. Form the linearly independent set of real solutions:

We find MAPLE will allow for an easily manipulation of these steps for us

```
> with(DEtools):with(plots):with(linalg):
> M:=array([[6,-1],[5,4]]);
> lambda:=eigenvects(M);
> homsol:=matrixDE(M,t);
```

$$M := \begin{bmatrix} 6 & -1 \\ 5 & 4 \end{bmatrix}$$

$$\lambda := [5+2I, 1, \{[1, 1-2I]\}], [5-2I, 1, \{[1, 1+2I]\}]$$

$$homsol := \left[\begin{bmatrix} e^{(5t)}\cos(2t) & e^{(5t)}\sin(2t) \\ e^{(5t)}\cos(2t)+2e^{(5t)}\sin(2t) & e^{(5t)}\sin(2t)-2e^{(5t)}\cos(2t) \end{bmatrix}, [0, 0]\right]$$

```
> phi:=homsol[1];
>
```

$$\phi := \begin{bmatrix} e^{(5t)}\cos(2t) & e^{(5t)}\sin(2t) \\ e^{(5t)}\cos(2t)+2e^{(5t)}\sin(2t) & e^{(5t)}\sin(2t)-2e^{(5t)}\cos(2t) \end{bmatrix}$$

We find the complimentary solution:

The matrix(2,1,[c1,c2]) means that we want a 2 row 1 columns matrix of $c_1$, $c_2$.

```
> Xc:=multiply(phi, matrix(2,1,[c1,c2]));
```

$$Xc := \begin{bmatrix} e^{(5t)}\cos(2t)\,c1 + e^{(5t)}\sin(2t)\,c2 \\ (e^{(5t)}\cos(2t)+2e^{(5t)}\sin(2t))\,c1 + (e^{(5t)}\sin(2t)-2e^{(5t)}\cos(2t))\,c2 \end{bmatrix}$$

Since we only had a homogeneous system, we will solve for $c_1$ and $c_2$ now using the initial conditions, x(0) = 1, y(0) = 2.

```
> eq1:=evalf(subs(t=0, Xc[1,1]));
```

$$eq1 := 1.\,c1$$

```
> eq2:=evalf(subs(t=0,Xc[2,1]));
```

$$eq2 := 1.\,c1 - 2.\,c2$$

```
> solve({1=eq1,2=eq2},{c1,c2});
```

$$\{c1 = 1.,\ c2 = -0.5000000000\}$$

```
> Xg:=multiply(phi, matrix(2,1,[1,-.5]));
```

$$Xg := \begin{bmatrix} e^{(5\,t)} \cos(2\,t) - 0.5\, e^{(5\,t)} \sin(2\,t) \\ 2.0\, e^{(5\,t)} \cos(2\,t) + 1.5\, e^{(5\,t)} \sin(2\,t) \end{bmatrix}$$

Again, we obtain plots (Figures 5.17 and 5.18) of $X_1$ and $X_2$ as functions of $t$.

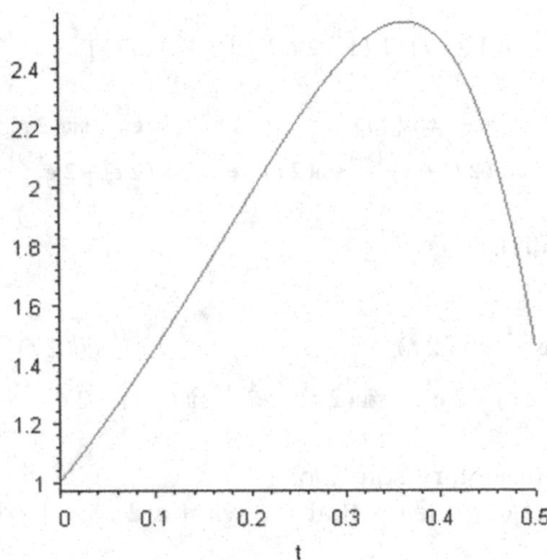


**FIGURE 5.17**
Plot of (exp(5*t)*cos(2*t)–.5*exp(5*t)*sin(2*t),t=0..0.5) in Maple.

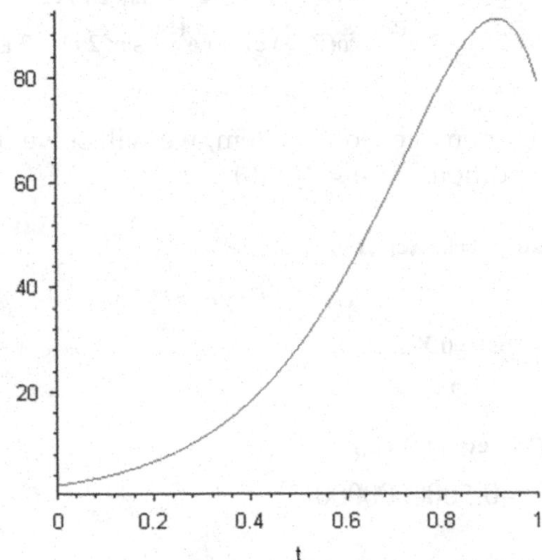


**FIGURE 5.18**
Plot of (2.0*exp(5*t)*cos(2*t)+1.5*exp(5*t)*sin(2*t),t=0..1) in Maple.

```
> plot(exp(5 * t) * cos(2 * t)-.5 * exp(5 * t) * sin(2 * t),t=0..0.5);
> plot(2.0 * exp(5 * t) * cos(2 * t)+1.5 * exp(5 * t) * sin(2 * t),t=0..1);
```

**Example 5.11:** Repeated Eigenvalues Solution

When eigenvalues are repeated, we must find a method to obtain independent solutions. The use the same rules as in the following for repeated real eigenvalues.

$$X_1 = Ke^{\lambda t}$$

and the second linearly independent solution is given by

$$X_2 = K\,t\,e^{\lambda t} + Pe^{\lambda t}$$

where the components of $P$ must satisfy the system

$$k_1 = (a - \lambda)p_1 + bp_2$$
$$k_2 = cp_1 + (d - \lambda)p_2$$

and

$$A = \begin{bmatrix} a & b \\ c & d \end{bmatrix}$$

Find the repeated eigenvalues, $\lambda_1 = \lambda_2, = \lambda$

One solution is

We now present an example of this method using Maple. Maple recognizes repeated eigenvalues and places the solution in the correct form.

```
> with(DEtools):with(plots):with(linalg):
> M:=array([[3,-18],[2,-9]]);
> lambda:=eigenvects(M);
> homsol:=matrixDE(M,t);
```

$$M := \begin{bmatrix} 3 & -18 \\ 2 & -9 \end{bmatrix}$$

$$\lambda := [-3, 2, \{[3, 1]\}]$$

$$homsol := \left[ \begin{bmatrix} e^{(-3\,t)} & e^{(-3\,t)}\, t \\ \frac{1}{3} e^{(-3\,t)} & \frac{1}{3} e^{(-3\,t)}\, t - \frac{1}{18} e^{(-3\,t)} \end{bmatrix}, [0, 0] \right]$$

**> phi:=homsol[1];**

$$\phi := \begin{bmatrix} e^{(-3\,t)} & e^{(-3\,t)}\, t \\ \frac{1}{3} e^{(-3\,t)} & \frac{1}{3} e^{(-3\,t)}\, t - \frac{1}{18} e^{(-3\,t)} \end{bmatrix}$$

We find the complimentary solution:

The matrix(2,1,[c1,c2]) means that we want a 2 row 1 columns matrix of $c_1$, $c_2$.

**> Xc:=multiply(phi, matrix(2,1,[c1,c2]));**

$$Xc := \begin{bmatrix} e^{(-3\,t)}\, c1 + e^{(-3\,t)}\, t\, c2 \\ \frac{1}{3} e^{(-3\,t)}\, c1 + \left( \frac{1}{3} e^{(-3\,t)}\, t - \frac{1}{18} e^{(-3\,t)} \right) c2 \end{bmatrix}$$

Since we only had a homogeneous system, we will solve for $c_1$ and $c_2$ now using the initial conditions, x(0) = 1, y(0) = 2 and plot the solutions (Figures 5.19 and 5.20).

**> eq1:=evalf(subs(t=0, Xc[1,1]));**

$eq1 := 1.\, c1$

**> eq2:=evalf(subs(t=0,Xc[2,1]));**

$eq2 := 0.3333333333\, c1 - 0.05555555556\, c2$

**> solve({1=eq1,2=eq2},{c1,c2});**

$\{c1 = 1.,\, c2 = -30.00000000\}$

**>Xg:=multiply(phi, matrix(2,1,[1,–30]));**

$$Xg := \begin{bmatrix} e^{(-3\,t)} - 30\, e^{(-3\,t)}\, t \\ 2\, e^{(-3\,t)} - 10\, e^{(-3\,t)}\, t \end{bmatrix}$$

**> plot(exp(–3 * t)–30 * exp(–3 * t) * t,t=0..0.5);**

**> plot(2 * exp(–3 * t)–10 * exp(–3 * t) * t,t=0..1);**

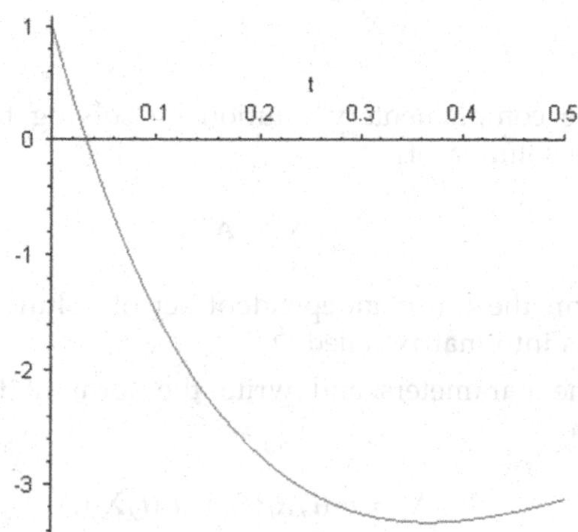


**FIGURE 5.19**
Plot of (exp(−3*t)−30*exp(−3*t)*t,t=0..0.5) in Maple.

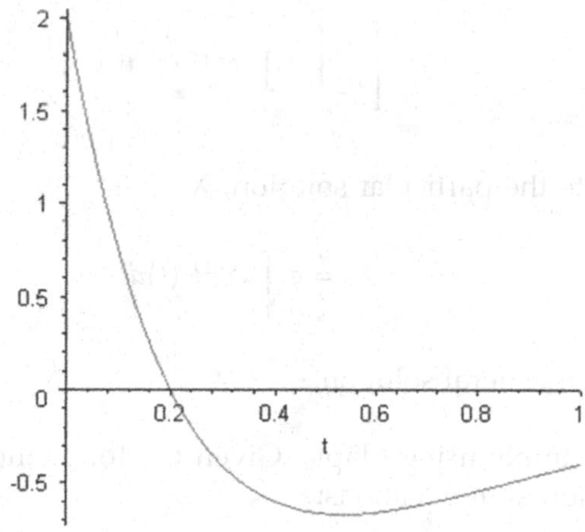


**FIGURE 5.20**
Plot of (2*exp(−3*t)−10*exp(−3*t)*t,t=0..1) in Maple.

### *5.4.2.1 Maple's Use in Nonhomogeneous Systems of Differential Equations for Closed Form Analytical Solutions*

In the nonhomogeneous form, we need both a complementary solution $X_c$ and a particular solution $X_p$. The complementary solution is found by solving the homogeneous part of the system. The particular solution is found using the variation of parameters. The following summary of the procedure is provided. Students should consult a differential equation textbook for a more detailed explanation and proof of the procedure.

$$X = X_c + X_p$$

Step 1. Find the complementary solution by solving for the homogeneous solution of

$$\mathbf{X'} = \mathbf{AX}$$

and form the linear independent set of solutions, put them as columns into matrix called $\Phi$.

Step 2. Vary the parameters and write the form of the particular solution

$$X_p = u_1(t)X_1(t) + u_2(t)X_2(t)$$

Step 3. Invert the matrix, $\Phi^{-1}$

Step 4. Determine the parameters $u_1$ and $u_2$.

$$\begin{bmatrix} u_1 \\ u_2 \end{bmatrix} = \int \phi^{-1}\mathrm{F}(\mathrm{t})\mathrm{dt}$$

Step 5. Calculate the particular solution, $X_p$

$$X_p = \phi \int \phi^{-1} F(t) dt$$

Step 6. Form the general solution.

We provide an example using Maple. Given the following system of non-homogeneous differential equations:

$$x' = 2x - y + 0$$
$$y' = 3x - 2y + 4t$$

$$x(0) = 1, \; y(0) = 2$$

Note that $g(t) = 0$ and $h(t) = 4t$.

### *5.4.2.2 Part 1: Homogeneous*

```
> with(DEtools):with(plots):with(linalg):
> M:=array([[2,-1],[3,-2]]);
```

```
> lambda:=eigenvects(M);
> homsol:=matrixDE(M,t);
```

$$M := \begin{bmatrix} 2 & -1 \\ 3 & -2 \end{bmatrix}$$

$$\lambda := [1, 1, \{[1, 1]\}], [-1, 1, \{[1, 3]\}]$$

$$homsol := \left[\begin{bmatrix} e^t & e^{(-t)} \\ e^t & 3\, e^{(-t)} \end{bmatrix}, [0, 0]\right]$$

```
> phi:=homsol[1];
```

$$\phi := \begin{bmatrix} e^t & e^{(-t)} \\ e^t & 3\, e^{(-t)} \end{bmatrix}$$

The Complimentary Solution:

The *matrix(2,1,[c1,c2])* means that we want a *2* row *1* columns matrix of $c_1$, $c_2$.

```
> Xc:=multiply(phi, matrix(2,1,[c1,c2]));
```

$$Xc := \begin{bmatrix} e^t\, c1 + e^{(-t)}\, c2 \\ e^t\, c1 + 3\, e^{(-t)}\, c2 \end{bmatrix}$$

If we only had a homogeneous systems, we would solve for $c_1$ and $c_2$ now, but since this is a nonhomogeneous system we wait until we get ***X*** = ***Xc*** + ***Xp***.

Finding *Xp*

phi ∗ int(phi^−1 ∗ F(t))dt

### 5.4.2.3 Nonhomogeneous Systems of Differential Equations

```
> A:=homsol[1];
```

$$A := \begin{bmatrix} e^t & e^{(-t)} \\ e^t & 3\, e^{(-t)} \end{bmatrix}$$

```
> A1:=inverse(A);
```

$$A1 := \begin{bmatrix} \frac{3}{2}\frac{1}{e^t} & -\frac{1}{2}\frac{1}{e^t} \\ -\frac{1}{2}\frac{1}{e^{(-t)}} & \frac{1}{2}\frac{1}{e^{(-t)}} \end{bmatrix}$$

```
> B:=matrix(2,1,[0,4*t]);
```

$$B := \begin{bmatrix} 0 \\ 4\,t \end{bmatrix}$$

```
> B1:=multiply(A1,B);
```

$$B1 := \begin{bmatrix} -\frac{2\,t}{e^t} \\ \frac{2\,t}{e^{(-t)}} \end{bmatrix}$$

```
> B2:=map(int,B1,t);
```

$$B2 := \begin{bmatrix} \frac{2\,(1+t)}{e^t} \\ \frac{2\,(t-1)}{e^{(-t)}} \end{bmatrix}$$

```
> B3:=multiply(A,B2);
```

$$B3 := \begin{bmatrix} 4\,t \\ 8\,t-4 \end{bmatrix}$$

```
> nB3:=simplify(%);
```

$$nB3 := \begin{bmatrix} 4\,t \\ 8\,t-4 \end{bmatrix}$$

```
> x:=evalm(multiply(A,matrix(2,1,[c1,c2]))+nB3);
```

$$x := \begin{bmatrix} e^t\,c1 + e^{(-t)}\,c2 + 4\,t \\ e^t\,c1 + 3\,e^{(-t)}\,c2 + 8\,t - 4 \end{bmatrix}$$

Now, use the initial conditions $x1(0) = 1$, $x2(0) = 2$ to find $c_1$ and $c_2$.

```
> solve({c1+c2=1,c1+3*c2-4=2},{c1,c2});
```

$$\{c2 = \frac{5}{2}, c1 = \frac{-3}{2}\}$$

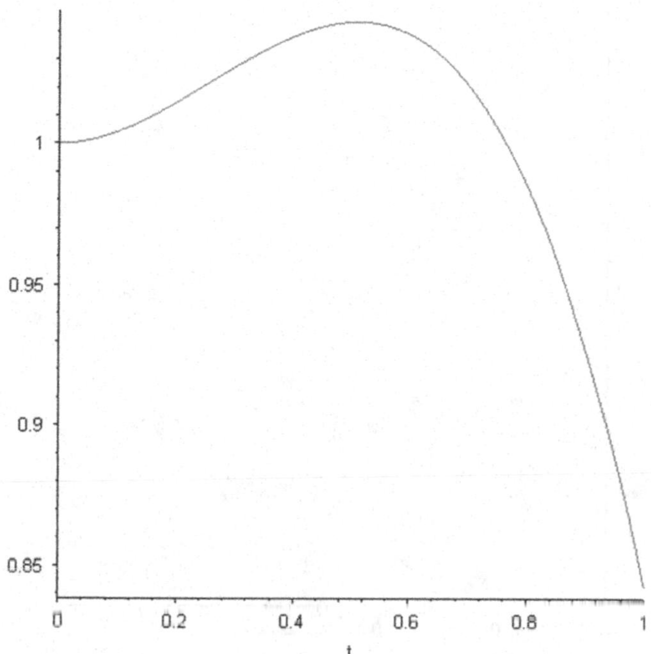


**FIGURE 5.21**
Plot of (x1,t=0..1) in Maple.

```
> x1:=subs({c2 = 2.5, c1 = −1.5},x[1,1]);
```

$$x1 := -1.5\, \mathbf{e}^{t} + 2.5\, \mathbf{e}^{(-t)} + 4\, t$$

```
> x2:=subs({c2 = 2.5, c1 = −1.5},x[2,1]);
```

$$x2 := -1.5\, \mathbf{e}^{t} + 7.5\, \mathbf{e}^{(-t)} + 8\, t - 4$$

We can plot each versus *t* (Figures 5.21 and 5.22).

```
> plot(x1,t=0..1);
> plot(x2,t=0..1);
```

### *5.4.2.4 Phase Portraits*

We now examine the phase portrait that provides the same information to us but in a slightly different format using Maple.

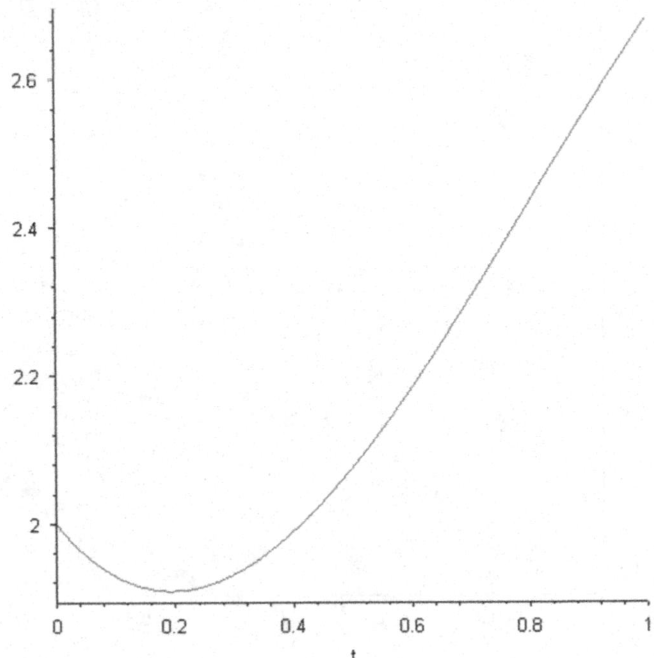


**FIGURE 5.22**
Plot of (x2,t=0..1) in Maple.

**restart; with(DEtools):with(linalg):**

*Phase Portrait in MAPLE*

> **> diffeq1:=diff(y(t),t)=4.5 * y(t)-.9 * y(t) * x(t);**

$$diffeq1 := \frac{d}{dt}\mathrm{y}(t) = 4.5\,\mathrm{y}(t) - 0.9\,\mathrm{y}(t)\,\mathrm{x}(t)$$

> **> diffeq2:=diff(x(t),t)=.24 * x(t)-.08 * y(t) * x(t);**

$$diffeq2 := \frac{d}{dt}\mathrm{x}(t) = 0.24\,\mathrm{x}(t) - 0.08\,\mathrm{y}(t)\,\mathrm{x}(t)$$

> **>DEplot({diffeq1,diffeq2},[y(t),x(t)],t=−3..3,y=0..5,x=0..5,title= 'Competing Species');**

The phase portrait (Figure 5.23) shows, from a starting point (the initial condition), who survives. The phase portrait traces out a possible solution curve.

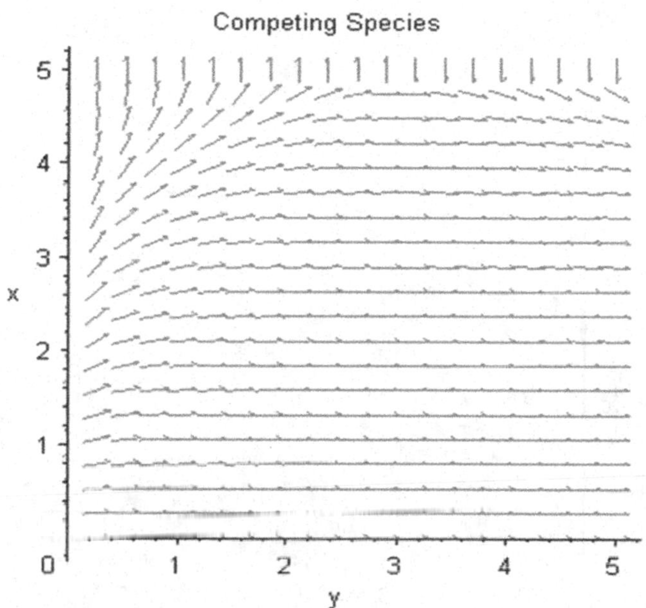


**FIGURE 5.23**
Plot of Example 5.11 phase portrait in Maple.

**Example 5.12:** The Fishpond

$$B(t) = \text{number of Bass fish after time } t$$
$$T(t) = \text{number of Trout after } t \text{ time}$$

Rate of Change of growth = rate in isolation + rate in competition for resources

$dB/dt = 0.7B - 0.02\,B * T$
$dT/dt = 0.5 - 0.01\,B * T$
Solve $dB/dt = 0$ and $dT/dt = 0$:
$dB/d = 0 = B(0.7 - 0.02T) = 0$ so either $B = 0$ or $T = 35.$
$dT/dt = 0 = T(0.5 - 0.01B) = 0$ so either $T = 0$ or $B = 5.0$

The equilibrium values are (0,0) and (50,35), and the phase portrait is in Figure 5.24.

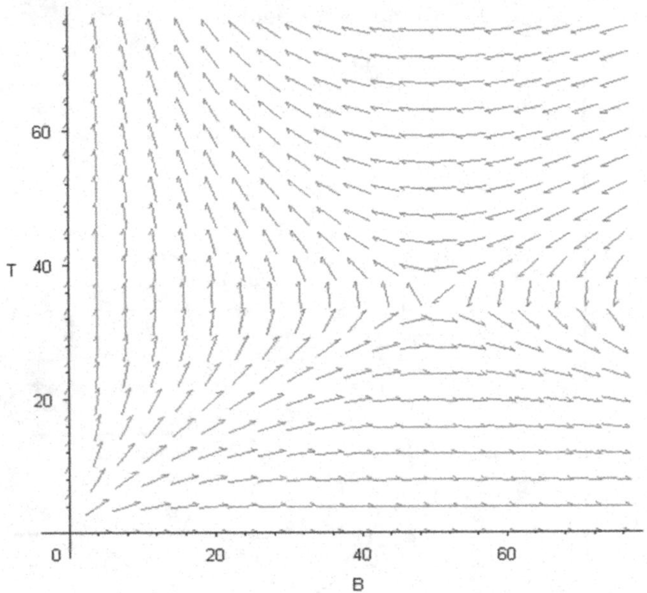


**FIGURE 5.24**
Plot of Example 5.12 phase portrait in Maple.

```
> with(plots):with(DEtools):
> eqn1:=diff(B(t),t)=.7 * B(t)-.02 * B(t) * T(t);
```

$$eqn1 := \frac{d}{dt} \mathrm{B}(t) = 0.7\, \mathrm{B}(t) - 0.02\, \mathrm{B}(t)\, \mathrm{T}(t)$$

```
> eqn2:=diff(T(t),t)=.5 * T(t)-.01 * B(t) * T(t);
```

$$eqn2 := \frac{d}{dt} \mathrm{T}(t) = 0.5\, \mathrm{T}(t) - 0.01\, \mathrm{B}(t)\, \mathrm{T}(t)$$

```
> DEplot([eqn1,eqn2], [B(t),T(t)], t=0..20, B=0..75,T=0..75);
```

Again both equilibrium values are not stable. Depending on the starting value one species will dominate over time. Thus, the phase portrait is useful to give us a sense of the possible solutions.

In Maple, we enter the system and initial conditions and then use the deSolve with classical numerical methods. Here is the command sequence to obtain the Euler estimates to our example.

```
> ode1 := diff(x(t), t) = 3·x(t) − 2·y(t);
```

$$ode1 := \frac{d}{dt}\, x(t) = 3\, x(t) - 2\, y(t)$$

```
> ode2 := diff(y(t), t) = 5·x(t) − 4·y(t);
```

$$ode2 := \frac{d}{dt}\, y(t) = 5\, x(t) - 4\, y(t)$$

```
> inits := x(0) = 3, y(0) = 6;
```

$$inits := x(0) = 3, y(0) = 6$$

```
    eulersol := dsolve({ode1, ode2, inits}, numeric,
        method = classical[foreuler],
        output = array([0, .1, .2, .3, .4, .5, .6, .7, .8, .9, 1, 1.1, 1.2,
        1.3, 1.4, 1.5, 1.6, 1.7, 1.8, 1.9, 2, 2.1, 2.2]), stepsize = 0.1);
>
```

eulersol := [[ t x(t) y(t) ]], [

| t | x(t) | y(t) |
|---|---|---|
| 0. | 3. | 6. |
| 0.1 | 2.70000000000000016 | 5.09999999999999964 |
| 0.2 | 2.49000000000000022 | 4.41000000000000014 |
| 0.3 | 2.35500000000000042 | 3.89100000000000044 |
| 0.4 | 2.28330000000000056 | 3.51210000000000022 |
| 0.5 | 2.26587000000000050 | 3.24891000000000042 |
| 0.6 | 2.29584900000000046 | 3.08228100000000050 |
| 0.7 | 2.36814750000000052 | 2.99729310000000070 |
| 0.8 | 2.47913313000000058 | 2.98244961000000064 |
| 0.9 | 2.62638314700000075 | 3.02903633100000080 |
| 1. | 2.80849082490000113 | 3.13061337210000090 |
| 1.1 | 3.02491539795000097 | 3.28261343571000142 |
| 1.2 | 3.27586733019300080 | 3.48202576040100098 |
| 1.3 | 3.56222237717070068 | 3.72714912133710108 |
| 1.4 | 3.88545926605448954 | 4.01740066138760987 |
| 1.5 | 4.24761691359331550 | 4.35317002985981106 |
| 1.6 | 4.65126798169934742 | 4.73571047471254403 |
| 1.7 | 5.09950628126664274 | 5.16706027567719950 |
| 1.8 | 5.59594611051119450 | 5.64998930603964044 |
| 1.9 | 6.14473208245662317 | 6.18796663887938080 |
| 2. | 6.75055837941773440 | 6.78514602455594052 |
| 2.1 | 7.41869668833186857 | 7.44636680444243292 |
| 2.2 | 8.15503233394294114 | 8.17716842683139332 |

]

The power of Euler's method is two-fold. First, it is easy to use, and second, as a numerical method it can be used to estimate a solution to a system of differential equations that does not have a closed form solution.

Assume we have the following predator-prey system that does not have a closed from analytical solution:

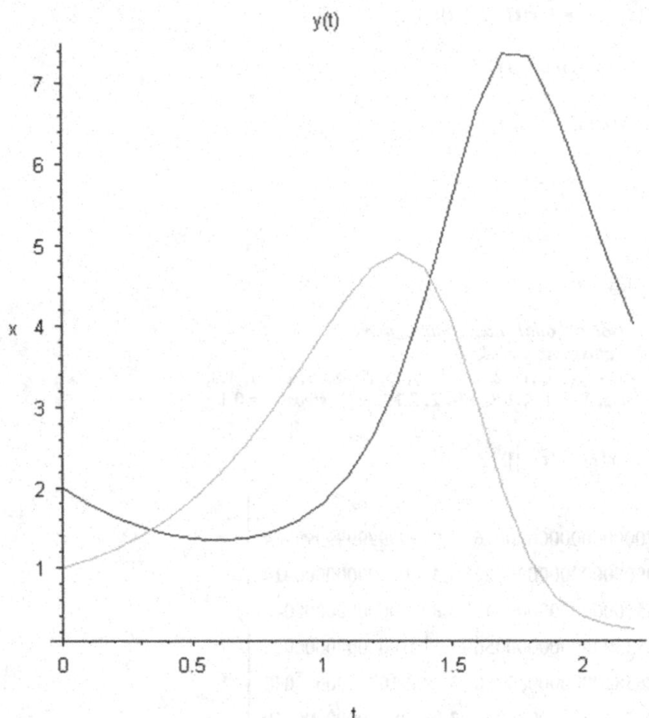


**FIGURE 5.25**
Plot of system of ODEs.

$$dx/dt = 3x - xy$$
$$dy/dt = xy - 2y$$
$$x(0) = 1,\ y(0) = 2$$
$$t_o = 1,\ \Delta t = 0.1$$

We will obtain an estimate *f* of the solution using Euler's method (Figure 5.25).

**> with(LinearAlgebra):with(DEtools):**

**> ode1:=diff(x(t),t)=3 * x(t)-x(t) * y(t);**

$ode1 := \frac{d}{dt}\mathrm{x}(t) = 3\,\mathrm{x}(t) - \mathrm{x}(t)\,\mathrm{y}(t)$

**> ode2:=diff(y(t),t)=x(t) * y(t)−2 * y(t);**

$ode2 := \frac{d}{dt}\mathrm{y}(t) = \mathrm{x}(t)\,\mathrm{y}(t) - 2\,\mathrm{y}(t)$

**> inits:=x(0)=1,y(0)=2;**

$inits := \mathrm{x}(0) = 1, \mathrm{y}(0) = 2$

```
>eulersol:=dsolve({ode1,ode2,inits},numeric,method=classical
[foreuler],

output=array([0,.1,.2,.3,.4,.5,.6,.7,.8,.9,1,1.1,1.2,1.3,1.4,1.5,1.6,1.7,1.8,1.9,2,
2.1,2.2]),stepsize=0.1);
```

*eulersol* :=

| $[t, x(t), y(t)]$ | | |
|---|---|---|
| 0. | 1. | 2. |
| 0.1 | 1.10000000000000008 | 1.80000000000000004 |
| 0.2 | 1.23200000000000021 | 1.63800000000000012 |
| 0.3 | 1.39979840000000010 | 1.51220160000000026 |
| 0.4 | 1.60806018198425638 | 1.42143901801574435 |
| 0.5 | 1.86190228798054114 | 1.36572716301158748 |
| 0.6 | 2.16618792141785876 | 1.34686678336611476 |
| 0.7 | 2.52428764205455636 | 1.36925008248155188 |
| 0.8 | 2.93593582846188727 | 1.44103817219427799 |
| 0.9 | 3.39363701700781206 | 1.57591009774806334 |
| 1. | 3.87692143779073328 | 1.79553476251787370 |
| 1.1 | 4.34388314781755014 | 2.13254253132470328 |
| 1.2 | 4.72069653578025861 | 2.63238558144231760 |
| 1.3 | 4.89423614699907182 | 3.34857781466911942 |
| 1.4 | 4.72363393293951717 | 4.31773530989456944 |
| 1.5 | 4.10118401049446124 | 5.49372835024256556 |
| 1.6 | 3.07846012684130698 | 6.64806176699554552 |
| 1.7 | 1.95541885784630654 | 7.36502872064382874 |
| 1.8 | 1.10187291030753887 | 7.33219458140772317 |
| 1.9 | 0.624520125164112150 | 6.67367032336186838 |
| 2. | 0.395092020148348210 | 5.75572040125449292 |
| 2.1 | 0.286215706118782388 | 4.83198024107766244 |
| 2.2 | 0.233781554289212323 | 4.00388305652733401 |

```
> with(plots):

> plot1:=odeplot(eulersol,[t,x(t)],0..10, color=green,title='x(t)'):

> plot2:=odeplot(eulersol,[t,y(t)],0..10,color=blue,title='y(t)'):

> display(plot1,plot2);
```

We experiment and find that when we plot $x(t)$ versus $y(t)$ we have an approximately closed loop (Figure 5.26).

```
> odeplot(eulersol,[x(t),y(t)],0..22, color=green,title='System');
```

Another method for numerical estimates is the Runge-Kutta 4 applied to systems. We illustrate with the same predator-prey example (Figures 5.27).

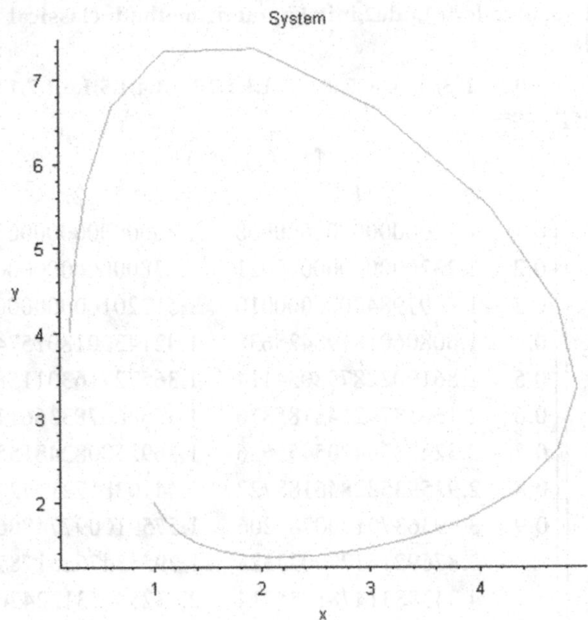


**FIGURE 5.26**
Plot of system of ODEs.

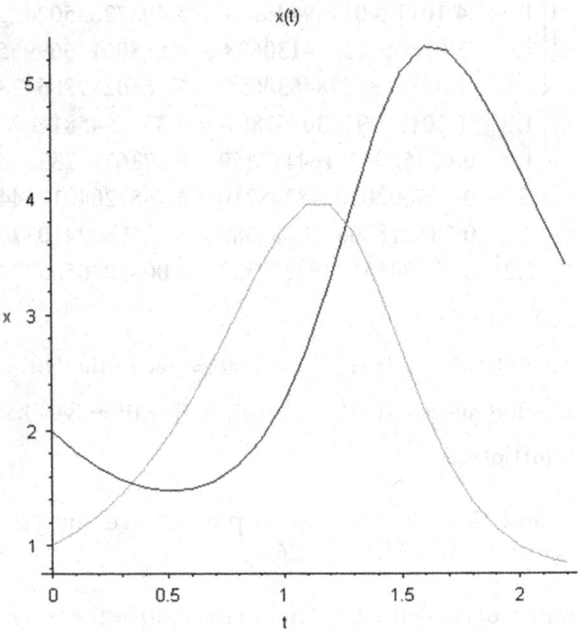


**FIGURE 5.27**
Plot of predator-prey system of ODEs.

```
> restart;
> with(LinearAlgebra):with(DEtools):
> ode1:=diff(x(t),t)=3*x(t)-x(t)*y(t);
```

$$ode1 := \frac{d}{dt}\,x(t) = 3\,x(t) - x(t)\,y(t)$$

```
> ode2:=diff(y(t),t)=x(t)*y(t)-2*y(t);
```

$$ode2 := \frac{d}{dt}\,y(t) = x(t)\,y(t) - 2\,y(t)$$

```
> inits:=x(0)=1,y(0)=2;
```

$$inits := x(0) = 1, y(0) = 2$$

```
> rk4sol:=dsolve({ode1,ode2,inits},numeric, method=classical[rk4],
output=array([0,.1,.2,.3,.4,.5,.6,.7,.8,.9,1,1.1,1.2,1.3,1.4,1.5,1.6,1.7,1.8,1.9,2,
2.1,2.2]),stepsize=0.1);
```

*rk4sol* :=

| $t$ | $x(t)$ | $y(t)$ |
|---|---|---|
| 0. | 1. | 2. |
| 0.1 | 1.11554071453956705 | 1.81968188493959970 |
| 0.2 | 1.26463746535620780 | 1.67761981669985926 |
| 0.3 | 1.45146389024689194 | 1.57278931221810914 |
| 0.4 | 1.68031037053259168 | 1.50542579227745321 |
| 0.5 | 1.95456986108324960 | 1.47762563817635396 |
| 0.6 | 2.27500034092209802 | 1.49412980191491540 |
| 0.7 | 2.63687047094120208 | 1.56336399876409616 |
| 0.8 | 3.02561346371586382 | 1.69866644464148696 |
| 0.9 | 3.41107833928229720 | 1.91916110241906601 |
| 1. | 3.74212348522216676 | 2.24852435676575446 |
| 1.1 | 3.94706090545572064 | 2.70772001337807611 |
| 1.2 | 3.95001172191016892 | 3.29658238745704324 |
| 1.3 | 3.70934802583802936 | 3.96638771088000476 |
| 1.4 | 3.25874674267819132 | 4.60727591934640213 |
| 1.5 | 2.70448435236999174 | 5.08419319694337712 |
| 1.6 | 2.16641586882149006 | 5.30768472071020270 |
| 1.7 | 1.71962027608752543 | 5.27274016506954890 |
| 1.8 | 1.38441109626295588 | 5.03717315896791806 |
| 1.9 | 1.14891557150367586 | 4.67751425643222784 |
| 2. | 0.991726731312949417 | 4.25984946993614156 |
| 2.1 | 0.893335577155144112 | 3.83076393278872685 |
| 2.2 | 0.839403883501402824 | 3.41909584046305914 |

```
> with(plots):
> plot1:=odeplot(rk4sol,[t,x(t)],0..10, color=green,title='x(t)'):
> plot2:=odeplot(rk4sol,[t,y(t)],0..10,color=blue,title='y(t)'):
> display(plot1,plot2);
> odeplot(rk4sol,[x(t),y(t)],0..22,color=green,title='System');
```

(Figure 5.28)

### 5.4.3 R for System of Ordinary Differential Equations

We provide R commands and code. R provides numerical results, plots, and phase portraits. You must install deSolve and phaseR from the CRAN library.

Let's resolve our predator-prey model in R. We will obtain numerical valuea and a portrait.

We provide R commands and code. R provides numerical results, plots, and phase portraits. You must install deSolve and phaseR from the CRAN library.

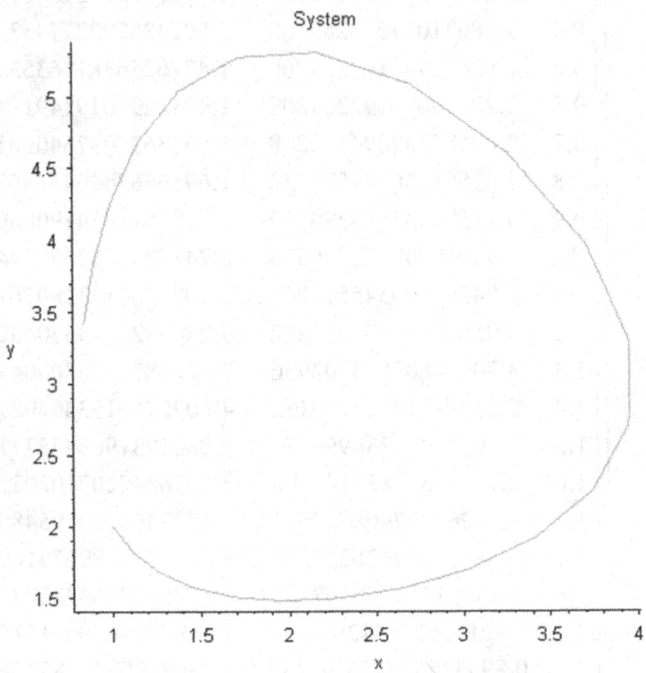


**FIGURE 5.28**
Plot of system of ODEs.

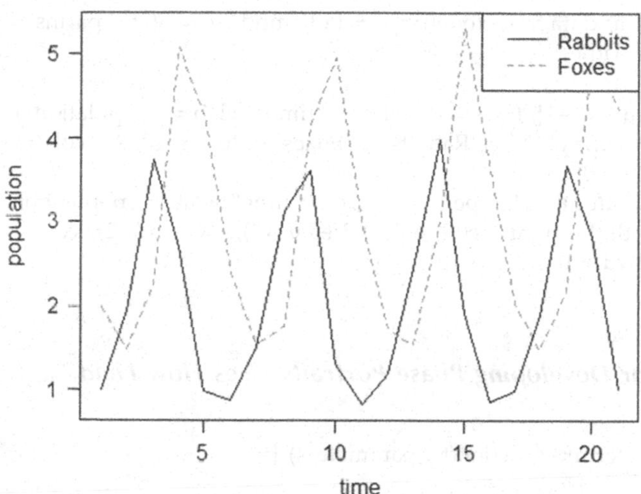


**FIGURE 5.29**
Screenshot of predator-prey system of ODEs in R.

Let's resolve our predator-prey model in R. We will obtain numerical valuea and a portrait.

The ODE we will solve is

$$\frac{dx}{dt} = 3 * x(t) - x(t)y(t)$$
$$\frac{dy}{dt} = x(t)y(t) - 2y(t)$$
$$x(0) = 1, \; y(0) = 2$$

The R code commands to obtain plots of *x(t)* and *y(t)* (Figure 5.29) are:

```
library(deSolve)

LotVmod <- function (Time, State, Pars) {
  with(as.list(c(State, Pars)), {
    dx = x * (alpha - beta * y)
    dy = -y * (gamma - delta * x)
    return(list(c(dx, dy)))
  })
}

Pars <- c(alpha = 3, beta = 1, gamma = 2, delta = 1)
State <- c(x = 1, y = 2)
Time <- seq(0, 10, by = 0.5)
```

```
out <- as.data.frame(ode(func = LotVmod, y = State, parms = Pars,
times = Time))

matplot(out[,-1], type = "l", xlab = "time", ylab = "population")
legend("topright", c("Rabbits", "Foxes"), lty = c(1,2), col = c(1,2),
box.lwd = 0)
  matplot(out[,-1], type = "l", xlab = "time", ylab = "population")
  legend("topright", c("Prey", "Predator"), lty = c(1,2), col = c(1,2),
  box.lwd = 0)
}
```

#### 5.4.3.1 R for Developing Phase Portraits Uses Flow Field

```
lotkavolterra<-function(t,y,parameters) {
x<-y[1]
y<-y[2]
lambda<-parameters[1]
epsilon<-parameters[2]
eta<-parameters[3]
delta<-parameters[4]
dy<-numeric(2)
dy[1]<-lambda * x-epsilon * x * y
dy[2]<-eta * x * y-delta * y
list(dy)
}
lotkavolterra.flowField <-flowField(lotkavolterra,
xlim=c(0,5),ylim=c(0,10), parameters =c(3,1,2,1),points=19, add=FALSE)
grid()
lotkavolterra.nullclines<-
nullclines(lotkavolterra, xlim=c(-1,5), ylim=c(-1,10),
parameters=c(3,1,2,1),points=500)
y0<-matrix(c(1,2,2,2,3,4),ncol=2,nrow=3, byrow=TRUE)
lotkavolterra.trajectory<-trajectory(lotkavolterra, y0=y0,t.end=10),
parameters=c(3,1,2,1), colour=rep("black",3))
```

Phase Portrait plot (Figure 5.30) from these commands

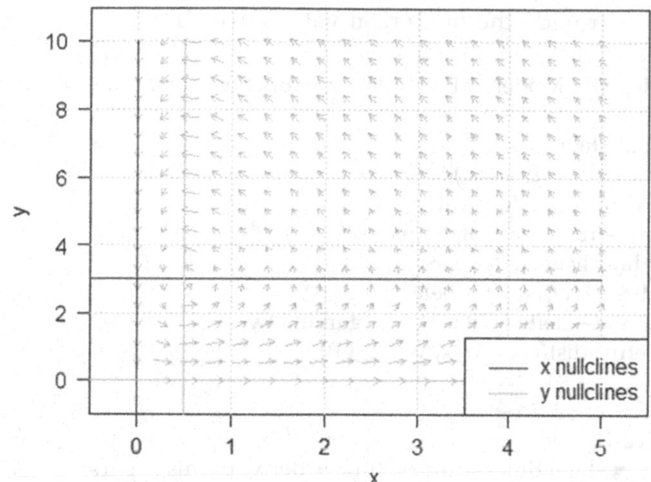


**FIGURE 5.30**
Screenshot of trajectories in phase portrait in R.

Following the trajectories, we see that the trajectories are elliptical.

We provide some additional R code and commands that allow you to see the numerical values.

```
# parameters
pars <- c(alpha = 3, beta = 1, delta = 1, gamma = 2)
# initial state
init <- c(x = 1, y = 2)
# times
times <- seq(0, 100, by = 1)
```

Next, we need to define a function that computes the derivatives in the ODE system at a given point in time.

```
deriv <- function(t, state, pars) {
  with(as.list(c(state, pars)), {
    d_x <- alpha * x - beta * x * y
    d_y <- delta * x * y - gamma * y
    return(list(c(x = d_x, y = d_y)))
  })
}
lv_results <- ode(init, times, deriv, pars)
```

The lv-results provide the numerical values over time.

```
lv_model XE "Model" XE "model" <- function(pars, times = seq(0, 50,
by = 1)) {
  # initial state
  state <- c(x = 1, y = 2)
  # derivative XE "derivative"
  deriv <- function(t, state, pars) {
    with(as.list(c(state, pars)), {
      d_x <- alpha * x - beta * x * y
      d_y <- delta * beta * x * y - gamma * y
      return(list(c(x = d_x, y = d_y)))
    })
  }
  # solve
  ode(y = state, times = times, func = deriv, parms = pars)
}
lv_results <- lv_model XE "Model" XE "model" (pars = pars, times =
seq(0, 50, by = 0.25))
```

We provide some of the numerical outputs as well.

```
   >lv_results
   time         x         y
1   0.00 1.0000000 2.000000
2   0.25 1.3530677 1.620569
3   0.50 1.9545794 1.477617
4   0.75 2.8293344 1.621674
5   1.00 3.7421685 2.248472
6   1.25 3.8602053 3.627084
7   1.50 2.7044130 5.084510
8   1.75 1.5381962 5.175234
9   2.00 0.9916079 4.259705
10  2.25 0.8259647 3.224968
11  2.50 0.8694643 2.409064
12  2.75 1.0854469 1.858638
13  3.00 1.5077400 1.551265
14  3.25 2.1955170 1.485853
15  3.50 3.1296117 1.747729
16  3.75 3.9144918 2.587278
17  4.00 3.6073011 4.144647
18  4.25 2.2865674 5.279742
19  4.50 1.3130843 4.951042
```

## 5.5 Exercises

1. Find and classify all the rest points. Then, sketch a few trajectories to indicate the motion.
   a. $dx/dt = x$, $dy/dt = y$
   b. $dx/dt = -x$, $dy/dt = 2y$
   c. $dx/dt = y$, $dy/dt = -2x$
   d. $dx/dt = -x + 1$, $dy/dt = -2y$
2. Given the following system of Linear 1st Order ODEs of species cooperation (symbiosis):

$$dx_1/dt = -0.5\, x_1 + x_2$$
$$dx_2/dt = 0.25\, x_1 - 0.5\, x_2$$
$$\text{and } x_1(0) = 200 \text{ and } x_2(0) = 500.$$

   a. Perform Euler's Method with step-size $h = 0.1$ to obtain graphs of numerical solutions for $x_1(t)$ and $x_2(t)$ versus $t$ and for $x_1$ versus $x_2$. You can put both $x_1(t)$ and $x_2(t)$ versus $t$ on one axis if you want.
   b. From the graphs discuss the long-term behavior of the system (discuss stability).
   c. Analytically using eigenvalues and eigenvectors solve the system of DEs to determine the population of each species for $t > 0$.
   d. Determine if there is a steady state solution for this system.
   e. Obtain real plots of $x_1(t)$ and $x_2(t)$ versus $t$ and for $x_1(t)$ versus $x_2(t)$. Compare to the numerical plots. Briefly discuss.
3. Given a competitive hunter model defined by the system:

$$dx/dt = 15x - x^2 - 2xy = x(15 - x - 2y)$$
$$dy/dt = 12y - y^2 - 1.5xy = y(12 - y - 1.5x)$$

   a. Perform a graphical analysis of this competitive hunter model in the $x$-$y$ plane.
   b. Identify all equilibrium points and classify their stability.
   c. Find the numerical solutions using Euler's Method with step size $h = 0.05$. Try it from two separate initial conditions: first, use $x(0) = 5$ and $y(0) = 4$, then use $x(0) = 3$, $y(0) = 9$. Obtain graphs of $x(t)$, $y(t)$ individually (or on the same axis) and then a plot of $x$

versus $y$ using your numerical approximations. Compare to your phase portrait analysis.

4. Since bass and trout both live in the same lake and eat the same food sources, they are competing for survival. The rate of growth for bass ($dB/dt$) and for trout ($dT/dt$) are estimated by the following equations:

$$dB/dt = (10\text{–}B\text{–}T)B$$
$$dT/dt = (15\text{–}B\text{–}3T)T$$

*Coefficients and values are in thousands.*

$$x_{n+1} = x_n + h\,f\,(x_n, y_n)\ \&\ y_{n+1} = y_n + h\,g\,(x_n, y_n)$$

   a. Obtain a "qualitative" graphical solution of this system. Find all equilibrium points of the system and classify each as unstable, stable, or asymptotically stable.
   b. If the initial conditions are $B(0) = 5$ and $T(0) = 2$, determine the long-term behavior of the system from your graph in part (a). Sketch it out.
   c. Using Euler's Method, $h = 0.1$ and the same initial conditions as above, obtain estimates for $B$ and $T$. Using these estimates determine a more accurate graph by plotting $B$ versus $T$ for the solution from $t = 0$ to $t = 7$.
   d. Euler's Method:
      Compare the graph in part (c) to the possible solutions found in (a) and (b). Briefly comment.
5. Find the equilibrium values for the SIR model presented.
6. In the predator-prey model, determine the outcomes with the following sets of parameters.
   a. Initial foxes are 200, and initial rabbits are 400.
   b. Initial foxes are 2,000, and initial rabbits are 10,000
   c. Birth rate of rabbits increases to 0.1
7. In the SIR model, determine the outcome with the following parameters changed.
   a. Initially 5 are sick and 10 the next week.
   b. The flu lasts 1 week.
   c. The flu lasts 4 weeks.
   d. There are 4,000 students in the dorm; 5 are initially infected, and 30 more the next week.

## 5.6 Projects

1. **Diffusion** – Diffusion through a membrane leads to a first-order system of ordinary linear differential equations. For example, consider the situation in which two solutions of substance are separated by a membrane of permeability $P$. Assume the amount of substance that passes through the membrane at any particular time is proportional to the difference between the concentrations of the two solutions. Therefore, if we let $x_1$ and $x_2$ represent the two concentrations, and $V_1$ and $V_2$ represent their corresponding volumes, then the system of differential equations is given by:

$$dx_1/dt = (P/V_1)(x_2 - x_1)$$
$$dx_2/dt = (P/V_2)(x_1 - x_2)$$

where the initial amounts of $x_1$ and $x_2$ are given.

Consider two salt concentrations of equal volume $V$ separated by a membrane of permeability $P$. Given that $P = V$, determine the amount of salt in each concentration at time $t$ if $x_1(0) = 2$ and $x_2(0) = 10$.

Write out the system of differential equations that models this behavior.

Using the methods described in Chapter 4, solve this system. Clearly indicate your eigenvalues and eigenvectors.

Plot the solutions for $x_1$ and $x_2$ on the same axis and label each. Comment about the plots.

Use a numerical method (Euler or Runge-Kutta 4) and iterate a numerical solution to predict $x_i(4)$, use a step size of 0.5. Obtain a plot of your numerical approach. Compare it to the analytical plot.

Comment about what you see in the plots.

Diffusion through a double-walled membrane, where the inner wall has permeability $P_1$ and the outer wall has permeability $P_2$ with $0 < P_1 < P_2$. Suppose the volume of the solution within the inner wall is $V_1$ and between the two walls is $V_2$. Let $x$ represent the concentration of the solution within the inner wall and $y$, the concentration between the two walls. This leads to the following system:

$$\frac{dx}{dt} = \frac{P_1}{V_1}(y - x)$$
$$\frac{dy}{dt} = \frac{1}{V_2}(P_2(C - y) + P_1(x - y))$$
$$x(0) = 2,\ y(0) = 1,\ C = 10$$

Also assume the following:

$$P_1 = 3$$
$$P_2 = 8$$
$$V_1 = 2$$
$$V_2 = 10.$$

Set up the system of ODEs with all coefficients.

Use the method of Variation of parameter for systems,

$$X = X_c + \phi \int \phi^{-1} F(t)dt$$

to find both $X_c$ and $X_p$.

Use the initial conditions to find the particular solution, find the coefficients for$X_c$ in the solution $X_c + X_p$.

Plot the solutions for $x(t)$ and $y(t)$ on the same axis. Comment about the solution.

2. **An Electrical Network** – An electrical network containing more than one loop also gives rise to a system of differential equations. For instance, in the electrical network displayed below, there are two resistors and two inductors. At branch point B in the network, the current $i_1(t)$ splits in two directions. Thus,

$$i_1(t) = i_2(t) + i_3(t)$$

Kirchhoff's law applies to each loop in the network. For loop ABEF, we find that

$$E(t) = i_1 R_1 + L_1 di_2/dt$$

The sum of the voltage drops across the loop ABCDEF is

$$E(t) = i_1R_1 + L_2di_3/dt + i_3R_3$$

Substituting, we find the following systems for equations:

$$\frac{di_1}{dt} = \frac{-(R_1 + R_2)}{L_1}i_1 + \frac{R_2}{L_2}i_2$$

$$\frac{di_2}{dt} = \left(\frac{R_2}{L_2} - \frac{1}{R_2C}\right)i_2 - \frac{(R_1 + R_2)}{L_2}i_1 + \frac{E(t)}{L_2}$$

$$i_1(0) = 0,\ i_2(0) = 0$$

Initially, let *E(t)* = 0 Volts, *L1* = 1 henry, *L2* = 1 henry, *R1* = 1 omhs, *R2* = 1 omhs, *C* = 3

Write out the system of differential equation that models this behavior.

Using the methods described in Chapter 4, solve this system. Clearly indicate your eigenvalues and eigenvectors.

Plot the solutions for $x_1$ and $x_2$ on the same axis and label each. Comment about the plots.

Use a numerical method (Euler or Runge-Kutta) and iterate a nu merical solution to predict $x_i(4)$, use a step size of 0.5. Obtain a plot of your numerical approach. Compare it to the analytical plot. Comment about the plots.

Now, let *E(t)* = $100 * \sin(t)$

Set up the system of ODEs with all coefficients.

Use the method of Variation of parameter for systems,

$$X = X_c + \phi\int\phi^{-1}F(t)dt$$

to find both $X_c$ and $X_p$.

Use the initial conditions to find the particular solution, find the coefficients for $X_c$ in the solution $X_c + X_p$.

Plot the solutions for *x(t)* and *y(t)* on the same axis. Comment about the solution.

3. **Interacting Species** – Suppose *x(t)* and*y(t)* represent respective populations of two species over time, t. One model might be

$$X' = R1\,X, \;\; X(0) = X_0$$
$$Y' = R2\,Y, \;\; Y(0) = Y_0$$

where *R1* and *R2* are intrinsic coefficients. Models involving competition between species or predator-prey models most often include interaction terms between the variables. These interactions terms, if included, will preclude any analytical solution attempts so we will simplify these models for this project.

Let's model bass and trout attempting to coexist in a small pond in South Carolina.

$$\mathbf{B' = -0.\,5\,B + T + H}$$
$$\mathbf{T' = 0.\,25B - 0.\,5\,T + K}$$

$$B(0) = 2000, \;\; T(0) = 5000$$

Initially, let $\boldsymbol{H = K = 0}$
Write out the system of differential equation that models this behavior.

Using the methods described in Chapter 8, solve this system. Clearly indicate your eigenvalues and eigenvectors.

Plot the solutions for $x_1$ and $x_2$ on the same axis and label each. Comment about the plots.

Use a numerical method (Euler or Runge-Kutta) and iterate a numerical solution to predict $x_i(10)$, use a step size of 0.5. Obtain a plot of your numerical approach. Compare it to the analytical plot.
Comment about the plots.

Now, let $\boldsymbol{H = 1{,}500; K = 1{,}000}$

Set up the system of ODEs with all coefficients.

Use the method of Variation of parameter for systems,

$$X = X_c + \phi \int \phi^{-1} F(t)\,dt$$

to find both $X_c$ and $X_p$.

Use the initial conditions to find the particular solution, find the coefficients for $X_c$ in the solution $X_c + X_p$.

Plot the solutions for *x(t)* and *y(t)* on the same axis. Comment about the solution.

Do these species coexist? Briefly explain. If any die out, determine when this happens.

4. **Trapezoidal Method** – The trapezoidal method is a more stable numerical method that is shown in Numerical Analysis textbooks (see Burden and Faires, Numerical Analysis, Brooks-Cole Publishers, pages 344–346). Find the trapezoidal algorithm and modify it for systems of ODEs. Write a Maple program to obtain the trapezoidal estimates and compare these to both Euler and Runge-Kutta estimates.

## References and Suggested Future Readings

Allman, E., and J. Rhodes (2004). *Mathematical Models in Biology: An Introduction*, Cambridge University Press, Cambridge, UK.

Abell, M. L., and J. P. Braselton (2000). *Differential Equation with MAPLE V*, 2nd Edition. Academic Press, San Diego, CA.

Barrow, D., A. Belmonte, A. Boggess, J. Bryant, T. Kiffe, J. Morgan, M. Rahe, K. Smith, and M. Stecher (1998). *Solving Differential Equations with MAPLE V Release 4*, Brooks-Cole Publisher, Pacific Grove, CA.

Fox, W. (2013) *Mathematical Modeling with MAPLE*, Brooks-Cole, Cengage Publisher, Boston, MA.

Fox, W., and W. Bauldry (2020). *Problem Solving with MAPLE*, Taylor and Francis, CRC Press, Boca Raton, FL.

Giordano, F. R., and Maurice D. Weir (1991). *Differential Equations: A Modeling Approach*, Addison-Wesley Publishers, Reading, MA.

Giordano, F. R., W. Fox, and S. Horton (2012). *A First Course in Mathematical Modeling*, 5th Edition. Brooks-Cole, Boston, MA.

Zill, D. (2015). *A first course in differential equations with modeling applications*, 10th Edition. Brooks-Cole, Cengage Publisher, Boston, MA.

# 6

# *Forecasting with Linear Programming and Machine Learning*

**OBJECTIVES**

1. Formulate a linear programming problem.
2. Solve and interpret a linear programming problem.
3. Perform sensitivity analysis and interpret that analysis.

## 6.1 Introduction to Forecasting

What is forecasting? We assume that forecasting is concerned with making predictions about future observations based upon past data. Mathematically, there are many algorithms that one might use to make these predictions: *regression (linear and non-linear), exponential smoothing* (time series) model, *autoregressive integrated moving average* (ARIMA) model to name a few that we will illustrate in this chapter.

Applications of forecasting might include the following:

- Operations management in business such as forecast of product sales or demand for services.
- Marketing: forecast of sales response to advertisement procedures, new item promotions, etc.
- Finance and risk management: forecast returns from investments or retirement accounts.
- Economics: forecast of major economic variables, e.g. GDP, population growth, unemployment rates, inflation; monetary and fiscal policy; budgeting plans and decisions.
- Industrial process control: forecasts of the quality characteristics of a production process.

DOI: 10.1201/9781003298762-6

- Demography: forecast of population; of demographic events (deaths, births, migration); forecasting language demand, forecasting crime or homicides
- To facilitate the process of deciding which forecasting process to use, we employ the following mathematical modeling approach within the decision process.

### 6.1.1 Steps in the Modeling Process

We adopt the framework for the mathematical modeling process (adapted from Giordano et al., 2013) which works very well in the decision-making contexts with a few minor adjustments, as shown in the following steps:

Step 1. Define the main objective for the forecasting tasks.
Step 2. Make assumptions and choose relevant variables.
Step 3. Collect necessary data.
Step 4. Choose the right forecasting model.
Step 5. Run the model.
Step 6. Perform model testing and sensitivity analysis.
Step 7. Perform a common sense test on the results.
Step 8. Consider both strengths and weaknesses to your modeling process.
Step 9. Present the results to the decision maker.

We discuss each of these nine steps in a little more depth.

Step 1. Understand the problem or the question asked. To make a good decision you need to understand the problem. Identifying the problem to study is usually difficult. In real life no one walks up to you and hands you an equation to be solved; usually, it is a comment like, "we need to make more money" or "we need to improve our efficiency". We need to be precise in our understanding of the problem to be precise in the formulation of the mathematics to describe the situation.

Step 2a. Make simplifying assumptions. Start by brainstorming the situation, making a list of as many factors, or variables, as you can. However, keep in mind that we usually cannot capture all these factors influencing a problem. The task is simplified by reducing the number of factors under consideration. We do this by making simplifying assumptions about the factors, such as holding certain factors as constants. We might then examine to see if relationships exist between the remaining factors (or variables). Assuming simple relationships might reduce the complexity of the problem.

Once you have a shorter list of variables, classify them as independent variables, dependent variables, or neither.

Step 2b. Define all variables and provide their respective units. It is critical to clearly define all your variables and provide the mathematical notation and units for each one.

Step 3. Acquire the data. We note that acquiring the data is not an easy process.

Step 4. Construct the model. Using the tools in this text and your own creativity to build a mathematical model that describes the situation and whose solution helps to answer important questions.

Step 5. Solve and interpret the model. We take the model constructed in Steps 1–4 and solve it. Often this model might be too complex so we cannot solve it or interpret it. If this happens, we return to Steps 2–4 and simplify the model further.

Step 6. Perform sensitivity analysis and model testing. Before we use the model, we should test it out. We must ask several questions. Does the model directly answer the question or does the model allow for the answer to the question(s) to be answered? During this step, we should review each of our assumptions to understand the impact on the mathematical model's solution if the assumption is incorrect.

Step 7. Passing the common sense test. Is the model useable in a practical sense (can we obtain data to use the model)? Does the model pass the common sense test? We will say that we "collaborate the reasonableness" of our model.

Step 8. Strengths and Weaknesses. No model is complete with self-reflection of the modeling process. We need to consider not only what we did right but what we did that might be suspect as well as what we could do better. This reflection also helps in refining models.

Step 9. Present results and sensitivity analysis to the decision maker. A model is pointless if we do not use it. The more user-friendly the model, the more it will be used. Sometimes the ease of obtaining data for the model can dictate its success or failure. The model must also remain current. Often this entails updating parameters used in the model.

Our plan stems from the modeling approach, which is an iterative approach as we move to refine models. Therefore, starting with least squares regression and "moving" models from time series we transition to the more formal machine learning methods.

## 6.2 Machine Learning

We have found and teach that ***machine learning*** has many definitions. Machine learning can range from almost any use of a computers (machine) performing calculations and analysis measures to the applications of artificial intelligence (AI) or ineural networks in the computer's calculations and analysis methods. In this chapter, we adapt the definition from Huddleston and Brown (2018),

> The defining characteristic of machine learning is the focus on using algorithmic methods to improve descriptive, predictive, and prescriptive performance in real-world contexts. An older, but perhaps more accurate, synonym for this approach from the statistical literature is *algorithmic modeling* (Breiman, 2001). This algorithmic approach to problem solving often entails sacrificing the interpretability of the resulting models. Therefore, machine learning is best applied when this trade-off makes business sense but is not appropriate for situations such as public policy decision-making, where the requirement to explain how one is making decisions about public resources is often essential.

Figure 6.1 provides a summary of the data needed to train the model, the format of the resulting model output, and a list of algorithms often used for this class of problem. Algorithms listed are addressed in this chapter, except Q Learning, which is omitted due to space limitations (Huddleston & Brown, 2018).

We concentrate in this chapter with the supervised learning aspect only and, those regression and time series models that we can readily apply to the forecasting tasks.

According to Huddleston and Brown (2018).

> Machine learning is an emerging field, with new algorithms regularly developed and fielded. Specific machine learning techniques are usually designed to address one of the three types of machine learning problems introduced here, but in practice, several methods are often combined for real-world application. As a general rule, unsupervised learning methods are designed to be *descriptive*, supervised learning methods are designed to be *predictive*, and reinforcement-learning methods are designed to be *prescriptive*.

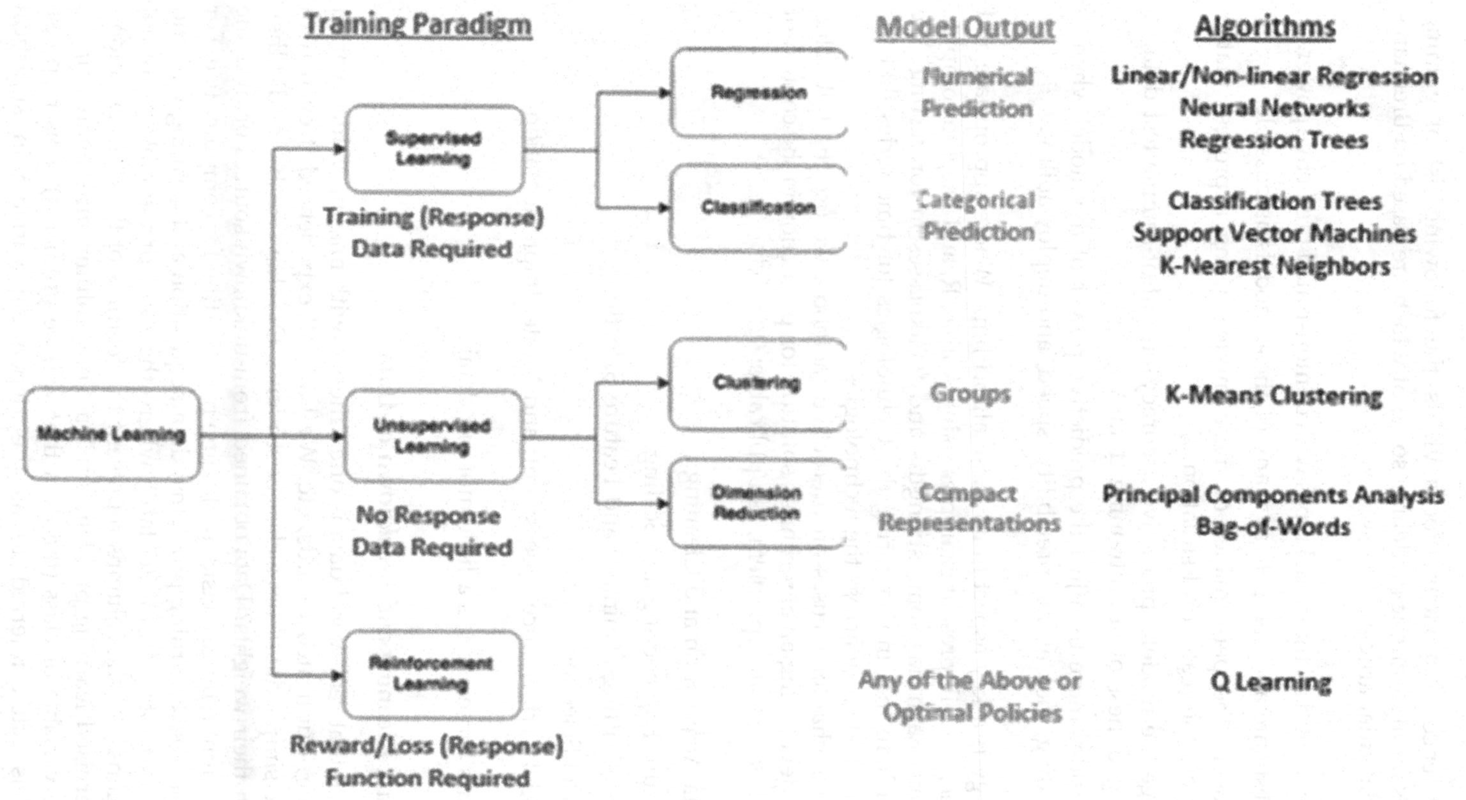


**FIGURE 6.1**
Overview of machine learning paradigms: supervised, unsupervised, and reinforcement learning.

This difference in modeling focus yields the following list of guiding principles for algorithmic modeling, some of which are based on Breiman's (2001) influential article:

- Understand all the model types with demonstrated predictive power.
- Investigate and compete as many of these models as possible.
- Measure the performance of these models using a process that mirrors the real-world situation.
- Judge the model's predictive accuracy on out-of-sample test data, not goodness of fit on training data.
- Consider more than just the predictive power of the model when deciding; look at interpretability, speed, and deploy ability.

Technology is a key aspect in all these algorithms. Where appropriate and applicable, we will present examples using Maple, R, and Excel to compute our output. Each has some strengths and weaknesses in forecasting. We point out here that in comparing the technologies for time series that the values differ slightly across the technologies.

Since this chapter focuses on predictive methods, we use the following steps for developing an algorithmic solution for prediction in the context of a supervised learning problem, as applicable:

- Data Acquisition and Cleaning.
- Feature Engineering and Scaling.
- Model Fitting (Training) and Feature Selection.
- Model Selection.
- Model Performance Assessment; and Model Implementation

We discuss these aspects a little more in detail.

### 6.2.1 Data Cleaning and Breakdown of Data

We realize that real-world data is often messy, with missing values, varied formatting, duplicated records, etc. We have all experienced it even with collecting simple data from a class of students. Did they use lbs., or did they use kg in their weights? Data obtained from unknown sources or over the Internet might require considerable time reformatting for computer use. For most supervised learning problems, data acquisition and cleaning will end with the data stored in a 2D table where the rows of the table represent observations and the columns represent the features of those observations. For supervised learning problems, at least one column must represent the response variable or class (the item that will be predicted). This response variable is often referred to as the dependent variable in statistical regression and as the target variable in machine learning. The forecasting model can be stated using the mathematical relationship:

$$y\,(the \quad response \quad variable) = f\,(x, \quad the \quad predictors).$$

In general, data needs to be divided into training, validation, and test sets. The training set is for model construction; the validation set is to tune the parameters, if any; and the test set is to check the model effectiveness.

### 6.2.2 Feature Engineering

Feature engineering is a heuristic process of creating new input features for the machine learning problem. It is considered as an effective approach to improving predictive models since it helps isolate key information; introduce valuable patterns via domain knowledge. However, this topic is quite open ended, and thus, it will not be explored in detail in this chapter. Note that for many problems, feature engineering might be worth exploring.

## 6.3 Model Fitting

Model fitting is a process of determining or measuring how well your model fits the data, and for machine learning models how well the model can generalize new data that was not used as part of the training process. Fitting typicall requires adjustment to the model's parameters to improve the model's accuracy in predicting the data. Typically model fitting follows a given algorithmic process.

In general, we suggest using the following adapted steps in regression analysis (Fox & Hammond, 2019).

- Step 1. Enter the data $(x, y)$, obtain a scatterplot of the data, and note the trends.
- Step 2. If necessary, transform the data into "$y$" and "$x$" components.
- Step 3. Build or compute the regression equation. Obtain all the output. Interpret the ANOVA output for $R^2$, *F-test, and P-values for coefficients.*
- Step 4. Plot the regression function and the data to obtain a visual fit.
- Step 5. Compute the predictions, the residuals, and percent relative error as described later.
- Step 6. Ensure the predictive results pass the common sense test.
- Step 7. Plot the residual versus prediction to determine model adequacy.

We discuss several methods to check for model adequacy. First, we suggest your predictions pass the "common sense" test. If not, return to your regression model, as we show with our exponential decay model. The residual plot is also very revealing. Figure 6.2 shows possible residual plot results where only random patterns indicate model adequacy from the

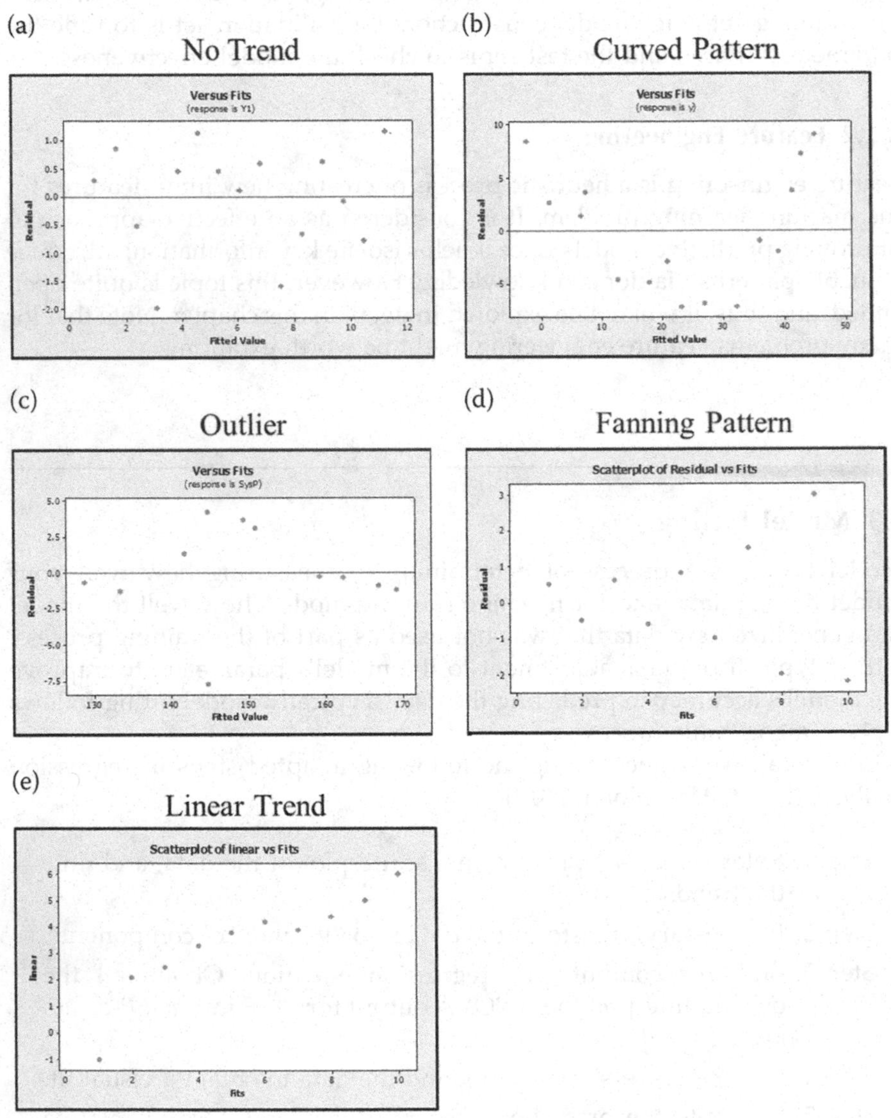


**FIGURE 6.2**
Patterns for residuals (a) No pattern (b) Curved pattern (c) Outliers (d) Fanning pattern (e) Linear trend.

residual plot perspective. Linear, curve, or fanning trends indicate a problem in the regression model (Figure 6.2b, d, & e). Affi and Azen (1979) have a good and useful discussion on corrective action based upon trends found in the residual analysis. Percent relative error also provides information about how well the model approximates the original values, and it provides insights into where the model fits well and where it might not fit well. We define the percent relative error with Equation (6.1),

$$\%RE = \frac{100\ |y_a - y_p|}{y_a} \tag{6.1}$$

where $y_a$ is the observed value and $y_p$ is the predicted value.

### 6.3.1 Simple Least Squares Regression

The method of least-squares curve fitting, also known as **ordinary least squares** and **linear regression**, is simply the solution to a model that minimizes the sum of the squares of the deviations between the observations and predictions. Least squares will find the parameters of the function, *f*(*x*) that will minimize the sum of squared differences between the real data and the proposed model, shown in Equation (6.2).

$$\text{Minimize } SSE = \sum_{i=1}^{n} (y_i - f(x_i))^2 \tag{6.2}$$

There are many technology packages, to include Excel, R, MINITAB, JUMP, Maple, and MATLAB that are capable of perform regression and model fitting.

**Example 6.1:** Regression of Recoil Data (adapted from Fox & Hammond, 2019)

We can then perform simple linear regression on this recoil data and produce tables presenting coefficient estimates and a range of diagnostic statistics to evaluate how well the model fits the data provided.

| | Estimate | Std. Error | t value | Pr(>\|t\|) |
|---|---|---|---|---|
| x | 0.001537 | 1.957e-05 | 78.57 | 4.437e-14 |
| **(Intercept)** | 0.03245 | 0.006635 | 4.891 | 0.0008579 |

*Fitting linear model: y ~ x*

| Observations | Residual Std. Error | $R^2$ | Adjusted $R^2$ |
|---|---|---|---|
| 11 | 0.01026 | 0.9985 | 0.9984 |

*Analysis of Variance Table*

| | Df | Sum Sq | Mean Sq | *F* value | *Pr(>F)* |
|---|---|---|---|---|---|
| x | 1 | 0.6499 | 0.6499 | 6173 | 4.437e-14 |
| **Residuals** | 9 | 0.0009475 | 0.0001053 | NA | NA |

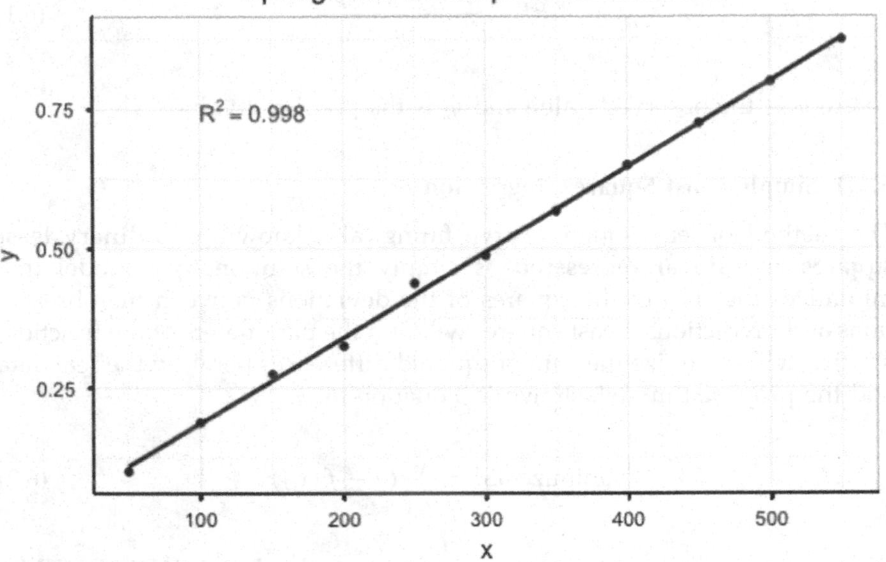


**FIGURE 6.3**
Regression plot of spring data (Example 6.1).

We visualize this estimated relationship by overlaying the fitted line to the spring data plot. This plot shows that the trend line estimated by the linear model fits the data quite well, as shown in Figure 6.3. The relationship between $R^2$ and the linear correlation coefficient ρ is that $R^2 = (\rho)^2$.

### 6.3.2 Exponential Decay Modeling

**Example 6.2:** Hospital Recovery Time

Introducing hospital recovery data from a typical hospital (data from Neter et al., 1996, Fox & Hammond, 2019). The data that we use is provided in Table 6.1.

Plotting the table of recovery data shows that once again, the structure of the data is amenable to statistical analysis. We have two columns, *T* (number of days in the hospital) and *Y* (estimated recovery index), and we

**TABLE 6.1**
Patient Recovery Time

| T | 2 | 5 | 7 | 10 | 14 | 19 | 26 | 31 | 34 | 338 | 45 | 52 | 53 | 60 | 65 |
|---|---|---|---|---|---|---|---|---|---|---|---|---|---|---|---|
| y | 54 | 50 | 45 | 37 | 35 | 25 | 20 | 16 | 18 | 13 | 8 | 11 | 8 | 4 | 6 |

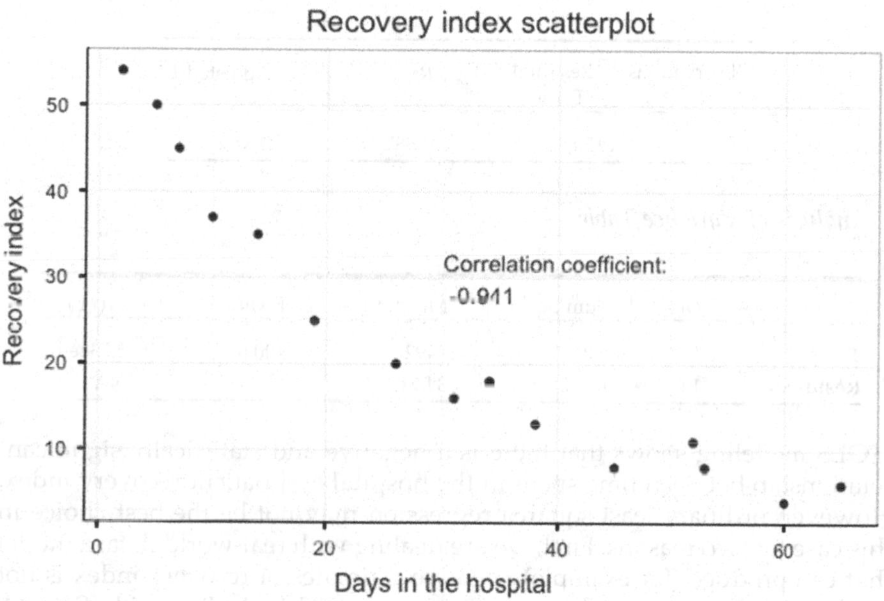


**FIGURE 6.4**
Scatterplot of days in the hospital and recovery index.

want to generate a model that predicts how well a patient will recover as a function of the time they spend in the hospital. Using Excel, we can compute the correlation coefficient of $\rho = -0.941$.

Once again, creating a scatter plot, Figure 6.4, of the data helps to visualize how closely the estimated correlation value matches the overall trend in the data.

In this example, we will demonstrate linear regression, polynomial regression, and then exponential regression in order to obtain a useful model.

### 6.3.3 Linear Regression of Hospital Recovery Data

It definitely appears that there is a strong negative relationship: the longer a patient spends in the hospital, the lower their recovery index. Next, we fit an OLS model to the data to estimate the magnitude of the linear relationship.

| | Estimate | Std. Error | t value | Pr(>\|t\|) |
|---|---|---|---|---|
| **T** | −0.7525 | 0.07502 | −10.03 | 1.736e-07 |
| **(Intercept)** | 46.46 | 2.762 | 16.82 | 3.335e-10 |

*Fitting linear model: Y ~ T*

| Observations | Residual Std. Error | $R^2$ | Adjusted $R^2$ |
|---|---|---|---|
| 15 | 5.891 | 0.8856 | 0.8768 |

*Analysis of Variance Table*

| | Df | Sum Sq | Mean Sq | F value | Pr(>F) |
|---|---|---|---|---|---|
| **T** | 1 | 3492 | 3492 | 100.6 | 1.736e-07 |
| **Residuals** | 13 | 451.2 | 34.71 | NA | NA |

OLS modeling shows that there is a negative and statistically significant relationship between time spent in the hospital and patient recovery index. However, ordinary least-squares regression may not be the best choice in this case for two reasons. First, we are dealing with real-world data: a model that can produce (for example) negative estimates of recovery index is not applicable to the underlying concepts our model is dealing with. Second, the assumption of OLS, like all linear models, is that the magnitude of the relationship between input and output variables stays constant over the entire range of values in the data. However, visualizing the data suggests that this assumption may not hold – in fact, it appears that the magnitude of the relationship is very high for low values of $T$ and decays somewhat for patients who spend more days in the hospital.

To test for this phenomenon, we examine the residuals of the linear model. Residual analysis can provide quick visual feedback about model fit and whether the relationships estimated hold over the full range of the data. We calculate residuals as the difference between observed values $Y$ and estimated values $Y^*$, or $Y_i - Y_i^*$. We then normalize residuals as percent relative error between the observed and estimated values, which helps us compare how well the model predicts each individual observation in the data set shown in Table 6.2:

These data can also be plotted to visualize how well the model fits over the range of our input variable. The residuals can be plotted and will show a curvilinear pattern, decreasing and then increasing in magnitude over the range of the input variable. This means that we can likely improve the fit of the model by allowing for non-linear effects. Furthermore, the current

**TABLE 6.2**

Residual Analysis

| T | Y | Index | Predicted | Residuals | Pct Relative Error |
|---|---|---|---|---|---|
| 2 | 54 | 1 | 44.96 | 9.05 | 16.75 |
| 5 | 50 | 2 | 42.70 | 7.30 | 14.61 |
| 7 | 45 | 3 | 41.19 | 3.81 | 8.46 |
| 10 | 37 | 4 | 38.94 | −1.94 | -5.23 |
| 14 | 35 | 5 | 35.93 | −0.92 | -2.64 |
| 19 | 25 | 6 | 32.16 | −7.16 | -28.65 |
| 26 | 20 | 7 | 26.90 | −6.90 | -34.48 |
| 31 | 16 | 8 | 23.13 | −7.13 | -44.58 |
| 34 | 18 | 9 | 20.88 | −2.88 | -15.97 |
| 38 | 13 | 10 | 17.87 | −4.87 | -37.42 |
| 45 | 8 | 11 | 12.60 | −4.60 | -57.47 |
| 42 | 11 | 12 | 14.86 | −3.86 | -35.05 |
| 53 | 8 | 13 | 6.58 | 1.42 | 17.78 |
| 60 | 4 | 14 | 1.31 | 2.69 | 67.25 |
| 65 | 6 | 15 | −2.45 | 8.45 | 140.88 |

model can make predictions that are substantively nonsensical, even if they were statistically valid. For example, our model predicts that after 100 days in the hospital, a patient's estimated recovery index value would be −29.79. This has no common sense, as the recovery index variable is always positive in the real world. By allowing for non-linear terms, perhaps we can also guard against these types of illogical predictions.

### 6.3.4 Quadratic Regression of Hospital Recovery Data

Note that including a quadratic term can be regarded as feature engineering, i.e. it modifies the model formula: $Y = \beta_0 + \beta_1 x + \beta_2 x^2$ Fitting this model to the data produces separate estimates of the effect of $T$ itself as well as the effect of $T^2$, the quadratic term.

| | Estimate | Std. Error | t value | Pr(>\|t\|) |
|---|---|---|---|---|
| **T** | −1.71 | 0.1248 | −13.7 | 1.087e-08 |
| **I(T^2)** | 0.01481 | 0.001868 | 7.927 | 4.127e-06 |
| **(Intercept)** | 55.82 | 1.649 | 33.85 | 2.811e-13 |

*Fitting linear model: Y ~ T + I(T^2)*

| Observations | Residual Std. Error | $R^2$ | Adjusted $R^2$ |
|---|---|---|---|
| 15 | 2.455 | 0.9817 | 0.9786 |

*Analysis of Variance Table*

| | Df | Sum Sq | Mean Sq | F value | Pr(>F) |
|---|---|---|---|---|---|
| **T** | 1 | 3492 | 3492 | 579.3 | 1.59e-11 |
| **I(T^2)** | 1 | 378.9 | 378.9 | 62.84 | 4.127e-06 |
| **Residuals** | 12 | 72.34 | 6.029 | NA | NA |

Including the quadratic term improves model fit as measured by $R^2$ from 0.88 to 0.98 – a sizable increase. To assess whether this new input variable deals with the curvilinear trend, as seen in the residuals from the first model, we now calculate and visualize the residuals from the quadratic model. The data are provided in Table 6.3.

Again, a residual plot can be obtained, and it will show that the trend has disappeared. This means that we can assume the same relationship holds whether $T = 1$ or $T = 100$. However, we are still unsure if the model produces numerical estimates that pass the common sense test. The simplest way to assess this is to generate predicted values of the recovery index variable using the quadratic model and plot them to see if they make sense.

To generate predicted values in R, we can pass the quadratic model object to the **predict()** function along with a set of hypothetical input values. In other words, we can ask the model what the recovery index would look like for a set of hypothetical patients who spend anywhere from 0 to 120 days in the hospital.

**TABLE 6.3**

Residual Analysis of Quadratic Model

| T | Y | Index | Predicted | Residuals | Relative Error |
|---|---|---|---|---|---|
| 2 | 54 | 1 | 52.46 | 1.54 | 2.85 |
| 5 | 50 | 2 | 47.64 | 2.36 | 4.72 |
| 7 | 45 | 3 | 44.58 | 0.42 | 0.94 |
| 10 | 37 | 4 | 40.20 | −3.20 | −8.65 |
| 14 | 35 | 5 | 34.78 | 0.22 | 0.62 |
| 19 | 25 | 6 | 28.68 | −3.68 | −14.71 |
| 26 | 20 | 7 | 21.37 | −1.37 | −6.86 |
| 31 | 16 | 8 | 17.04 | −1.04 | −6.52 |
| 34 | 18 | 9 | 14.80 | 3.20 | 17.78 |
| 38 | 13 | 10 | 12.23 | 0.77 | 5.96 |
| 45 | 8 | 11 | 8.86 | −0.86 | −10.75 |
| 42 | 11 | 12 | 10.12 | 0.88 | 7.96 |
| 53 | 8 | 13 | 6.79 | 1.21 | 15.11 |
| 60 | 4 | 14 | 6.54 | −2.54 | −63.40 |
| 65 | 6 | 15 | 7.24 | −1.24 | −20.70 |

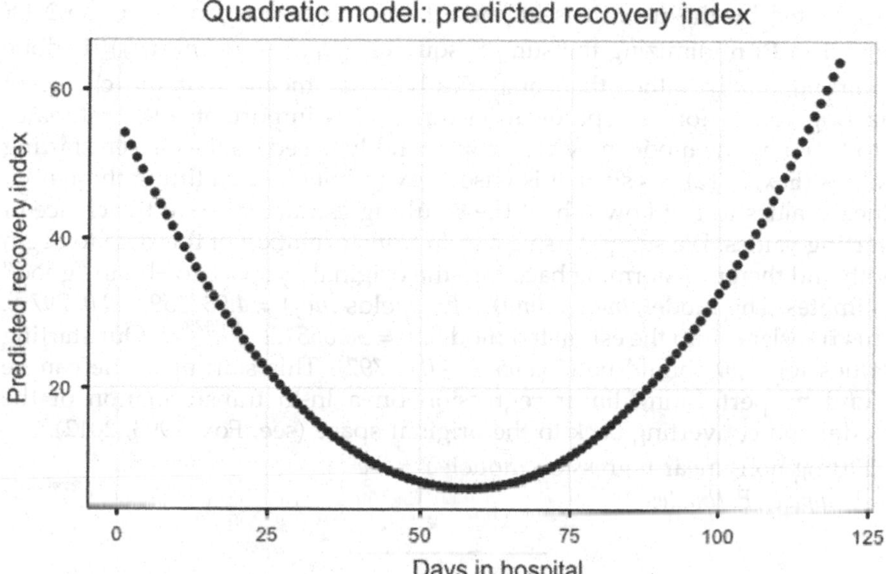


**FIGURE 6.5**
Polynomial regression plot (quadratic polynomial).

We can then plot these estimates to conveniently access whether they pass the common sense test for real-world predictive values as shown in Figure 6.5.

The predicted values curve up toward infinity (see Figure 6.5); clearly, this is a problem. The quadratic term we included in the model leads to unrealistic estimates of recovery index at larger values of $T$. Not only is this unacceptable for the context of our model, but it is unrealistic on its face. After all, we understand that people generally spend long periods in the hospital for serious or life-threatening conditions such as severe disease or major bodily injury. As such, we can assess that someone who spends six months in the hospital probably should not have a higher recovery index than someone who was only hospitalized for a day or two.

### 6.3.5 Exponential Decay Modeling of Hospital Recovery Data

We build a model that both accurately fits the data and produces estimates that pass the common sense test by using an exponential decay model. This modeling approach lets us model relationships that vary over time in a non-linear fashion – in this case, we want to accurately capture the strong correlation for lower ranges of $T$, but allow the magnitude of this relationship to decay as $T$ increases, as the data seem to indicate.

Generating non-linear models in **R** is done using the non-linear least squares or NLS function, appropriately labeled **nls()**. This function automatically fits a wide range of non-linear models based on a functional form designated by the user. It is important to note that when fitting an NLS model in **R**, minimizing the sum of squares $\sum_{i=1}^{n} (y_i - a(exp(bx_i)))^2$ is done computationally rather than analytically. That means that the choice of starting values for the optimization function is important – the estimates produced by the model may vary considerably based on the chosen starting values (Fox, 2012). As such, it is wise to experiment when fitting these non-linear values to test how robust the resulting estimates are to the choice of starting values. We suggest using a ln-ln transformation of this data to begin with and then transforming back into the original $xy$ space to obtain "good" estimates. The model, $ln(y) = ln(a) + bx$, yields $ln(y) = 4.037159 - 0.03797\,x$. This translates into the estimated model: $y = 56.66512e^{(-0.03797x)}$. Our starting values for $(a,b)$ should be $(56.66512, -0.03797)$. This starting value can be found by performing linear regression on a ln-ln transformation of the model and converting back to the original space (see, Fox, 1993, 2012).

Fitting non-linear regression model: $Y \approx a \cdot e^{bT}$

*Parameter Estimates*

| a | b |
|---|---|
| 58.61 | −0.03959 |

residual sum-of-squares: 1.951

The final model is $y = 58.61e^{-0.03959x}$. Overlaying the trend produced by the model on the plot of observed values (see Figure 6.6), we observe that the NLS modeling approach fits the data very well.

Once again, we can visually assess model fit by calculating and plotting the residuals. Figures 6.7a and b, show the same residuals plotted along both days in the hospital $T$ and recovery index $Y$. The data are displayed in Table 6.4.

In both cases we see that there is no easily distinguishable pattern in residuals. Finally, we apply the common sense check by generating and plotting estimated recovery index values for a set of values of $T$ from 1 to 120.

The predicted values generated by the exponential decay model make intuitive sense. As the number of days a patient spends in the hospital increases, the model predicts that their recovery index will decrease at a decreasing rate. This means that while the recovery index variable will continuously decrease, it will not take on negative values (as predicted by the linear model) or explosively large values (as predicted by the quadratic model). It appears that the exponential decay model not only fits the data best from a purely statistical point of view, but also generates values that pass the common sense test to an observer or analyst shown in Figure 6.6.

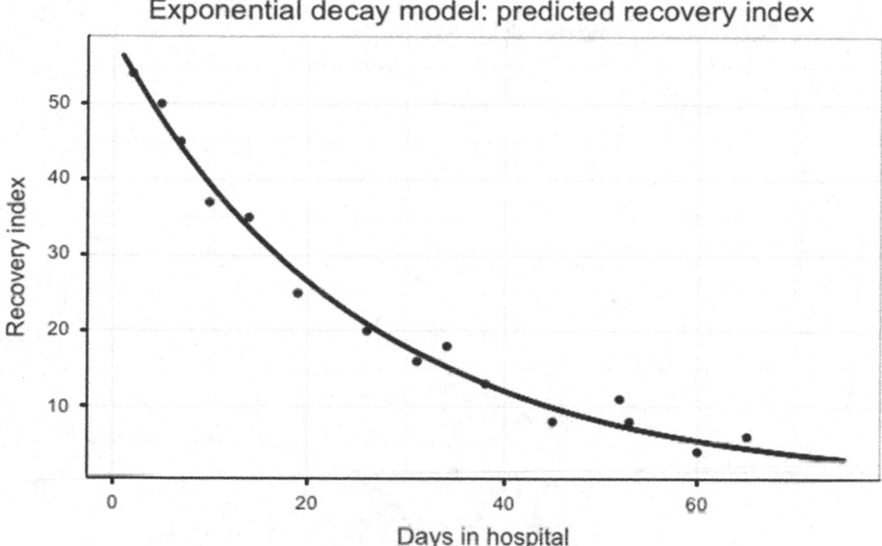


**FIGURE 6.6**
Exponential regression model and data.

### 6.3.6 Sinusoidal Regression with Demand Data

Therefore, using regression techniques is much more than just running the model. Analysis of the results needs to be accomplished **by analysts involved in the problem.**

This involves a multi-step "contest" for choosing a model from the many (perhaps thousands when considering all the different combinations of features) available:

- Model Fitting: We fit and perform *feature selection* and parameter optimization for each of the modeling methods (algorithms) under consideration on a *training dataset*. The output of this step is a list of "best of breed" models that we will compete against each other in the next step.
- Model (Algorithm) Selection: We compete the "best of breed" models against each other on an out-of-sample *validation dataset*. The best performing algorithm (on a range of criteria) is chosen for implementation.
- Model Performance Assessment: We assess the performance of our selected approach on an out-of-sample *test dataset*. This gives us an unbiased estimate for how well the algorithm will perform in practice.
- Model (Algorithm) Implementation: The selected algorithm is applied to the full dataset (i.e. training, validation, and test datasets, now combined) so that the model we deploy uses all available information.

(a)

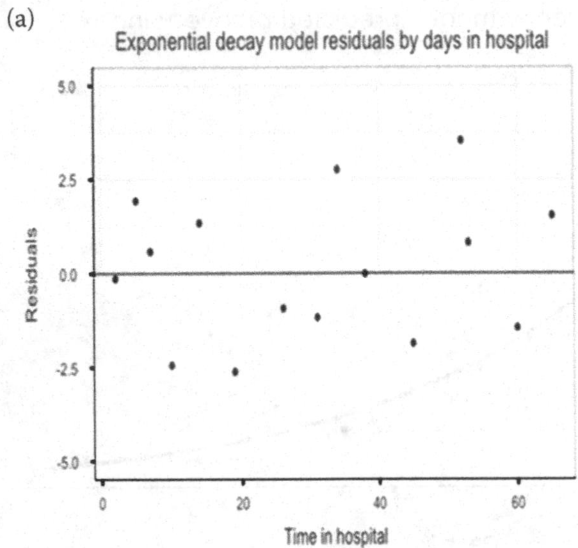


(b)

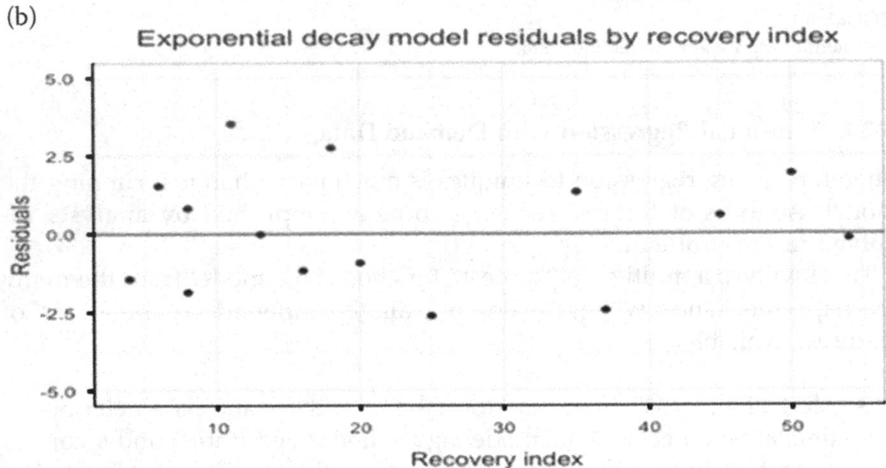


**FIGURE 6.7**
(a) and (b) Residual plot as functions of the time and the model.

Training, testing, and validation are key aspects as well as error analysis.

According to Huddleston and Brown (2018) "In the context of time series forecasting, we make a clear distinction between a prediction and a forecast. A prediction is defined as an assertion (often probabilistic) that a specific event will take place, whereas a forecast is an assertion of how much of something will occur over a specified geographic area and period of time. In this context, the nightly weather "forecast" might include a prediction about

**TABLE 6.4**
Residual Analysis of Exponential Model

| T | Y | Index | Predicted | Residuals | Relative Error |
|---|---|---|---|---|---|
| 2 | 54 | 1 | 54.15 | −0.15 | −0.27 |
| 5 | 50 | 2 | 48.08 | 1.92 | 3.83 |
| 7 | 45 | 3 | 44.42 | 0.58 | 1.28 |
| 10 | 37 | 4 | 39.45 | −2.45 | −6.62 |
| 14 | 35 | 5 | 33.67 | 1.33 | 3.80 |
| 19 | 25 | 6 | 27.62 | −2.62 | −10.50 |
| 26 | 20 | 7 | 20.94 | −0.94 | −4.69 |
| 31 | 16 | 8 | 17.18 | −1.18 | −7.36 |
| 34 | 18 | 9 | 15.25 | 2.75 | 15.26 |
| 38 | 13 | 10 | 13.02 | −0.02 | −0.15 |
| 45 | 8 | 11 | 9.87 | −1.87 | −23.36 |
| 42 | 11 | 12 | 11.11 | −0.11 | −1.03 |
| 53 | 8 | 13 | 7.19 | 0.81 | 10.13 |
| 60 | 4 | 14 | 5.45 | −1.45 | −36.23 |
| 65 | 6 | 15 | 4.47 | 1.53 | 25.49 |

the high temperature for the following day and a forecast for the amount of rain. We provide a brief discussion of some frequently used approaches here and recommend forecasting techniques."

## 6.4 Time Series Models

Time series model fitting is similar in many ways to predictive regression modeling but is unique in that previous observations of the response variable are often the most important predictive feature for the forecast. The statistical term for this use of previous observations of the response variable is *auto-regression* (Box & Jenkins). Fitting time series models often involves modeling four components of a time series model. These include the trend, seasonality effects, cycles, and noise. Fitting and evaluating time series models also require the use of rolling horizon design due to the time dependency inherent in these problems.

The three most common methods used for time series forecasting are *time series regression, exponential smoothing,* and *Auto-Regressive Integrated Moving Average* (ARIMA) models. Time series regression extends basic regression to include auto-regression against previous observations of the response variables. Time series regression also facilitates the use of other variables (i.e. another time series) as predictive features. For example, various studies have related temperature and weather effects to crime occurrence and so predicted temperature over the next week could be incorporated into a forecasting model using time series regression. ARIMA models extend basic auto-regressive modeling to account for trends and other effects.

Measure of effectiveness, especially for exponential smoothing in time series may be defined by the following, any of which can be minimized in an optimization procedure:

### 6.4.1 Mean Absolute Percentage Error

Mean absolute percentage error (MAPE) measures the accuracy of fitted time series values. MAPE expresses accuracy as a percentage.

#### 6.4.1.1 Formula

$$\frac{\sum \frac{y_t - \widehat{y_t}}{y_t} x \quad 100\% \quad (y \neq 0)}{n}$$

#### 6.4.1.2 Notation

| Term | Description |
|---|---|
| $y_t$ | actual value at time $t$ |
| $\hat{y}_t$ | fitted value |
| $n$ | number of observations |

### 6.4.2 MAD

Mean absolute deviation (MAD) measures the accuracy of fitted time series values. MAD expresses accuracy in the same units as the data, which helps conceptualize the amount of error.

#### 6.4.2.1 Formula

$$\frac{\sum_{t=1}^{n} |y_t - \widehat{y_t}|}{n}$$

#### 6.4.2.2 Notation

| Term | Description |
|---|---|
| $y_t$ | actual value at time $t$ |
| $\hat{y}_t$ | fitted value |
| $n$ | number of observations |

### 6.4.3 MSD

Mean squared deviation (MSD) is always computed using the same denominator, $n$, regardless of the model. MSD is a more sensitive measure of an unusually large forecast error than MAD.

#### 6.4.3.1 Formula

$$\frac{\sum_{t=1}^{n} |Y_t - \widehat{Y_t}|^2}{n}$$

#### 6.4.3.2 Notation

| Term | Description |
|---|---|
| $y_t$ | actual value at time $t$ |
| $\hat{y}_t$ | fitted value |
| $n$ | number of observations |

### 6.4.4 Exponential Smoothing

Exponential smoothing is a non-parametric technique that develops a forecast for the next period by using the immediately previous forecast and the immediately previous observation as the predictor variables according to the following formula:

$$\text{New Forecast} = \alpha(\text{Previous Forecast}) + (1 - \alpha)(\text{Observed Value}).$$

The parameter $\alpha$ is a tuning parameter that places more weight either on the previous forecast or the previous observation, taking on values between 0 and 1. This model form results in a recursive relationship with previous forecasts, with the effects of previous forecasts decaying exponentially backwards in time; hence, the name for the algorithm. Fitting an exponential smoothing model is relatively simple and straightforward as it requires only the optimization of the weighting parameter *a*. Basic exponential smoothing has been extended to account for trend and seasonal effects in an algorithm, but our examples will only be basic single-term exponential smoothing. Good discussion is also in Box-Jenkins (1976).

We choose the best value for α so the value results in the smallest MAPE, MAD, or MSE.

**Example 6.3:** Exponential Smoothing

Consider the following data set consisting of 12 observations taken over time (Table 6.5)

The sum of the squared errors (SSE) = 208.94. The mean of the squared errors (MSE) is the SSE /11 = 19.0.

The MSE was again calculated for α = 0.5 and turned out to be 16.29, so in this case we would prefer an α of 0.5 or an α of 0.1. Can we do better? We could apply the proven trial-and-error method. This is an iterative procedure beginning with a range of α between 0.1 and 0.9. We determine the best initial choice for α and then search between α – Δ and α + Δ. We could repeat this perhaps one more time to find the best α to three decimal places. However, we can employ optimization on α.

In this case, α = 0.9 is best with a MSE value of 12.9506; MSD is the Minitab plot in Figure 6.8.

But there are better search methods, such as the Leven-Marquardt or a gradient search procedure. These are non-linear optimizing schemes that minimize the sum of squares of residuals (Fox, 2012). In general, most well-designed

**TABLE 6.5**

Observation Data

| Time | yt | S(*α* = 0.1) | Error | Error$^2$ |
|---|---|---|---|---|
| 1 | 71 | | | |
| 2 | 70 | 71.00 | –1.0 | 1 |
| 3 | 69 | 70.90 | –1.9 | 3.61 |
| 4 | 68 | 70.71 | –2.71 | 7.34 |
| 5 | 64 | 70.44 | –6.44 | 41.47 |
| 6 | 65 | 69.80 | –4.80 | 23.04 |
| 7 | 72 | 69.32 | 2.68 | 7.18 |
| 8 | 78 | 69.58 | 8.42 | 79.90 |
| 9 | 75 | 70.43 | 4.57 | 20.88 |
| 10 | 75 | 70.88 | 4.12 | 16.97 |
| 11 | 75 | 71.29 | 3.71 | 13.76 |
| 12 | 70 | 71.67 | –1.67 | 2.79 |

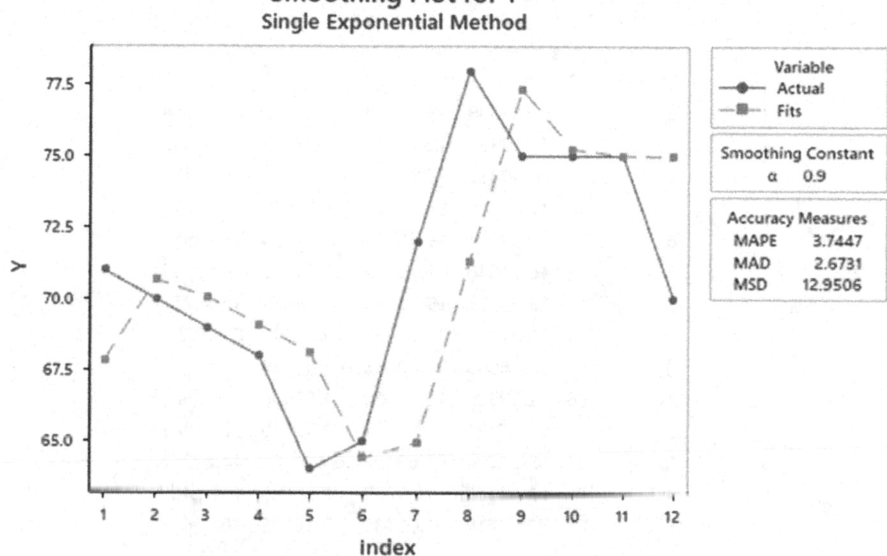


**FIGURE 6.8**
Exponential smoothing plots with $\alpha = 0.9$.

statistical software programs should be able to find the value of α that minimizes the MSE. The plots in Figure 6.8 show the visual comparison.

**Example 6.4:** Optimizing α

We optimize α for another example. Using a program to optimize α involves machine learning techniques in this simple example.

We find the best α is 0.65621 with a MAD of 10.0824, seen from Figure 6.9. We graphically depict this using MINITAB and note that the MAD value is slightly different. We see a good fit of the data in Figure 6.10.

### 6.4.5 Auto-regressive Integrated Moving Average (ARIMA)

ARIMA stands for auto-regressive integrated moving average and is specified by these three order parameters: *(p, d, q)*. The process of fitting an ARIMA model is sometimes referred to as the Box-Jenkins method. For more discussion, see Box-Jenkins (1976).

An auto-regressive (AR(p)) component is referring to the use of past values in the regression equation for the series *Y*. The auto-regressive parameter *p* specifies the number of lags used in the model. For example, AR(2) or, equivalently, ARIMA(2,0,0), is represented as

Such a model has hyperparameters *p*, *q*, and *d*.

| A | B | C | D | E | F |
|---|---|---|---|---|---|
| Exponential Smoothing | | | | | |
| t | y | y_pred | ny | € | e^2 |
| 1 | 3 | 3 | 3 | 0 | 0 |
| 2 | 5 | 3.687589 | 3 | 2 | 4 |
| 3 | 9 | 5.513968 | 4.312411 | 4.687589 | 21.97349 |
| 4 | 20 | 10.49419 | 7.388432 | 12.61157 | 159.0517 |
| 5 | 12 | 11.01188 | 15.66421 | 3.66421 | 13.42643 |
| 6 | 17 | 13.07056 | 13.25974 | 3.740264 | 13.98958 |
| 7 | 22 | 16.14046 | 15.71412 | 6.285883 | 39.51232 |
| 8 | 23 | 18.49873 | 19.83895 | 3.161053 | 9.992257 |
| 9 | 51 | 29.67249 | 21.91325 | 29.08675 | 846.0392 |
| 10 | 41 | 33.56683 | 41.00013 | 0.000129 | 1.67E-08 |
| 11 | 56 | 41.27923 | 41.00004 | 14.99996 | 224.9987 |
| 12 | 75 | 52.87225 | 50.8431 | 24.1569 | 583.5561 |
| 13 | 60 | 55.32274 | 66.69498 | 6.694985 | 44.82282 |
| 14 | 75 | 62.08767 | 62.3017 | 12.6983 | 161.2468 |
| 15 | 88 | 70.99619 | 70.63439 | 17.36561 | 301.5643 |
| | | | | 141.1532 | 2424.174 |
| | | | | 10.08237 | 173.1553 |
| alpha | 0.656205 | | | | |

**FIGURE 6.9**
Screenshot from Excel.

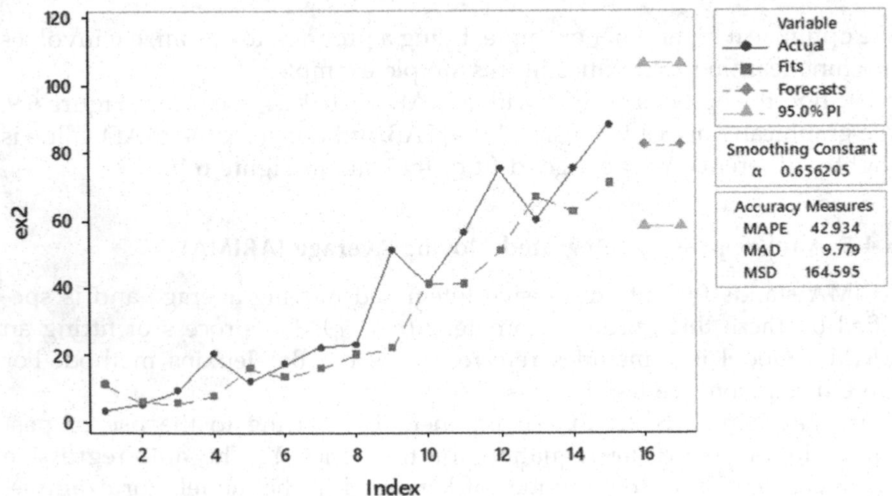


**FIGURE 6.10**
Smoothing plot for Example 6.4.

- $p$ is the order of the AR model
- $q$ is the order of the MA model
- $d$ is the differencing order (how often we difference the data)

The ARMA and ARIMA combination is defined as

$$X_t = c + \varepsilon_t + \sum_{t=1}^{p} \varphi_i X_{t-i} + \sum_{i=1}^{q} \theta_i \epsilon_{t-i}$$

where $\varphi_1$, $\varphi_2$ are parameters for the model.

The $d$ represents the degree of differencing in the **integrated** (*I(d)*) component. Differencing a series involves simply subtracting its current and previous values $d$ times. Often, differencing is used to stabilize the series when the stationarity assumption is not met, which we will discuss below.

ARIMA methodology does have its limitations. These models directly rely on past values and therefore work best on long and stable series. Also note that ARIMA simply approximates historical patterns and therefore does not aim to explain the structure of the underlying data mechanism

## 6.5 Case Studies of Time Series Data

The case studies provided in this section to reinforce the concepts presented in this chapter take advantage of several publicly available data sources. Specifically, one small data set, stock prices (https://support.minitab.com/en-us/minitab/18/help-and-how-to/graphs/how-to/time-series-plot/before-you-start/example/), and one larger data set, Walmart sales from Kaggle.com (https://www.kaggle.com/c/walmart-recruiting-store-sales-forecasting/data).

**CASE STUDY 6.1 TYPICAL SUPPLY/DEMAND SHIPPING DATA (FOX & HAMMOND, 2019)**

Consider a situation where we have shipping data (Table 6.6) that we need to model to estimate future results.

**TABLE 6.6**

Shipping Data for Case Study 6.1

| Month | Usage |
|---|---|
| 1 | 20 |
| 2 | 15 |
| 3 | 10 |
| 4 | 18 |
| 5 | 28 |
| 6 | 18 |
| 7 | 13 |
| 8 | 21 |
| 9 | 28 |
| 10 | 22 |
| 11 | 19 |
| 12 | 25 |
| 13 | 32 |
| 14 | 26 |
| 15 | 21 |
| 16 | 29 |
| 17 | 35 |
| 18 | 28 |
| 19 | 22 |
| 20 | 32 |

First, we can use Excel to obtain the correlation $\rho = 0.6726$.

Once again, we can visualize the data in a scatter plot to assess whether this positive correlation is borne out by the overall trend.

Visualizing the data, Figure 6.11, we see that there is a clear positive trend over time in shipping usage. However, examining the data in more detail suggests that a simple linear model may not be best-suited to capturing the variation in these data. One way to plot more complex patterns in data is through the use of a trend line using polynomial or non-parametric smoothing functions.

Plotting a trend line generated via a spline function (discussed in Chapter 3) shows that there seems to be an oscillating pattern, Figure 6.12, with a steady increase over time in the shipping data.

1. **Sinusoidal regression of shipping data**

R, as well as other software, treats sinusoidal regression models as part of the larger family of non-linear least-squares (NLS) regression models. This

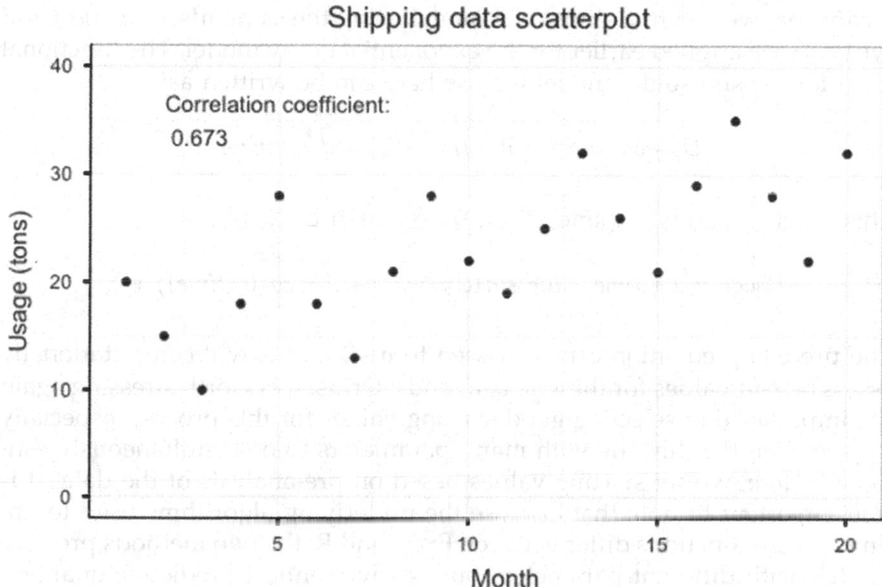


**FIGURE 6.11**
Scatterplot of shipping data.

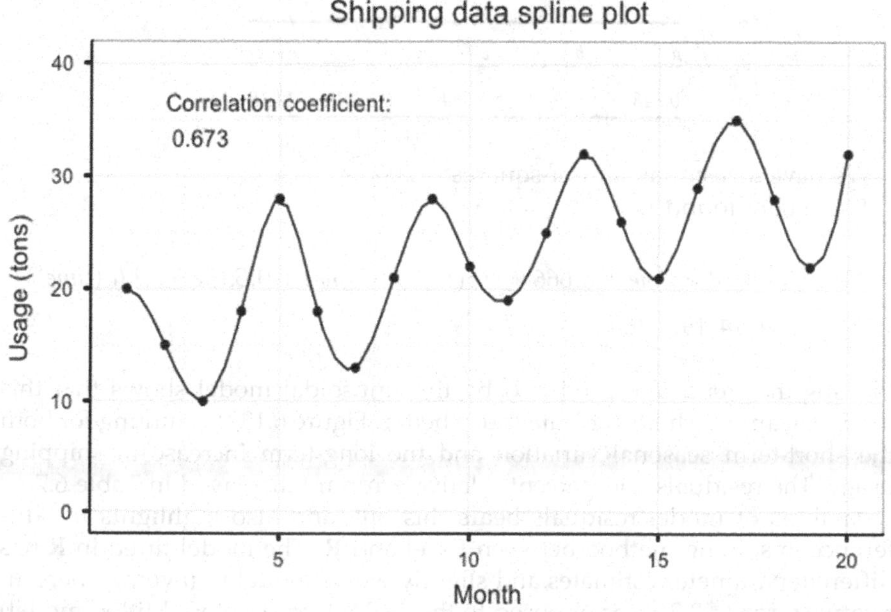


**FIGURE 6.12**
Shipping data with data points connected show an oscillating trend.

means that we can fit a sinusoidal model using the same **nls()** function and syntax as we applied earlier for the exponential decay model. The functional form for the sinusoidal model we use here can be written as:

$$Usage = a * sin(b * time + c) + d * time + e$$

This function can be expanded out trigonometrically as:

$$Usage = a * time + b * sin(c * time) + d * cos(c(time)) + e$$

The preceding equation can be passed to **nls()** and **R** will computationally assess best-fit values for the $a$, $b$, $c$, $d$, and $e$ terms. It is worth stressing again the importance of selecting good starting values for this process, especially for a model like this one with many parameters to be simultaneously estimated. Here, we set starting values based on pre-analysis of the data. It is also important to note that because the underlying algorithms used to optimize these functions differ between Excel and R, the two methods produce models with different parameters but nearly identical predictive qualities. The model can be specified in **R** as follows.

Fitting non-linear regression model: *UsageTons~a·Month* + *b*·sin (*c·Month* + +*d*·cos(*c·Month*) + *e*

*Parameter Estimates*

| *a* | *b* | *c* | *d* | *e* |
|---|---|---|---|---|
| *0.848* | *6.666* | *1.574* | *0.5521* | *14.19* |

We have a residual sum-of-squares: 1.206

The model found is:

$$Usage = 0.848 * time + 6.666 * sin(1.574 * time) + 0.5521 * cos(c(time)) + 14.19.$$

Plotting the trend line produced by the sinusoidal model shows that this modeling approach fits the data much better, Figure 6.13, accounting for both the short-term seasonal variation and the long-term increase in shipping usage. The residuals and percent relative error are displayed in Table 6.7.

Analysis of model residuals bears this out, and also highlights the difference in solving method between Excel and **R**. The model fitted in **R** has different parameter estimates and slightly worse model fit (average percent relative error of 3.26% as opposed to the 3.03% from the Excel-fitted model) but the overall trend identified in the data is virtually identical.

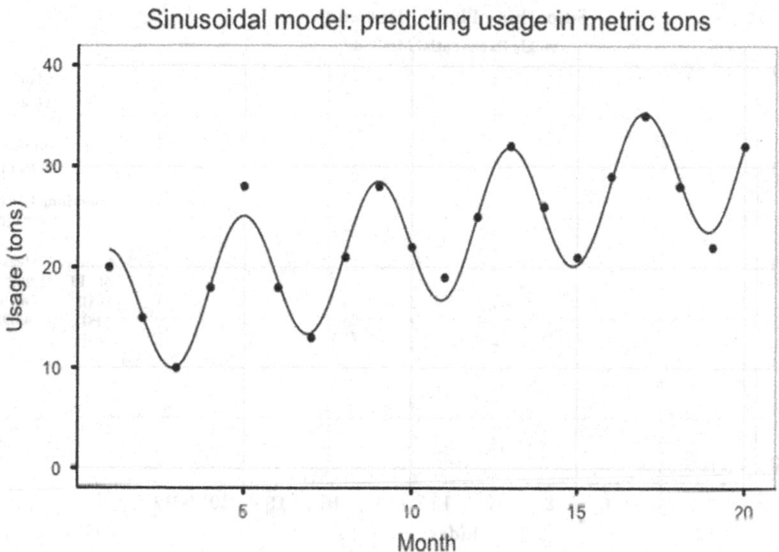


**FIGURE 6.13**
Overlay of regression model and data.

**TABLE 6.7**
Residual Analysis for Case Study 6.1

| Month | Usage | Predicted | Residuals | % Rel. Error |
|---|---|---|---|---|
| 1 | 20 | 21.70 | −1.70 | −8.51 |
| 2 | 15 | 15.29 | −0.29 | −1.94 |
| 3 | 10 | 10.07 | −0.07 | −0.74 |
| 4 | 18 | 18.22 | −0.22 | −1.22 |
| 5 | 28 | 25.09 | 2.91 | 10.41 |
| 6 | 18 | 18.60 | −0.60 | -3.32 |
| 7 | 13 | 13.47 | −0.47 | -3.65 |
| 8 | 21 | 21.70 | −0.70 | -3.32 |
| 9 | 28 | 28.47 | −0.47 | −1.68 |
| 10 | 22 | 21.90 | 0.10 | 0.43 |
| 11 | 19 | 16.88 | 2.12 | 11.18 |
| 12 | 25 | 25.17 | −0.17 | −0.70 |
| 13 | 32 | 31.85 | 0.15 | 0.46 |
| 14 | 26 | 25.21 | 0.79 | 3.03 |
| 15 | 21 | 20.28 | 0.72 | 3.44 |
| 16 | 29 | 28.65 | 0.35 | 1.20 |
| 17 | 35 | 35.23 | −0.23 | −0.66 |
| 18 | 28 | 28.52 | −0.52 | −1.85 |
| 19 | 22 | 23.68 | −1.68 | −7.65 |
| 20 | 32 | 32.13 | −0.13 | −0.40 |

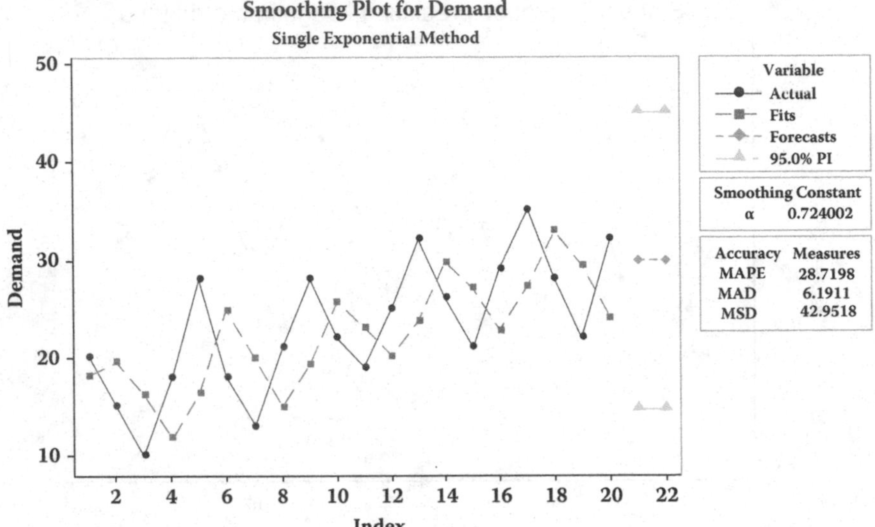


**FIGURE 6.14**
Exponential smoothing of supply/demand data.

### 6.5.1 Time Series with Exponential Smoothing

We apply the exponential smoothing model for a single exponential smoothing. We optimize for α by using the Solver to find the optimal α value is 0.7240. We find the best α is 0.7240 and the optimal MAD is 5.10945, optimizing is done in Excel. Figure 6.14 from MINITAB shows the relative accuracy. Again, we note the slight differences between Excel's measures and MINITAB's.

We estimate the next time permit and obtain a value of 27.9051. After the next time period arrives, we can compare to our estimates to determine which model does better and (perhaps) use that model for the next time period estimate.

#### CASE STUDY 6.2 HOMICIDES IN HAMPTON VA FUTURE PREDICTIONS (SOURCE: DAILY PRESS, HAMPTON, VIRGINIA, MARCH 4, 2018)

Data were provided in a story in the *Daily Press* concerning homicides in Hampton, VA, and Newport News, VA. The purpose was to provide comparative information and not to predict anything. However, let's analyze the homicide data in Hampton and try to forecast the number of homicides, at least in 2018.

First, we plot the time series, and then we plot the data from 2003–2017, as displayed in Figure 6.15.

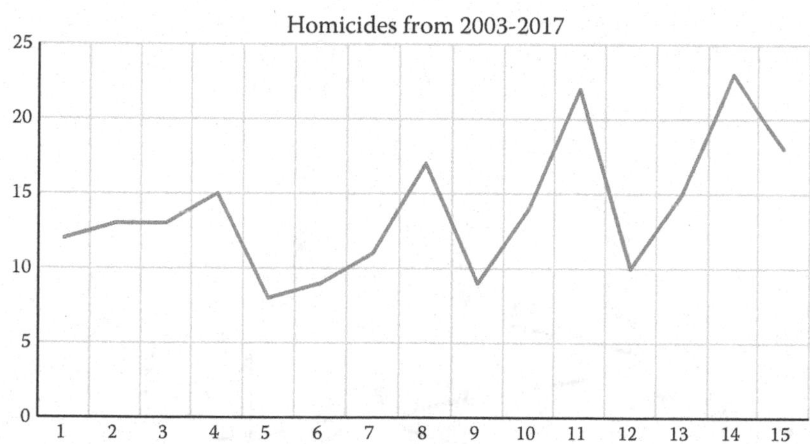


**FIGURE 6.15**
Time series plot of homicide data.

We first use moving average ARIMA model to forecast the future. We train with the data from 2003–2016 and test with the data point from 2017 since the 2018 data are not yet available. Our ARIMA model predicts 20 homicides, and we had only 17 in Hampton. The percent relative error is 17.6%.

Next, we use sine regression in Maple to model and predict. Even varying our initial input estimates we obtain the same sine regression model (see Figure 6.16) as:

$$y = -3.97687527582931 * sin(4.33262380863333 * x + 0.514250245580131) + 0.524461138685776 * x + 9.47706296518979$$

Using sine regression our percent relative absolute error is 8.55%. In this case, the regression model was slightly better in predicting over the ARIMA model.

## 6.6 Summary and Conclusions

As has been shown in the many examples in this chapter, training machines (i.e. computers) to "learn" via the application of algorithmic modeling has a wide variety of very diverse applications. This chapter has also

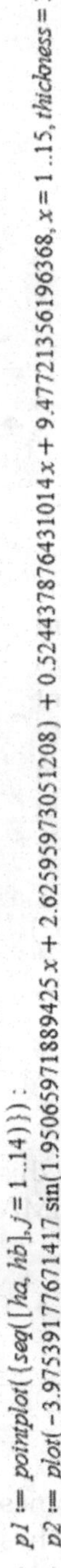

```
> p1 := pointplot( [seq( [ha, hb], j = 1..14)]) :
> p2 := plot( -3.975391776714 17 sin(1.95065971889425 x + 2.62595973051208) + 0.524437876431014 x + 9.47721356196368, x = 1..15, thickness = 3) :
> display( p1, p2, title = "Sine Regression Homicides");
```

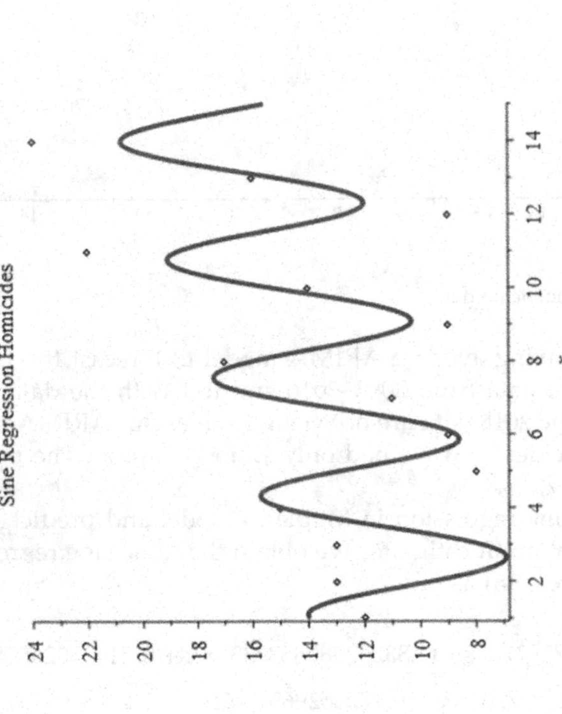


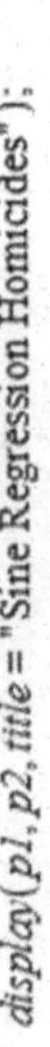

```
> 
> evalf( subs(x = 15, -3.97539177671417 sin(1.95065971889425 x + 2.62595973051208) + 0.524437876431014 x + 9.47721356196368));
                                15.54363304
> 
```

**FIGURE 6.16**
Screenshot from Maple's output.

demonstrated that even though the algorithmic procedures employed to "fit" these models can be automated to a certain extent, the machines still require significant input from analysts for their "training". While individual machine learning algorithms provide a framework for approaching a particular class of problem, choosing the right machine learning algorithm for any particular problem is a highly complex and iterative process that requires significant expertise, judgment, and often the active participation of domain experts and users of your results. Often, for best results, multiple machine learning algorithms, as well as best practices for data storage, data engineering, and computing will be needed. Practitioners are well advised to algorithmically model in teams that incorporate statisticians, operations research analysts, computer scientists, data engineers, data scientists, and domain experts to form a comprehensive unit dedicated to training the machines to "learn" to solve the right problems the best way.

In modeling the analyst must know the purpose of the model. Is it to predict the short term, predict the long term, interpolate, or just estimate? In other words, how accurate must our model be? All these answers affect the modeling choice we make for analyzing the data in any forecasting situations.

## References and Suggested Further Readings

Affi, A., and S. Azen (1979). *Statistical Analysis*, 2nd Edition. Academic Press, London, UK, pp. 143–144.

Box, G., and G. Jenkins (1976). *Time Series Analysis: Forecast and Control*, John Wiley & Sons, Hoboken NJ.

Breiman, L. (2001). Statistical Modeling: The Two Cultures. *Statistical Science* 16(3): 199–231.

Devore, J. (2012). *Probability and Statistics for Engineering and the Sciences*, 8th Edition. Cengage Publisher, Belmont, CA, pp. 211–217.

Fox, W. P. (2012). Importance of "Good" Starting Points in Nonlinear Regression in Mathematical Modeling in Maple. *JCMST* 31(1): 1–16.

Fox, W. P. (2012). Issues and Importance of "Good" Starting Points for Nonlinear Regression for Mathematical Modeling with Maple: Basic Model Fitting to Make Predictions with Oscillating Data. *Journal of Computers in Mathematics and Science Teaching* 31(1): 1–16.

Fox, W. P. (2018). *Mathematical Modeling for Business Analytics*, CRC Press, Boca Raton, Fl.

Fox, W. P. (2012). *Mathematical Modeling with Maple*, Cengage Publishers, Boston, MA.

Fox, W. P. (1993). The Use of Transformed Least Squares in Mathematical Modeling. *Computers in Education Journal* III(1): 25–31.

Fox, W. P. (2011). Using Excel for Nonlinear Regression. *COED Journal* 2(4): 77–86.

Fox, W. P. (2011 October–December). Using the EXCEL Solver for Nonlinear Regression. *Computers in Education Journal (COED)* 2(4): 77–86.

Fox, W. P., and F. Christopher (1996). Understanding Covariance and Correlation. *Primus* VI(3): 235–244.

Fox, W. P., and J. Hammond (2019). Advanced Regression Models: Least Squares, Nonlinear, Poisson and Binary Logistics Regression Using R. In: Gausto and Levy (Eds.) *Data Science and Digital Business*, Springer, Switzerland, AG, pp. 221–262.

Giordano, F., W. Fox, and S. Horton (2013). *A First Course in Mathematical Modeling*, 5th Edition. Cengage Publishers, Boston, MA.

Huddleston, S., and G. Brown (2018). Chapter 7, *INFORMS Analytics Body of Knowledge*, John Wiley & Sons and Naval Postgraduate School updated notes. MAPLE ©, Version 18.

Hyndman, R., and A. Athanasopoulos (2018). *Forecasting: Principles and Practices*, 2nd Edition. OTexts, Melbourne, Australia. (https://otexts.com/fpp2/ets.html (accessed May 2019).

Johnson, I. (2012). An Introductory Handbook on Probability, Statistics, and Excel. http://records.viu.ca/~johnstoi/maybe/maybe4.htm (accessed July 11, 2012).

MINITAB ©, (2019). https://support.minitab.com/en-us/minitab/18/help-and-how-to/graphs/how-to/time-series-plot/before-you-start/example/ (accessed May 2019). Introduction to Forecasting with Arima in R-Learning Data Science Tutorials. https://www.datascience.com/blog/introduction-to-forecasting-with-arima-in-r-learn-data-science-tutorials) (accessed May 20, 2019).

Neter, J., M. Kutner, C. Nachtsheim, and W. Wasserman (1996). *Applied Linear Statistical Models*, 4th Edition. Irwin Press, Chicago, Il, pp. 531–547.

# 7

# *Stochastic Models and Markov Chains*

**OBJECTIVES**

1. Formulate a nonlinear optimization problem.
2. Solve and interpret the nonlinear optimization problem.
3. Perform sensitivity analysis and interpret that analysis.

## 7.1 Introduction

Uncertainty and randomness are all around us, and we must take both of them in account during the modeling process. Many real applications will require us to explicitly represent uncertainty in our mathematical models. In this chapter, we introduce some of the most important and commonly used stochastic models.

A **Markov chain** is a mathematical system that experiences transitions from one state to another state according to certain probabilistic rules. The defining characteristic of a Markov chain is that no matter *how* the process arrived at its present state, the possible future states are fixed. In other words, the probability of transitioning to any particular state is dependent solely on the current state and the amount of time that has elapsed. The **state space**, the set of all possible states, can be anything: letters, numbers, weather conditions, baseball scores, or stock performances.

The finite state, also known as stages, is a common modeling approach for Markov chains, and random walks provide a prolific example of their usefulness in mathematics. They arise broadly in statistical and information-theoretical contexts and are widely employed in economics, game theory, queuing theory, genetics, and finance. While it is possible to discuss Markov chains with any size of state space, the initial theory and most applications are focused on cases with a finite (or countably infinite) number of states.

DOI: 10.1201/9781003298762-7

Many uses of Markov chains require proficiency with common matrix methods. In some cases, we will refer back to the discrete dynamical systems (DDS) discussed in Chapter 2 to iterate some of our examples.

In the language of conditional probability and random variables, a Markov chain is a sequence $X_0, X_1, X_2, \ldots$ of random variables satisfying the rule of conditional independence.

Conditional independence implies $P(A \mid B) = \frac{P(A \cap B)}{P(B)} = P(A)$.

### 7.1.1 The Markov Property

For any positive integer $n$ and possible states $i_0, i_1, \ldots, i_n$ of the random variables,

$$P(X_n = i_n \mid X_{n-1} = i_{n-1}) = P(X_n = i_n \mid X_0 = i_0, X_1 = i_1, \ldots i_{n-1}).$$

This definition implies that knowledge of the previous state is all that is necessary to determine the probability distribution of the current state. This definition is broader than the one explored above, as it allows for *non-stationary transition probabilities* and therefore *time-inhomogeneous Markov chains;* that is, as time goes on (steps increase), the probability of moving from one state to another may change.

**Example 7.1:** Simple Markov Chain

We want to start by describing the behavior of the following Markov chain, where the state variables are: $X_n \in \{1, 2, 3\}$.

If $X_n = 1$, then $X_{n+1} = 1, 2$, or $3$ with equal probability $p_{11} = p_{12} = p_{13} = 1/3$.
If $X_n, = 2$, then $X_{n+1} = 1$, with $p_{21} = 0.7$ and $X_{n+1} = 2$ with probability, $0.3$.
If $X_n = 3$, then $X_{n+1} = 1$ with probability $1$.

All other probabilities among the states are zero. This can be written as a matrix, $P$.

$$\begin{pmatrix} p_{11} & \cdots & p_{1m} \\ \vdots & \ddots & \vdots \\ p_{m1} & \cdots & p_{mm} \end{pmatrix}$$

For our example, we have

$$\begin{bmatrix} 1/3 & 1/3 & 1/3 \\ 0.7 & 0.3 & 0 \\ 1 & 0 & 0 \end{bmatrix}$$

We can create a state transition diagram (Figure 7.1) to make it easier to visualize the Markov chain as a sequence of random jumps.

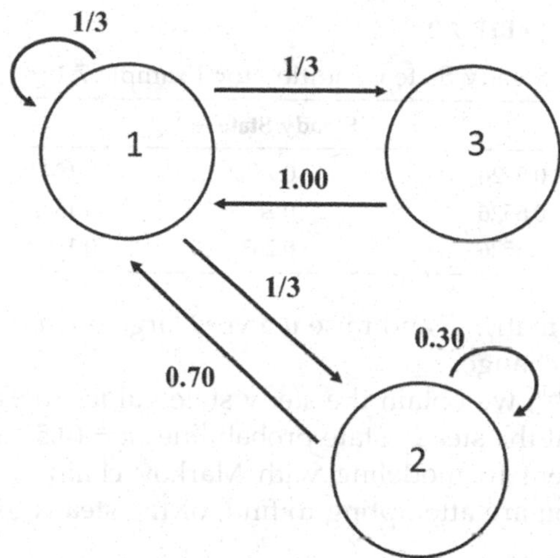


**FIGURE 7.1**
Transition diagram for Example 7.1.

We can define $\pi_n(i) = p(X_n = i)$ and then $\pi_{n+1}(j) = \Sigma_i p_{ij}\pi_n(i)$.

We denote $\pi_n$ as a vector with entries $\pi_n(1)$, $\pi_n(2)$, …, and let $P$ denote the matrix. We can write this now as

$$\pi_{n+1} = \pi_n P$$

For example, we have $\pi_2 = \pi_1 P$, or **(1/3, 1/3, 1/3)**$*$

$$\begin{bmatrix} 1/3 & 1/3 & 1/3 \\ 0.7 & 0.3 & 0 \\ 1 & 0 & 0 \end{bmatrix} = (0.677, \quad 0.211, \quad 0.111).$$

Similarly, we can move through the sequence as we do in Table 7.1.

Our objective here is to find the steady state values or after a long period of time the probabilistic results. We might either transfer to a DDS and

**TABLE 7.1**
Outcomes for Some $\pi$'s

| | State | | |
|---|---|---|---|
| π1 | 0.333 | 0.333 | 0.333 |
| π2 | 0.678 | 0.211 | 0.111 |
| π3 | 0.485 | 0.289 | 0.226 |
| π4 | 0.590 | 0.248 | 0.162 |

**TABLE 7.2**

Steady State Condition for Example 7.1

| | Steady State | |
|---|---|---|
| 0.5526 | 0.2632 | 0.1842 |
| 0.5526 | 0.2632 | 0.1842 |
| 0.5526 | 0.2632 | 0.1842 |

iterate or take the matrix $P$ and raise it a very large power, where the matrix entries no longer change.

So, if we take $P^{20}$, we obtain the stedy state values in Table 7.2.

This implies that the steady state probabilities $\pi = (0.5526, 0.2632, 0.1842)$.

A key component to modeling with Markov chains is to know which transition state you are attempting to find, or the steady state probabilities.

**Example 7.2:** Inventory Analysis, adapted from Meerschaert (1999).

A local pet store sells special-sized dog pens for different breeds of dogs. To ensure customer happiness, the store manager takes inventory of the number of pens still on the shelves and if necessary, places an order to replenish the stock. His current policy is to order three replacement pens if the current inventory has been sold by the end of the week. This means that if even one pen is left, then the order is not placed. This policy was built on historic observations that showed the store only sells, on average, one pen per week. We would like to know if this policy is adequate.

The distribution of demand based upon Poisson probabilities is (Table 7.3).

We can rewrite these simply as

$$P(D_n = 0) = 0.368$$
$$P(D_n = 1) = 0.368$$
$$P(D_n = 2) = 0.184$$
$$P(D_n = 3) = 0.061$$
$$P(D_n > 0) = 0.019$$

**TABLE 7.3**

Distribution of Demand for Example 7.2

| k | p(k) |
|---|---|
| 0 | 0.3679 |
| 1 | 0.3679 |
| 2 | 0.1839 |
| 3 | 0.0613 |
| 4 | 0.0153 |
| 5 | 0.0031 |

If $X_n = 3$ then

$$P(X_{n+1} = 1) = P(D_n = 2) = 0.184$$
$$P(X_{n+1} = 2) = P(D_n = 1) = 0.368$$
$$P(X_{n+1} = 3) = 1 - 0.184 - 0.368 = 0.448$$

We put this information into the matrix $P$

$$\mathrm{P} = \begin{bmatrix} 0.368 & 0 & 0.632 \\ 0.368 & 0.368 & 0.264 \\ 0.184 & 0.368 & 0.448 \end{bmatrix}$$

If we raise $P$ to a large power, we can obtain the steady state probabilities of (0.285, 0.263, 0.452) for states 1, 2, 3.

### 7.1.2 Maple for Markov Chains

```
s := {Pi1 = 0.368 * Pi1 + 0.368 * Pi2 + 0.184 * Pi3, Pi2 = 0.368 * Pi2 + 0.368 * Pi3,
Pi1 + Pi2 + Pi3 = 1};

s := {Pi1 = 0.368 Pi1 + 0.368 Pi2 + 0.184 Pi3,

Pi2 = 0.368 Pi2 + 0.368 Pi3, Pi1 + Pi2 + Pi3 = 1}
f solve (s, {Pi1, Pi2, Pi3});

{Pi1 = 0.2848348783, Pi2 = 0.2631807648, Pi3 = 0.4519843569}
```

Putting this together with our demand information about $D_n$, we can compute $P(D_n > S_n)$ = (0.264)(0.285) + (0.080)(0.263) + (0.019)(0.452) = 0.105. Thus, for large $n$ or the long run, demand will exceed supply by a little over 10% of the time. As a manger, you can decide if this is too great.

## 7.2 Transition Matrices

A transition matrix, sometimes referred to as a stochastic matrix or probability matrix, is a square matrix that succinctly describes probability of transitions from one state to another in a Markov chain. Each entry in the matrix is a nonnegative number that represents the probability of moving from one state to another state.

For the time-independent Markov chain described in the transition diagram (Figure 7.2), what is its two-step transition matrix? What is the steady state probability?

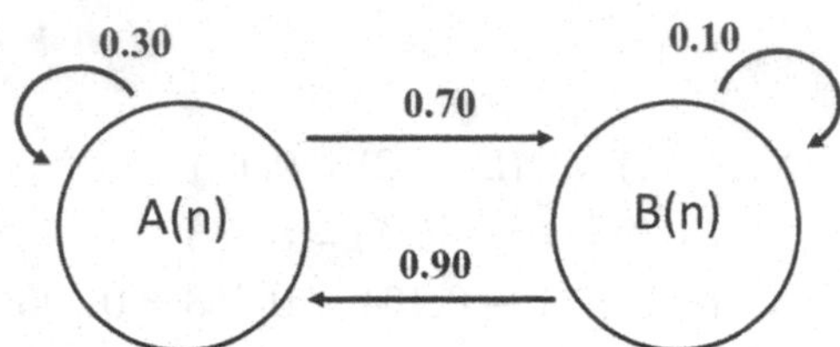


**FIGURE 7.2**
Transition diagram.

Note the transition matrix is written as:

$$P = \begin{bmatrix} 0.3 & 0.7 \\ 0.9 & 0.1 \end{bmatrix}$$

The two-step transition matrix is

$$P^2 = \left( \begin{bmatrix} 0.3 & 0.7 \\ 0.9 & 0.1 \end{bmatrix}^2 = \begin{bmatrix} 0.72 & 0.28 \\ 0.36 & 0.64 \end{bmatrix} \right)$$

The steady state probabilities are 0.5625 and 0.4375.

## 7.3 Markov Chains and Bayes' Theorem

We begin with Markov chains. A Markov chain is a discrete-time stochastic model. Markov chains also involve change. We may address these as a dynamical system.

A Markov chain is a stochastic model describing a sequence of possible events in which the probability of each event depends only on the state attained in the previous event. A countably infinite sequence, in which the chain moves state at discrete time steps, gives a discrete-time Markov chain (DTMC).

**Example 7.3:** Fast Food. Simple Markov Process

The local restaurants Fast Pizza ($P$) and Fast Chicken (C) are in direct competition for the same set of customers. The state transition diagram (Figure 7.3) provides the probabilities of customers staying with a particular restaurant. With this identified Markov process, we want to seek the steady state probabilities of this system.

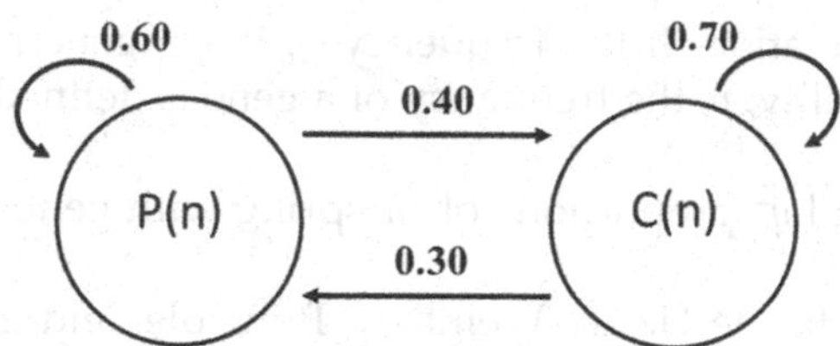


**FIGURE 7.3**
Fast food transition diagram.

Recall in Chapter 2, we presented the DDSs, and we can use them now to create our system.

$$P(n+1) = P(n) - k_1 P(n) + k_2 C(n)$$

$$C(n+1) = C(n) - (1 - k_2)C(n) + (1 - k_1)P(n)$$

We can write in matrix form as

$$\begin{bmatrix} P(n+1) \\ C(n+1) \end{bmatrix} = \begin{bmatrix} .6 & .3 \\ .4 & .7 \end{bmatrix} \begin{bmatrix} P(n) \\ C(n) \end{bmatrix}$$

We can raise the coefficient matrix to a large power to obtain the steady state probabilities as we also could iterate the DDS until we become steady state.

The values 0.429 and 0.571 are our steady state probabilities.

**Example 7.4:** Genetic Transfer as a Markov Chain

In a sexually reproducing organism, the unit of evolutionary change in nature is a population composed of interbreeding individuals. Population genetics is the branch of genetics that seeks to explain the genetic structure of such populations. How this structure changes over time in response to other natural or unnatural forces helps define the underlying evolution of the population. Thus, we will say genetics is the study of mechanisms of heredity transmissions and variation of organism characteristics. Genetic characteristics are determined by the gene received from the mother and the gene received from the father. *A* and *a* represent two possible genes for a characteristic. The possible combinations are *AA*, *aa*, *Aa*, and *aA*. We state that *Aa* and *aA* are equivalent and interchangeable. The combinations *AA*, *aa*, and *Aa* are called genotypes.

If the *A* is the dominant trait, then *Aa* tends to follow the *AA* traits. If left undisturbed, the frequency of *A* and *a* in the population remain unchanged in future generations, and the proportion of these genotypes will not change after the first generation. If the frequency of the *A* gene in a population is defined as probability, *p*, the frequency of a gene is defined with frequency, $q = 1 - p$.

Then $p^2$, $2pq$, and $q^2$ proportions of offspring with genotype *AA*, *Aa*, and *aa*, respectively.

This is referred to the Hardy-Weinberg Principle, and interested readers are invited to see https://www.nature.com/scitable/definition/hardy-weinberg-equilibrium-122/ for more information.

Suppose a parent population has a random mating with initial values.

$p(0) = p_0$ of *A* and $q(0) = q_0$ of *a* but the proportions of the genotypes, *AA*, *Aa*, and *aa* are not known. If there is no outside influences, such as mutation, migration, or selection, then the proportion of *A* and *a* are in equilibrium. The proportions of *AA*, *Aa*, and *aa* would be $p^2$, $2pq$, and $q^2$, respectively.

For another example, suppose the gene type in a particular flower determines the color of the flower and that it comes in only two forms: *A*, which gives the flower a red color and which gives the flower a white color. Each flower has a pair of genes. Suppose initially that proportion of *A* in the generation is $P(0) = 0.7$ and the proportion of *a* in the generation is $q(0) = 0.3$. Then, according to the Hardy-Weinberg principle the proportion of flowers for the next generation is: $AA = 0.7^2 = 0.49$, $aa = 0.3^2 = 0.09$ and for $Aa = 2(0.7)(0.3) = 0.42$.

If we let $u = p(1)$, $w = q(1)$, and $v = 2p(0)q(0)$, then the sum $u + w + v = 1$. As an aside, if the number of flowers in generation 1 is *N*, then the total plants by genotype is a fraction of this total: $u = 0.49\ N$, $w = 0.09\ N$, and $v = 0.42\ N$. Recall that we previously stated that the proportion of gene type *A* in a population must remain the same from one generation to the next generation. In our example since each *A* has two *AA* and each *Aa* has one *A*, then the proportion of *A* is $2u + v = 0.98\ N + 0.42\ N = 1.4\ N$. To get the proportion in generation 1, we divide the total number of *A* genes by the total number of genes, $1.4\ N/2\ N = 0.7 = p(1)$.

**Example 7.5:** Genotype Frequencies

An allele is one of the two or more alternative forms of a given gene. Alleles differ from each other in the precise DNA sequence in the phenotypes they confer. Consider the genotype frequencies in two different populations with two different alleles (Table 7.4).

**TABLE 7.4**

Genotype Frequency of Two Different Populations

| | AA | Aa | aa |
|---|---|---|---|
| Population 1 | 0.09 | 0.42 | 0.49 |
| Population 2 | 0.20 | 0.20 | 0.60 |

We want to determine the allele frequencies of *A* and *a* in the populations. We want to know if either population is in Hardy-Weinberg equilibrium. We also want to find the frequencies of alleles and genotypes after one random mating in each population.

**Solution:**
We want to determine the allele frequencies of *A* and *a* in the populations.

Population 1: *P(A)* = 0.09 + (0.50)(0.42) = 0.30, *P(a)*= 1 – 0.30 = 0.70
Population 2: *P(A)* = 0.20 + (0.50)(0.20) = 0.30, *P(a)* = 1 – 0.30 = 0.70

We want to know if either population is in Hardy-Weinberg equilibrium.

$$P(AA) = 0.09, P(Aa) = 0.42, \quad P(aa) = 0.70^2 = 0.49$$

Only Population 2 is in Hardy-Weinberg equilibrium.

We also want to find the frequencies of alleles and genotypes after one random mating in each population.

$$P(AA) = 0.09, P(Aa) = 0.42, \text{ and } P(aa) = 0.49$$

$$P(A) = p = 0.30 \text{ and } P(a) = q - 0.70$$

They become $P(AA) = p^2 = 0.09$; $P(Aa) = 2pq = 0.42$; and $P(aa) = q^2 = 0.049$

**Example 7.6:** Migration of Genes Across Border as a Markov Chain

Assume the border a country has been open for *n* years. No one know the proportions of residents, and hence genes, will migrate each generation, Let's consider a simplified version between two countries, **A** and **B**. Assume the migrations from **A** to **B** is 0.09 and the migration from **B** to **A** is 0.02. We want to find the equilibrium migration of the genes.

**Solution:**
Let

$$A(n) = \text{gene type in country A after period n.}$$

$$B(n) = \text{gene type in country B after period n}$$

Using DDS our model is

$$A(n+1) = 0.91A(n) + 0.02B(n)$$

$$B(n + 1) = 0.09A(n) + 0.98B(n)$$

We may iterate this DDS and obtain the steady state probabilities for $A(n)$ and $B(n)$ of 0.1819 and 0.8181.

We may also solve for the eigenvalues and eigenvectors of the matrix.

$$\begin{bmatrix} 0.91 & 0.02 \\ 0.09 & 0.98 \end{bmatrix}$$

We do this by solving the determinant of

$$\begin{bmatrix} 0.91 - \lambda & 0.02 \\ 0.09 & 0.98 - \lambda \end{bmatrix}$$

equal to zero. The eignevalues are 0.89 and 1 with eigenvectors [1, −1] and [1, 4.5], respectively.

**Example 7.7:** Extension to Bayes' Theorem

We extend the previous example (Example 7.6) to determine two probability questions: *P*(gene changes county during one generation) and *P*(gene stays the same during one generation).

**Solution:**
Let

$P(A)$ = probability that gene was originally in country A
$P(B)$ = probability that gene was originally in country B
$S$ = gene stays in country
$S'$ = gene does not stay in country

Then, we have

$P(A \mid S') = 0.09$
$P(B \mid S') = 0.02$
$P(A \cap S) = 0.16544$
$P(A \cap S') = 0.01636$
$P(B \cap S) = 0.8018$
$P(B \cap S') = 0.01636$

Then, we use the law of total probability,

$P(S) = 0.16544 + 0.8018 = 0.96724$
$P(S') = 0.01636 + 0.01636 = 0.03276$

**TABLE 7.5**
Three Countries

| Source | A | B | C |
|---|---|---|---|
| A | 0.02 | 0.90 | 0.08 |
| B | 0.01 | 0.01 | 0.98 |
| C | 0.95 | 0.02 | 0.03 |

**Example 7.8:** Extending Migration

We can now extend Example 7.7 beyond two countries to include three countries: A, B, and C (Table 7.5).

We want to find the steady state probabilities of migration between the three countries.

**Solution:**
We find the steady state probabilities as 0.1866, 0.1081, and 0.7053 for *A(n)*, *B(n)*, and *C(n)*, respectively.

Markov chains are a common, and relatively simple, way to statistically model random processes. They have been used in many different domains, ranging from text generation to financial modeling. Overall, Markov Chains are conceptually quite intuitive, and are very accessible in that they can be implemented without the use of any advanced statistical or mathematical concepts. They are a great way to start learning about probabilistic modeling and data science techniques.

**Conclusion:**
Now that you know the basics of Markov chains, you should now be able to easily implement them in a language of your choice. If coding is not your forte, there are also many more advanced properties of Markov chains and Markov processes to dive into. In my opinion, the natural progression along the theory route would be toward Hidden Markov Processes or MCMC. Simple Markov chains are the building blocks of other, more sophisticated, modeling techniques, so with this knowledge, you can now move onto various techniques within topics such as belief modeling and sampling.

## 7.4 Markov Processes

A Markov processes is the continuous-time analog of the Markov chains in Sections 7.1 and 7.2. We might also say it is the stochastic analog to the dynamical systems.

We will start the process by first providing some basic definitions and notation for Markov processes.

Let {X(t) : t ∈ [0, ∞)} be a collection of discrete random variables with values in a finite or infinite set, {1, 2, ..., *N*} or {0, 1, 2, ...}. The index set is continuous $t \in [0, \infty)$.

Continuous-Time Markov Chain (CTMC) is the stochastic process {X(t) : t ∈ [0, ∞)} is a CTMC if it satisfies the following conditions for any sequence of real numbers satisfying $0 \le t_0 \le t_1 \le \ldots \le t_n \le t_{n+1}$.

$$\begin{aligned} Prob\{X(t_{n+1}) &= i_{n+1} \mid X(t_0) = i_0,\ X(t_1) = i_1, \ldots, X(t_n) = i_n\} \\ &= Prob\{X(t_{n+1}) = i_{n+1} X(t_n) = i_n\} \end{aligned}$$

Each random variable *X(t)* has an associated probability distribution {pi(t)}∞ = 0 where pi(t) = Prob{X(t) = i}

Let $p(t) = (p_0(t), p_1(t), \ldots)$ *T* be the vector of probabilities A (Peace 2017).

The transition probabilities define the relation between the random variables *X(s)* and *X(t)* for $s < t$ I transition probabilities are defined below for $i, j = 0, 1, 2, \ldots$: pjj(t, s) = Prob{X(t) = j | X(s) = i}, $s < t$.

If the transition probabilities do not explicitly depend on *s* or *t* but only depend on the length of the time interval $t - s$, they are called stationary or homogeneous; otherwise, they are nonstationary or nonhomogeneous.

We will assume the transition probabilities are stationary, unless stated otherwise (Peace 2017).

Stationary transition probabilities *pji(t − s)* = Prob{X(t) = j | X(s) = i} = Prob {X(t − s) = j | X(0) = i} for $s < t$.

The transition matrix is *P(t) = (pji(t))* where in most cases, *pji(t)* ≥ 0 and X∞ j = 0 *pji(t)* = 1 for $t \ge 0$. *P(t)* is a stochastic matrix for all $t \ge 0$.

The transition probabilities are solutions of the Chapman-Kolmogorov equations: X∞ *k* = 0 pjk (s)pki(t) = pji(t + s) or in matrix form: *P(s)P(t)* =*P(s + t)* for all *s*, $t \in [0, \infty)$.

DTMC: There is a jump to a new state at discrete times: 1, 2, ..., I CTMC: The jump can occur at any time +$t \ge 0$ I consider a CTMC beginning at state *X(0)* I the process stays in state *X(0)* for a random amount of time: W1 I it then jumps to a new state: *X(W1)* I is stays in state *X(W1)* for a random amount of time: W2 I it then jumps to a new state *X(W2)* I Wi is a random variable for the time of the *i*-th jump.

**Example 7.9:** Continuous Time Markov Chain

Consider the continuous-time Markov chain (CTMC) with jumps shown in Figure 7.4. We will assume that $\lambda_1 = 2$, $\lambda_2 = 3$, $\lambda_3 = 4$.

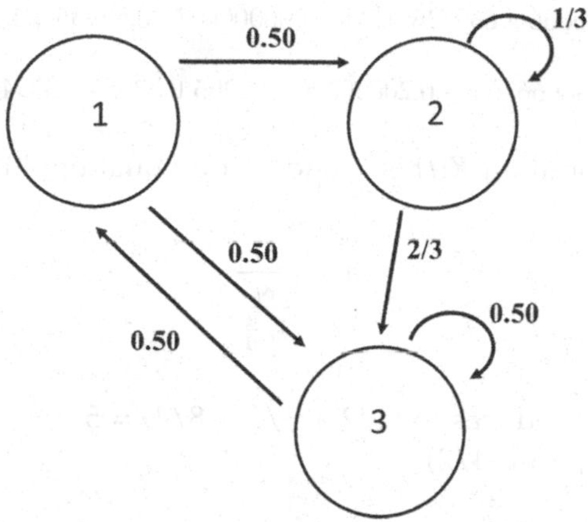


**FIGURE 7.4**
Transition diagram for Example 1 CTMC.

We want to find both the stationary (steady state) probabilities, and then using those probabilities, find the stationary distribution for $X(t)$.

**Solution:**
We find the matrix $P$ as

$$P = \begin{bmatrix} 0 & 0.5 & 0.5 \\ 0 & 0.333 & 0.667 \\ 0.5 & 0 & 0.5 \end{bmatrix}$$

As before, we raise $P$ to a very large power. The steady state probabilities for this system are (4/15. 3/15. 8/15).

We can use R to quickly determine the approximations for these steady state probabilities.

```
with(LinearAlgebra);

    A := Matrix(3, 3, [0, 0.5, 0.5, 0, 1/3, 2/3, 0.5, 0, 0.5]);
                          [ 0 0.5 0.5]
                          [          ]
                          [    1   2 ]
                     A := [ 0  -   - ]
                          [    3   3 ]
                          [          ]
                          [0.5  0  0.5]

    SS := A^20;
        [0.266666666682417 0.199999999990212 0.533333333327370]
        [                                                     ]
```

SS := [0.266666666669726 0.200000000006931 0.533333333323343]
[ ]
[0.266666666657644 0.200000000002295 0.533333333340061]

The limiting distribution ***X(t)*** is found using Equation 7.1.

$$\pi_j = \frac{\frac{\tilde{\pi}_j}{\lambda_j}}{\sum_k \frac{\tilde{\pi}_k}{\lambda_k}} \tag{7.1}$$

The denomination reduces to (4/2 + 3/3 + 8/4) = 5
Numerator of $\pi_1$ will (4/2) = 2

$$\pi_1 = 2/5$$

Continuing to $\pi_2$: we have 1/5 and then $\pi_1 = 2/5$.
The limiting distribution is 1/5 [2,1,2] for ***X****(t)*.

## 7.5 Exercises

1. Reconsider the inventory problem of Example 7.2, but now suppose that the store policy is to order more pens if there are fewer than two left in stock at the end of the week. We reorder if zero or one pen is remaining; we order enough pens to bring the inventory back to three.
   a. Find the probability that demand exceeds supply.
   b. Perform some sensitivity analysis on the demand rate $\lambda$ and determine the steady state probabilities assuming $\lambda$ is 0.50, 0.75, 1.0, 1.2, and 1.25.
2. Five locations are connected by radio. The radio link is active 20% of the time, and there is no radio activity the remaining 80% of the time. The main location sends a radio message with an average duration of 30 seconds and the remaining four locations send radio messages that average 10 seconds in length. Half of all messages originate from the main location and the remaining equally divided by the other four stations. For each location, compute the steady state probability.
3. Consider our fast food problem in Example 7.3. Assume now that the transition diagram is as follows (Figure 7.5).

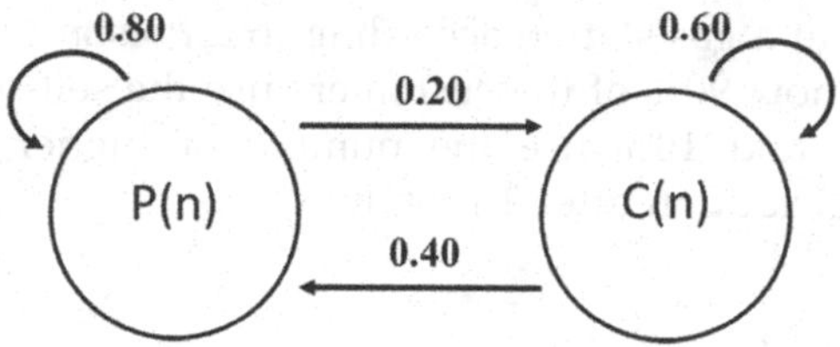


**FIGURE 7.5**
Fast food transition diagram.

Determine the probabilities after two steps and then determine the steady state probabilities.

4. Consider the genetic transfer example extended to include three new countries: A, B, and C (Example 7.8). Given new migration probabilities between countries (Table 7.6), find the steady state probabilities.
5. Consider a CTMC that has a jump chain as shown in Figure 7.6 and assume $\lambda_1 = 2, \lambda_2 = 1, \lambda_3 = 3$. Find the steady state probabilities and the limiting distribution for $\mathbf{X}(t)$.

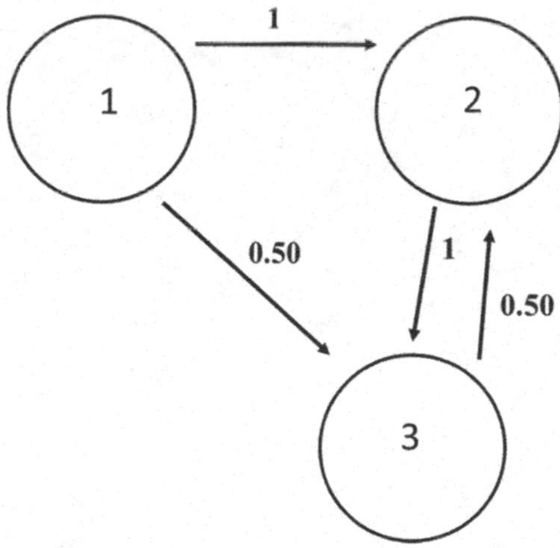


**FIGURE 7.6**
Transition diagram for Exercise 6 concerning CTMC.

**TABLE 7.6**
Updated Migration Probabilities

| Source | A | B | C |
|---|---|---|---|
| A | 0.04 | 0.88 | 0.08 |
| B | 0.02 | 0.02 | 0.96 |
| C | 0.93 | 0.03 | 0.04 |

Customers arrive at a gas station according to a Poison process with a rate of 20/hour. We know 90% of the customers use the self-service pumps for regular unleaded and 10% use the pumps for higher octane gasoline. Determine the distribution after 1 month.

## Reference

Meerschaert, M. (1999). *Mathematical Modeling,* 2nd Edition. Academic Press, New York.

# 8

# Linear Programming

**OBJECTIVES**

1. Apply modeling to continuous optimization problems.
2. Understand both unconstrained and constrained optimization in three or more variables.
3. Understand the application of numerical procedures to optimization problems.

## 8.1 Introduction

Linear programming (LP) is simply a mathematical modeling technique, which can trace its roots to the 1940s, where a linear function is either maximized or minimized subjected to a set of constraints. LP has been used to support many decisions in business, industrial engineering, and the social and physical sciences.

Typically, LP reduces a problem to seeking an optimum solution to an objective function (linear regression)

$$f(x) = c_1x_1 + c_2x_2 + c_3x_3 + \ldots + c_nx_n$$

based on a set of inequality constraints.

$$a_{11}x_1 + a_{12}x_2 + \ \ldots + a_{1n}x_n \le b_1$$

$$\forall\ x_i \ge 0$$

To help make the points, we will start by considering the Emergency Service Coordinator (ESC) for a county that is interested in locating the county's three ambulances to maximize the residents that can be reached within 8 minutes in emergency situations. The county is divided into 6 zones, and

DOI: 10.1201/9781003298762-8

**TABLE 8.1**

Average Travel Times from Zone *i* to Zone *j* in Perfect Conditions

| Zones | 1 | 2 | 3 | 4 | 5 | 6 |
|---|---|---|---|---|---|---|
| 1 | 1 | 8 | 12 | 14 | 10 | 16 |
| 2 | 8 | 1 | 6 | 18 | 16 | 16 |
| 3 | 12 | 18 | 1.5 | 12 | 6 | 4 |
| 4 | 16 | 14 | 4 | 1 | 16 | 12 |
| 5 | 18 | 16 | 10 | 4 | 2 | 2 |
| 6 | 16 | 18 | 4 | 12 | 2 | 2 |

the average time required to travel from one region to the next under semi-perfect conditions are summarized in the following Table 8.1.

The population in zones 1, 2, 3, 4, 5, and 6 are given by the Table 8.2 below:

**Problem Identification:** Determine the location for placement of the ambulances to maximize coverage within the allotted time.

**Assumptions:** Time travel between zones is negligible. Times in the data are averages under ideal circumstances.

Here, we further assume that employing an optimization technique would be worthwhile. We will start with assuming a linear model, and then we will enhance the model with integer programming.

Perhaps, consider planning the shipment of needed items from the warehouses where they are manufactured and stored to the distribution centers where they are needed.

There are three warehouses at different cities: Detroit, Pittsburgh, and Buffalo. They have 250, 130, and 235 tons of paper accordingly. There are four publishers in Boston, New York, Chicago, and Indianapolis. They ordered 75, 230, 240, and 70 tons of paper to publish new books. The following are costs in dollars of transportation of one ton of paper (Table 8.3).

**TABLE 8.2**

Populations in Each Zone

| Zone | Population |
|---|---|
| 1 | 50,000 |
| 2 | 80,000 |
| 3 | 30,000 |
| 4 | 55,000 |
| 5 | 35,000 |
| 6 | 20,000 |
| Total | 270,000 |

**TABLE 8.3**
Transportation Costs between Cities

| From/To | Boston (BS) | New York (NY) | Chicago (CH) | Indianapolis (IN) |
|---|---|---|---|---|
| **Detroit (DT)** | 15 | 20 | 16 | 21 |
| **Pittsburgh (PT)** | 25 | 13 | 5 | 11 |
| **Buffalo (BF)** | 15 | 15 | 7 | 17 |

Management wants you to minimize the shipping costs while meeting demand. This problem involves the allocation of resources and can be modeled as a LP problem, as we will discuss.

In engineering management, the ability to optimize results in a constrained environment is crucial to success. Additionally, the ability to perform critical sensitivity analysis, or "what if analysis", is extremely important for decision making. Consider starting a new diet, which you need to be healthy. You go to a nutritionist who gives you lots of information on foods. They recommend sticking to six different foods – bread, milk, cheese, fish, potatoes, and yogurt and provides you a table of information, including the average cost of the items (Table 8.4):

We go to a nutritionist, and they recommend that our diet contains no less than **150** calories, no more than **10 g** of protein, no less than **10** g of carbohydrates, and no less than **8** g of fat. Also, we decide that our diet should have **minimal cost**. In addition, we conclude that our diet should include at least **0.5** g of fish and not more than **1** cup of milk. Again, this is an allocation of recourses problem where we want the optimal diet at minimum cost. We have six unknown variables that define weight of the food. There is a lower bound for Fish as 0.5 g. There is an upper bound for Milk as 1 cup. To model and solve this problem, we can use LP.

Modern LP was the result of a research project undertaken by the U.S. Department of Air Force under the title of Project SCOOP (Scientific Computation of Optimum Programs). As the number of fronts in the Second World War increased, it became more and more difficult to

**TABLE 8.4**
Costs for Food Items

| | Bread | Milk | Cheese | Potato | Fish | Yogurt |
|---|---|---|---|---|---|---|
| **Cost, $** | 2.0 | 3.5 | 8.0 | 1.5 | 11.0 | 1.0 |
| **Protein, g** | 4.0 | 8.0 | 7.0 | 1.3 | 8.0 | 9.2 |
| **Fat, g** | 1.0 | 5.0 | 9.0 | 0.1 | 7.0 | 1.0 |
| **Carbohydrates, g** | 15.0 | 11.7 | 0.4 | 22.6 | 0.0 | 17.0 |
| **Calories, Cal** | 90 | 120 | 106 | 97 | 130 | 180 |

coordinate troop supplies effectively. Mathematicians looked for ways to use the new computers being developed to perform calculations quickly. One of the SCOOP team members, George Dantzig, developed the simplex algorithm for solving simultaneous LP problems. The simplex method has several advantageous properties: it is very efficient, allowing its use for solving problems with many variables; it uses methods from linear algebra, which are readily solvable.

In January 1952, the first successful solution to a LP problem was found using a high-speed electronic computer on the National Bureau of Standards SEAC machine. Today, most LPs are solved via high-speed computers. Computer specific software, such as LINDO, EXCEL SOLVER, GAMS, have been developed to help in the solving and analysis of LP problems. We may use the power of LINDO to solve our LP problems in this chapter.

To provide a framework for our discussions, we offer the following basic model:

Maximize (or minimize) *f(X)*

Subject to

$$g_i(x)\begin{Bmatrix} \geq \\ = \\ \leq \end{Bmatrix} b_i \text{ for all } i.$$

The various component of the vector X are called the decision variables of the model. These are the variables that can be controlled or manipulated. The function, *f(X)*, is called the objective function. By subject *to*, we connote that there are certain side conditions, resource requirement, or resource limitations that must be met. These conditions are called constraints. The constant $b_i$ represents the level that the associated constraint *g (Xi)* and is called the right-hand side in the model.

LP is a method for solving linear problems, which occur very frequently in almost every modern industry. In fact, areas using LP are as diverse as defense, health, transportation, manufacturing, advertising, and telecommunications. The reason for this is that in most situations, the classic economic problem exists – you want to maximize output, but you are competing for limited resources. The "linear" in linear programming means that in the case of production, the quantity produced is proportional to the resources used and also the revenue generated. The coefficients are constants, and no products of variables are allowed.

In order to use this technique, the company must identify a number of constraints that will limit the production or transportation of their goods; these may include factors such as labor hours, energy, and raw materials. Each constraint must be quantified in terms of one unit of output, as the problem-solving method relies on the constraints being used.

An optimization problem that satisfies the following five properties is said to be a LP problem.

- There is a unique objective function, *f(X)*.
- Whenever a decision variable, *X*, appears in either the objective function or a constraint function, it must appear with an exponent of 1, possibly multiplied by a constant.
- No terms contain products of decision variables.
- All coefficients of decision variables are constants.
- Decision variables are permitted to assume fractional as well as integer values.

Linear problems, by the nature of the many unknowns, are very hard to solve by human inspection, but methods have been developed to use the power of computers to do the hard work.

## 8.2 Formulating Linear Programming Problems

A LP problem is a problem that requires an objective function to be maximized or minimized subject to resource constraints. The key to formulating a LP problem is recognizing the decision variables. The objective function and all constraints are written in terms of these decision variables.

The conditions for a mathematical model to be a linear program were:

- All variables are continuous (i.e. can take fractional values).
- There is a single objective (minimize or maximize).
- The objective and constraints are linear i.e. any term is either a constant or a constant multiplied by an unknown.
- The decision variables must be non-negative.

LP's are important – this is because:

- Many practical problems can be formulated as LPs.
- There exists an algorithm (called the *simplex* algorithm) that enables us to solve LPs numerically relatively easily.

We will return later to the simplex algorithm for solving LPs but for the moment we will concentrate upon formulating LP's.

Some of the major application areas to which LP can be applied are:

- Blending
- Production planning
- Oil refinery management
- Distribution
- Financial and economic planning
- Manpower planning
- Blast furnace burdening
- Farm planning

We consider below some specific examples of the types of problem that can be formulated as LPs. Note here that the key to formulating LPs is ***practice***. However, a useful hint is that common objectives for LPs are ***minimize cost/ maximize profit***.

**Example 8.1:** Manufacturing

Consider the following problem statement: A company wants to can two new different drinks for the holiday season. It takes 2 hours to can one gross of Drink A, and it takes 1 hour to label the cans. It takes 3 hours to can one gross of Drink B, and it takes 4 hours to label the cans. The company makes $10 profit on one gross of Drink A and a $20 profit of one gross of Drink B. Given that we have 20 hours to devote to canning the drinks and 15 hours to devote to labeling cans per week, how many cans of each type drink should the company package to maximize profits?

Problem Identification: Maximize the profit of selling these new drinks.

Define variables:

$X_1$ = the number of gross cans produced for Drink A per week
$X_2$ = the number of gross cans produced for Drink B per week

Objective Function:

$$Z = 10X_1 + 20X_2$$

Constraints:

1. Canning with only 20 hours available per week

$$2X_1 + 3X_2 \leq 20$$

2. Labeling with only 15 hours available per week

$$X_1 + 4X_2 \leq 15$$

3. Non-negativity restrictions
   $X_1 \geq 0$ (non-negativity of the production items)
   $X_2 \geq 0$ (non-negativity of the production items)

The Complete FORMULATION:

$$MAXIMIZE\ Z = 10X_1 + 20X_2$$

*subject to*

$$2X_1 + 3X_2 \leq 20$$

$$X_1 + 4X_2 \leq 15$$

$$X_1 \geq 0$$
$$X_2 \geq 0$$

We will see in the next section how to solve these two-variable problems graphically.

**Example 8.2:** Financial Planning
A bank makes four kinds of loans to its personal customers, and these loans yield the following annual interest rates to the bank:

- First mortgage 14%
- Second mortgage 20%
- Home improvement 20%
- Personal overdraft 10%

The bank has a maximum foreseeable lending capability of $250 million and is further constrained by the policies:

1. First mortgages must be at least 55% of all mortgages issued and at least 25% of all loans issued (in $ terms).
2. Second mortgages cannot exceed 25% of all loans issued (in $ terms).
3. Third, to avoid public displeasure and the introduction of a new windfall tax the average interest rate on all loans must not exceed 15%.

Formulate the bank's loan problem as an LP so that the bank can maximize interest income while satisfying the policy limitations.

Note here that these policy conditions, while potentially limiting the profit that the bank can make, also limit its exposure to risk in a particular area. It is a fundamental principle of risk reduction that risk is reduced by spreading money (appropriately) across different areas.

- ***Financial planning Formulation***

Note here that as in *all* formulation exercises, we are translating a verbal description of the problem into an *equivalent* mathematical description.

A useful tip when formulating LP's is to express the variables, constraints, and objective in words before attempting to express them in mathematics.

- ***Variables***

Essentially, we are interested in the amount (in dollars) the bank has loaned to customers in each of the four different areas (not in the actual number of such loans). Hence, let

$x_i$ = amount loaned in area $i$ in millions of dollars (where $i = 1$ corresponds to first mortgages, $I = 2$ to second mortgages etc) and note that each $x_i \geq 0$ (i = 1, 2, 3, 4).

Note here that it is conventional in LP's to have all variables ≥ 0. Any variable (X, say) which can be positive *or* negative can be written as $X_1 - X_2$ (the difference of two new variables) where $X_1 \geq 0$ and $X_2 \geq 0$.

- ***Constraints***

restricts the amount that will be loaned

$$x_1 + x_2 + x_3 + x_4 < 250$$

a. policy condition 1

$$x_1 \geq 0.55(x_1 + x_2)$$

i.e. first mortgages ≥ 0.55(total mortgage lending) and also

$$x_1 \geq 0.25(x_1 + x_2 + x_3 + x_4)$$

i.e. first mortgages ≥ 0.25(total loans)

b. policy condition 2

$$x_2 < 0.25(x_1 + x_2 + x_3 + x_4)$$

c. policy condition 3 – we know that the total annual interest is $0.14x_1 + 0.20x_2 + 0.20x_3 + 0.10x_4$ on total loans of $(x_1 + x_2 + x_3 + x_4)$. Hence, the constraint relating to policy condition (3) is

$$0.14x_1 + 0.20x_2 + 0.20x_3 + 0.10x_4 \leq 0.15(x_1 + x_2 + x_3 + x_4)$$

- ***Objective Function***

To maximize interest income (which is given above) i.e.

$$\text{Maximize } Z = 0.14x_1 + 0.20x_2 + 0.20x_3 + 0.10x_4$$

**Example 8.3:** Blending and Formulation

Consider the example of a manufacturer of animal feed who is producing feed mix for dairy cattle. In our simple example, the feed mix contains two active ingredients. One kg of feed mix must contain a minimum quantity of each of four nutrients in Table 8.5.

The ingredients have the following nutrient values and cost (Table 8.6):

What should be the amounts of active ingredients in one kg of feed mix that minimizes cost?

- ***Blending problem solution***
- ***Variables***

In order to solve this problem, it is best to think in terms of one kilogram of feed mix. That kilogram is made up of two parts – ingredient 1 and ingredient 2:

**TABLE 8.5**

Animal Feed Mix

| Nutrient | A | B | C | D |
|---|---|---|---|---|
| Grams | 90 | 50 | 50 | 2 |

**TABLE 8.6**

Nutrient Values and Cost

| Ingredient (gram/kg) | A | B | C | D | Cost/kg |
|---|---|---|---|---|---|
| 1 | 100 | 80 | 40 | 10 | 40 |
| 2 | 200 | 150 | 20 | 0 | 60 |

$x_1$ = amount(kg) of ingredient 1 in one kg of feed mix
$x_2$ = amount(kg) of ingredient 2 in one kg of feed mix

where $x_1 \geq 0$, $x_2 \geq 0$

Essentially these variables ($x_1$ and $x_2$) can be thought of as the recipe telling us how to make up one kilogram of feed mix.

$$x_1 + x_2 = 1$$

- ***Objective function***
- ***Constraints***
  - nutrient constraints

$$100x_1 + 200x_2 >= 90\,(nutrient\ A)$$
$$80x_1 + 150x_2 >= 50\,(nutrient\ B)$$
$$40x_1 + 20x_2 >= 20\,(nutrient\ C)$$
$$10x_1 >= 2\,(nutrient\ D)$$

  - balancing constraint (an *implicit* constraint due to the definition of the variables)

Presumably to minimize cost, i.e.

$$\text{Minimize } Z = 40x_1 + 60x_2$$

This gives us our complete LP model for the blending problem.

**Example 8.4:** Production Planning Problem

A company manufactures four variants of the same table and in the final part of the manufacturing process there are assembly, polishing and packing operations. For each variant the time required for these operations is in minutes as is the profit per unit sold (Table 8.7).

**TABLE 8.7**

Production Planning Data

| Variant | Assembly | Polish | Pack | Profit ($) |
|---|---|---|---|---|
| 1 | 2 | 3 | 2 | 1.5 |
| 2 | 4 | 2 | 3 | 2.5 |
| 3 | 3 | 3 | 2 | 3 |
| 4 | 7 | 4 | 5 | 4.5 |

- ***Variables***
  - Given the current state of the labor force the company estimate that, each year, they have 100,000 minutes of assembly time, 50,000 minutes of polishing time and 60,000 minutes of packing time available How many of each variant should the company make per year and what is the associated profit?

Let: $x_i$ be the number of units of variant $i$ ($i$ = 1,2,3,4) made per year where $x_i \geq 0$ $i = 1,2,3,4$

- ***Constraints***

Resources for the operations of assembly, polishing, and packing.

$$2x_1 + 4x_2 + 3x_3 + 7x_4 <= 100{,}000\,(assembly)$$
$$3x_1 + 2x_2 + 3x_3 + 4x_4 <= 50{,}000\,(polishing)$$
$$2x_1 + 3x_2 + 2x_3 + 5x_4 <= 60{,}000\,(packing)$$

- ***Objective function***

$$\text{Maximize } Z = 1.5x_1 + 2.5x_2 + 3.0x_3 + 4.5x_4$$

**Example 8.5:** Shipping

Consider planning the shipment of needed items from the warehouses, where they are manufactured and stored, to the distribution centers, where they are needed, as shown in the introduction. There are three warehouses at different cities: Detroit, Pittsburgh and Buffalo. They have 250, 130 and 235 tons of paper accordingly. There are four publishers in Boston, New York, Chicago and Indianapolis. They ordered 75, 230, 240 and 70 tons of paper to publish new books.

These are the costs in dollars of transportation of one ton of paper (Table 8.8):

**TABLE 8.8**

Transportation Costs

| From/To | Boston (BS) | New York (NY) | Chicago (CH) | Indianapolis (IN) |
|---|---|---|---|---|
| **Detroit (DT)** | 15 | 20 | 16 | 21 |
| **Pittsburgh (PT)** | 25 | 13 | 5 | 11 |
| **Buffalo (BF)** | 15 | 15 | 7 | 17 |

Management wants you to minimize the shipping costs while meeting demand.

We define $x_{ij}$ to be the travel from city $i$ (1 is Detroit, 2 is Pittsburg, 3 is Buffalo) to city $j$ (1 is Boston, 2 is New York, 3 is Chicago, and 4 is Indianapolis).

*Minimize* $Z =$

$$15x_{11} + 20x_{12} + 16x_{13} + 21x_{14} + 25x_{21} + 13x_{22} + 5x_{23} + 11x_{24} + 15x_{31} + 15x_{32} + 7x_{33} + 17x_{34}$$

*Subject to:*

$$x_{11} + x_{12} + x_{13} + x_{14} \le 250 \text{ (availability in Detroit)}$$
$$x_{31} + x_{32} + x_{33} + x_{34} \le 235 \text{ (availability in Buffalo)}$$
$$x_{11} + x_{21} + x_{31} \ge 75 \text{ (demand Boston)}$$
$$x_{12} + x_{22} + x_{32} \ge 230 \text{ (demand New York)}$$
$$x_{13} + x_{23} + x_{33} \ge 240 \text{ (demand Chicago)}$$
$$x_{14} + x_{24} + x_{34} \ge 70 \text{ (demand Indianapolis)}$$
$$x_{ij} \ge 0$$

### 8.2.1 Integer Programming

For Integer programming, we will take advantage of technology. We will not present the branch and bound technique but suggest a thorough review of the topic can be found in Winston (2005) or other similar math programming books.

For mixed integer programming, we show the formulations and the use of Excel.

### 8.2.2 Nonlinear Programming

Often, we have nonlinear objective functions or constraints. Suffice it to say, we will recognize these and use technology to assist in the solution. Excellent nonlinear programming can be read for additional information.

## 8.3 Technology Examples for Linear Programming

### 8.3.1 Memory Chips for CPUs

Let's start with a manufacturing example. Suppose a small business wants to know how many of two types of high-speed computer chips to manufacturer

**TABLE 8.9**
Assembly Requirements and Profit

| | Chip A | Chip B | Quantity Available |
|---|---|---|---|
| **Assembly time (hours)** | 2 | 4 | 1,400 |
| **Installation time (hours)** | 4 | 3 | 1,500 |
| **Profit (per unit)** | 140 | 120 | |

weekly to maximize their profits. First, we need to define our decision variables. Let,

$x_1$ = *number of high-speed chip type A to produce weekly*
$x_2$ = *number of high-speed chip type B to produce week*

The company reports a profit of \$140 for each type *A* chip and \$120 for each type *B* chip sold. The production line reports the information in Table 8.9.

The constraint information from the table becomes inequalities that are written mathematical as:

Maximize $Z = 140x_1 + 120x_2$
Subject to:

$$2x_1 + 4x_2 \leq 1400$$
$$4x_1 + 3x_2 \leq 1500$$
$$x1,\ x2,\ \geq 0$$

$$2x_1 + 4x_2 \leq 1400 \text{ (assembly time)}$$
$$4x_1 + 3x_2 \leq 1500 \text{ (installation time)}$$
$$x_1 \geq 0,\ x_2 \geq 0$$

The profit equation is:

$$Profit\ Z = 140x_1 + 120x_2$$

We will illustrate this example with technology.

### 8.3.2 Linear Programing in Excel

Technology is critical to solving, analyzing, and performing sensitivity analysis on LP problems. Technology provides a suite of powerful, robust routines for solving optimization problems, including linear programs. Excel and other technology such as LINDO, and LINGO often appear in often in the engineering industry. We have tested all of these software packages and found them useful but will use Excel to solve our problem.

First we present a "How to" in Excel.

**Example 8.6:** Linear Programing in Excel

Consider the following example:

Maximize: $2X1 + 6X2 + 5X3$

Subject to:

$$X1 + X2 + X3 = 3$$
$$X1 + 2X2 + 3X3 \leq 10$$
$$2X1 + 6X2 + X3 \geq 5$$
$$X1,\ X2,\ X3 \geq 0$$

Using two variable models, we can graph it on paper and easily solve for the solution. With more than two variables (X1, X2, X3), LP must be used and this example will demonstrate how to use Excel in that capacity.

First, open Excel, and type in/set-up an LP that looks exactly like the one below (Figure 8.1):

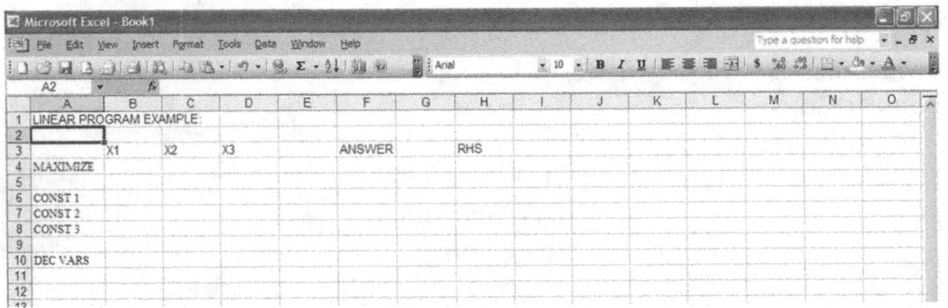


**FIGURE 8.1**
Excel screenshot of Example 8.6.

**Key: RHS = Right Hand Side (for constraints)**

**Const 1 = Constraint #1 (same for Const 2 and 3)**

**Dec vars = Decision Variables (the values for X1, X2, and X3 we seek)**

To this point, the Linear Program is all "just typing", with no blocks identified with formulas in them. A title is input in cell A1 – call it whatever you want, in this case "Linear Program Example" – and the spacings between cells are merely to make it an "easy to read" format up to this point. In future problems, you can label the constraints and Decision Variables any way you would like – Instead of "Const 1", you can use "Labor Hours", or "Amount of Material Available", etc... instead of Dec Vars, you can use "Number of fixtures to produce", etc...You can use any terminology you like, as long as YOU understand it – just remember that Dec Vars in Row 10 are the values of X1, X2 and X3 that you seek.

Next, add in the numbers for your problem in the correct columns and rows. In Row G, add in the = and/or < = constraints, then the RHS values in Column H.

At this point, there are two things you can do to make your program look better and make it easier to read and understand: first, highlight the block of cells B3 to B10 to H3 to H10 and Click on "Center", for aesthetics and ease of data presentation; second, color in the Objective Function row, the Constraints rows, and the Dec Vars with different colors to offset them. It is starting to look more like a model now (Figure 8.2)!

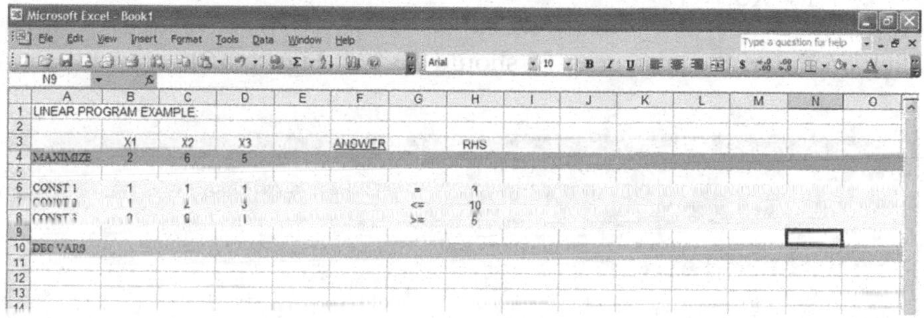

**FIGURE 8.2**
Excel screenshot of update model for Example 8.6

Now, we need to define cells. In this example, cells F4, F6, F7, and F8 need to be defined. Also, in cells B10, C10, and D10, place the number "1" in each for reasons that will become clear later (briefly, they are to test that the answer cells F4, F6, F7, and F8 have been defined properly). In cells F4, F6, F7, and F8, the "=" sign before the equation tells the cell that this is a mathematical formula to be computed.

Cell F4: This cell will contain the final answer to the Value of the Objective Function. Highlight the cell F4, and type in the following: = B4*B10 + C4***C10** + **D4***D10, which signifies 2*X1 + 6*X2 + 5*X3. After completion, the answer in F4 should be "13" at this point, since the Dec Vars are set to 1,1,1, ensuring your coding is correct. If your answer is not 13, re-check your coding in F4.

Cell F6: This cell contains the answer to the first constraint. Highlight cell F6, and type in the following:

$$= B6 * B10 + C6 * C10 + D6 * D10 \text{ which signifies } 1 * X1 + 1 * X2 + 1 * X3.$$

Click on

Enter – the answer in F6 should be "3". Don't forget the "=" sign!

Cell F7: This cell contains the answer to the second constraint. Highlight cell F7, and type in the following:

$= B7 * B10 + C7 * C10 + D7 * D10$ *which signifies* $1 * X1 + 2 * X2 + 3 * X3$. *Click on*

Enter: the answer in F7 should be "6".

Cell F8: This cell contains the answer to the third constraint. Highlight cell F8, and type in the following:

$= B8 * B10 + C8 * C10 + D8 * D10$ *which signifies* $2 * X1 + 6 * X2 + 1 * X3$.

Click on Enter: the answer in F8 should be "9" (Figure 8.3).

| | A | B | C | D | E | F | G | H |
|---|---|---|---|---|---|---|---|---|
| 3 | | X1 | X2 | X3 | | ANSWER | | RHS |
| 4 | MAXIMIZE | 2 | 6 | 5 | | 13 | | |
| 5 | | | | | | | | |
| 6 | CONST 1 | 1 | 1 | 1 | | 3 | = | 3 |
| 7 | CONST 2 | 1 | 2 | 3 | | 6 | <= | 10 |
| 8 | CONST 3 | 2 | 6 | 1 | | 9 | >= | 5 |
| 9 | | | | | | | | |
| 10 | DEC VARS | 1 | 1 | 1 | | | | |

**FIGURE 8.3**
Excel screenshot of completed model for Example 8.6.

Now, we are ready to use the Solver (Figure 8.4). Save what you have, then go to "Tools", and then select "Solver". (If you can't find Solver, it will be in Add-ins – extract it from there).

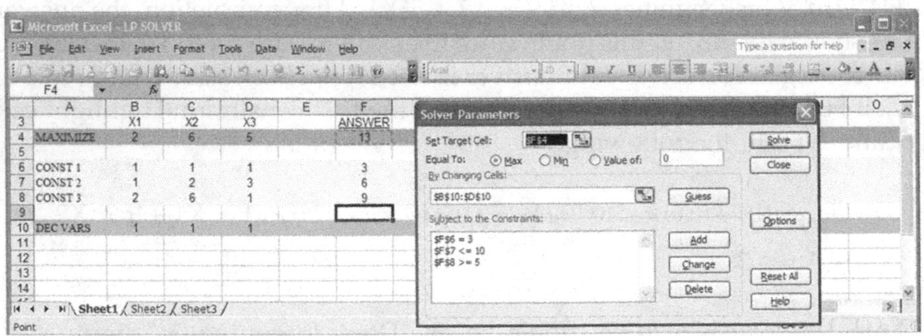


**FIGURE 8.4**
Solver parameters dialogue box for Example 8.6.

1. Set Target Cell: Input F4.
2. Select the "Max" button.
3. In the "By Changing Cells" area, merely highlight all three cells in B10, C10, and D10.
4. Now, move to the constraints. Click on "Add".
   In the first constraint: Type in F6, select =, and type in the number 3, in the three areas. Click on OK,
   then Add .....
   In the second constraint: Type in F7, select <=, type in 10. Click on OK, then Add.....
   In the third constraint: Type in F8, select >=, type in 5. Click on OK
5. Now, go to Options: Select "Assume Linear Model", "Assume Non-Negative", "Use Auto Scaling" (Figure 8.5)

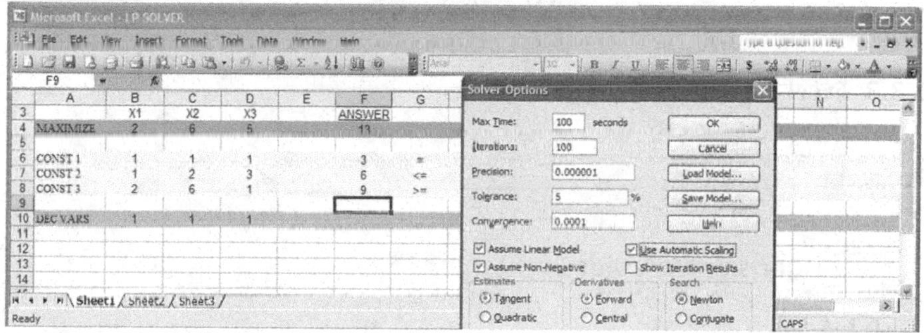


**FIGURE 8.5**
Solver parameters dialogue box with "Auto Scaling" for Example 8.6.

6. Click on OK, then click on Solve (Figure 8.6).

Microsoft Excel - LP SOLVER

| | A | B | C | D | E | F | G | H |
|---|---|---|---|---|---|---|---|---|
| 3 | | X1 | X2 | X3 | | ANSWER | | RHS |
| 4 | MAXIMIZE | 2 | 6 | 5 | | 18 | | |
| 5 | | | | | | | | |
| 6 | CONST 1 | 1 | 1 | 1 | | 3 | = | 3 |
| 7 | CONST 2 | 1 | 2 | 3 | | 6 | <= | 10 |
| 8 | CONST 3 | 2 | 6 | 1 | | 18 | >= | 5 |
| 9 | | | | | | | | |
| 10 | DEC VARS | 0 | 3 | 0 | | | | |
| 11 | | | | | | | | |
| 12 | | | | | | | | |
| 13 | | | | | | | | |
| 14 | | | | | | | | |

Sheet1 / Sheet2 / Sheet3

Ready CAPS

**FIGURE 8.6**
Final model and solution for Example 8.6.

Answer should now be revealed: Objective Function = **18** in Cell F4
Constraints are Satisfied: **3 = 3**
**6 <= 10**
**18 >= 5**
Decision Variables: **X1 = 0, X2 = 3, X3 =0**

Thus, the problem is maximized by making three *X2*s only, with an objective value profit of 18. Now, we present our previous example (Example 8.5) solved via each technology.

Maximize $Z = 140x_1 + 120x_2$

Subject to:

$$2x_1 + 4x_2 \leq 1400$$
$$4x_1 + 3x_2 \leq 1500$$
$$x1, \; x2, \geq 0$$

### 8.3.3 Excel for Linear Programing

| Problem | | | | | |
|---|---|---|---|---|---|
| | | | | | |
| | Decision Variables | | | | |
| | x1 | | 0 | | |
| | x2 | | 0 | | |
| | | | | | |
| | Objective Function | | | 0 | |
| | | | | | |
| | Constraints | | | | |
| | | | 0 | 1400 | |
| | | | 0 | 1500 | |

**Solver** (Figure 8.7)

**Constraints into solver** (Figure 8.8)

We now have the Full Set UP. We can click on Solve for the results (Figure 8.9).

Obtain the answers as *x1 = 180, x2 = 260, Z = 56400.*

Additionally, we can obtain reports from Excel. Two key reports are the answer report and the sensitivity report.

**Answer Report** (Figure 8.10).

**Sensitivity Report** (Figure 8.11).

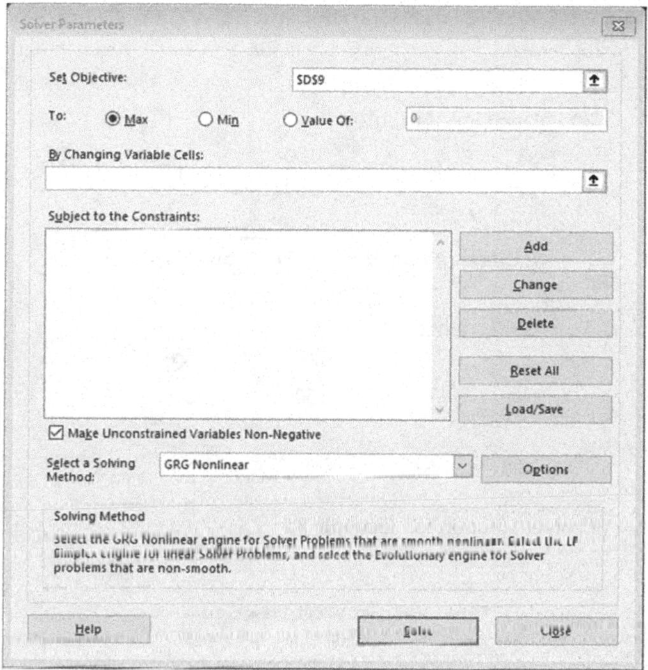


**FIGURE 8.7**
Solver dialogue box.

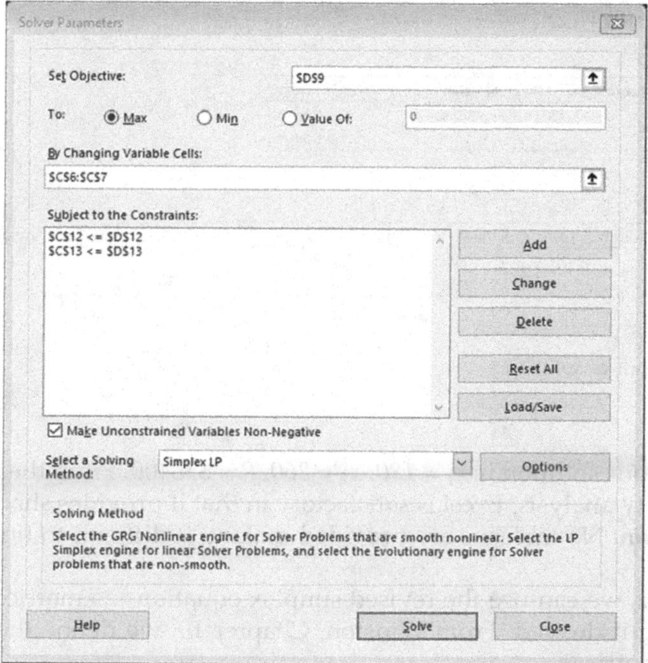


**FIGURE 8.8**
Solver constraints added.

| Problem | | | |
|---|---|---|---|
| | Decision Variables | | |
| | x1 | 180 | |
| | x2 | 260 | |
| | Objective Function | | 56400 |
| | Constraints | | |
| | | 1400 | 1400 |
| | | 1500 | 1500 |

**FIGURE 8.9**
Excel screenshot of solver solution for Example 8.5.

**Solver Options**
Max Time Unlimited, Iterations Unlimited, Precision 0.000001, Use Automatic Scaling
Max Subproblems Unlimited, Max Integer Sols Unlimited, Integer Tolerance 1%, Assume NonNegative

Objective Cell (Max)

| Cell | Name | Original Value | Final Value |
|---|---|---|---|
| $D$9 | Objective Function | 0 | 56400 |

Variable Cells

| Cell | Name | Original Value | Final Value | Integer |
|---|---|---|---|---|
| $C$6 | x1 | 0 | 180 | Contin |
| $C$7 | x2 | 0 | 260 | Contin |

Constraints

| Cell | Name | Cell Value | Formula | Status | Slack |
|---|---|---|---|---|---|
| $C$12 | | 1400 | $C$12<=$D$12 | Binding | 0 |
| $C$13 | | 1500 | $C$13<=$D$13 | Binding | 0 |

**FIGURE 8.10**
Excel screenshot of solution.

We find our solution is $x_1 = 180$, $x_2 = 260$, $P = \$56400$. From the standpoint of sensitivity analysis, Excel is satisfactory in that it provides shadow prices.

**Limitation**: No tableaus are provided making it difficult to find alternate solutions.

However, we can use the revised simplex equations as matrices to create the tableau if desired. From Winston, Chapter 10, we define the following terms as matrices.

Microsoft Excel 16.0 Sensitivity Report
Worksheet: [Book1]Sheet1
Report Created: 1/25/2021 9:28:57 AM

Variable Cells

| Cell | Name | Final Value | Reduced Cost | Objective Coefficient | Allowable Increase | Allowable Decrease |
|---|---|---|---|---|---|---|
| $C$6 | x1 | 180 | 0 | 140 | 20 | 80 |
| $C$7 | x2 | 260 | 0 | 120 | 160 | 15 |

Constraints

| Cell | Name | Final Value | Shadow Price | Constraint R.H. Side | Allowable Increase | Allowable Decrease |
|---|---|---|---|---|---|---|
| $C$12 | | 1400 | 6 | 1400 | 600 | 650 |
| $C$13 | | 1500 | 32 | 1500 | 1300 | 450 |

**FIGURE 8.11**
Excel sensitivity analysis.

***BV*** = matrix of basic variables
***B*** = right hand side column matrix
$A_j$ = column matrix for $x_j$ in the *j*th constraints
***B*** = $m \times m$ matrix
$C_j$ = row matrix of costs.
$C_{BV}$ = original costs of basic variables

Summarizing formulas that we can use,

$$B^{-1}a_j$$

$$C_{BV}B^{-1}a_j - C_j$$

$$B^{-1}b$$

Maximize Z = $140x_1 + 120x_2$
Subject to:

$$2x_1 + 4x_2 \le 1400$$
$$4x_1 + 3x_2 \le 1500$$
$$x1,\ x2, \ge 0$$

Set up the LP

| Z | X1 | X2 | S1 | S2 | Rhs |
|---|---|---|---|---|---|
| 1 | −140 | −120 | 0 | 0 | 0 |
| 0 | 2 | 4 | 1 | 0 | 1400 |
| 0 | 4 | 3 | 0 | 1 | 1500 |

Step 1. Place this matrix in Excel, as shown below in Standard form (Figure 8.12).

| BV | z | x1 | x2 | s1 | s2 | rhs |
|---|---|---|---|---|---|---|
| z | 1 | -140 | -120 | 0 | 0 | 0 |
| s1 | 0 | 2 | 4 | 1 | 0 | 1400 |
| s2 | 0 | 4 | 3 | 0 | 1 | 1500 |

**FIGURE 8.12**
Excel screenshot of matrix.

Step 2. Determine the entering variables as the cost in row 1 that is most negative. In this case, it is *x1* at −140.

Step 3. Determine who leaves by performing the minimum positive ratio test. In this case, we compare 1400/2 = 700 and 1500/4 = 375. We find 375 is smaller so *S2* departs.

Step 4. Write the new BV matrix (Figure 8.13), replacing column for S2 0 0,1,0 with the original column of x1 −140, 2, 4.

| NewBV | | |
|---|---|---|
| 1 | 0 | -140 |
| 0 | 1 | 2 |
| 0 | 0 | 4 |

**FIGURE 8.13**
New BV matrix in Excel.

Step 5. Find the matrix inverse of the new Bbv matrix (Figure 8.14).

| | | |
|---|---|---|
| 1 | 0 | 35 |
| 0 | 1 | -0.5 |
| 0 | 0 | 0.25 |

**FIGURE 8.14**
Matrix inverse in Excel.

Step 6. Multiple the new Bbv and the original B matrix to obtain the updated tableau (Figure 8.15).

| BV | z | x1 | x2 | s1 | s2 | rhs |
|---|---|---|---|---|---|---|
| z | 1 | 0 | -15 | 0 | 35 | 52500 |
| s1 | 0 | 0 | 2.5 | 1 | -0.5 | 650 |
| x1 | 0 | 1 | 0.75 | 0 | 0.25 | 375 |

**FIGURE 8.15**
Updated tableau in Excel.

Step 7. Check for optimality conditions, all *Cj* >= 0. If not, return to Step 2 and repeat Steps 2–7 again.

We are not optimal as the coefficient of *x2* in the *Cj* row is –15.

We repeat Steps 2–7.

Ratio test, BV, and New Inverse (Figure 8.16).

| ratio test | | | | |
|---|---|---|---|---|
| | | 1 | -140 | -120 |
| 260 | Min | 0 | 2 | 4 |
| 500 | | 0 | 4 | 3 |
| | | | | |
| | | 1 | 6 | 32 |
| | | 0 | -0.3 | 0.4 |
| | | 0 | 0.4 | -0.2 |

**FIGURE 8.16**
New inverse matrix in Excel.

We obtain a new tableau that is optimal (Figure 8.17).

| BV | z | x1 | x2 | s1 | s2 | rhs |
|---|---|---|---|---|---|---|
| z | 1 | 0 | -2.84217E-14 | 6 | 32 | 56400 |
| x2 | 0 | 1 | 0 | -0.3 | 0.4 | 180 |
| x1 | 0 | 0 | 1 | 0.4 | -0.2 | 260 |

**FIGURE 8.17**
Optimal matrix in Excel.

We find we are now optimal as all $cj \geq 0$. Our solution is $Z = 56{,}400$ when $x_1 = 260$, $x_2 = 180$.

### 8.3.4 Maple for Linear Programing

Maple provides a suite of powerful, robust routines for solving optimization problems, including linear programs. Using Maple's flexible mathematical programming language, it is simple to conduct thorough sensitivity studies on solutions to optimization problems. Maple has intrinsic built-in features that solve LP problems.

This method evokes the use of the new Optimization package in Maple and requires us to enter the same inputs as in Method I but input them in the LP Solver.

We present the LPSolve command.

- Optimization[LPSolve] – solve a linear program
- Calling Sequence

LPSolve(**obj, constr, bd, opts**)

- Parameters\

obj – **algebraic**; linear objective function

constr – (optional) **set(relation)** or **list(relation)**; linear constraints

bd – (optional) sequence of **name = range**; bounds for one or more variables

opts – (optional) equation(s) of the form **option = value** where **option** is one of **assume, binaryvariables, depthlimit, feasibilitytolerance, infinitebound, initialpoint, integertolerance, integervariables, iterationlimit, maximize, nodelimit** or **output**; specify options for the **LPSolve** command

```
> with(Optimization) : with(simplex);
[basis, convexhull, cterm, define_zero, display, dual, feasible, maximize, minimize, pivot,
    pivoteqn, pivotvar, ratio, setup, standardize]
> objfunc := 140·x1 + 120·x2;
                                objfunc := 140 x1 + 120 x2
> constr := {2 x1 + 4 x2 ≤ 1400, 4 x1 + 3 x2 ≤ 1500};
                        constr := {2 x1 + 4 x2 ≤ 1400, 4 x1 + 3 x2 ≤ 1500}
> LPSolve(objfunc, constr, assume = nonnegative, maximize);
                  [56400., [x1 = 180.000000000000, x2 = 260.000000000000]]
>
```

Maximize Z = $140x_1 + 120x_2$

Subject to:

$$2x_1 + 4x_2 \leq 1400$$
$$4x_1 + 3x_2 \leq 1500$$
$$x1,\ x2, \geq 0$$

## 8.4 Transportation and Assignment Problems

Transportation and assignment problems are special cases of generic LP. The typical transportation problem requires determining how to move products or goods across a network of routes (roads, highways, tail, etc.) connecting cities or destinations. These cities or destinations are referred to as either a source, origin of the product, or a sink, final destination of the product. The sources have a given supply of the product, and the sources have a specified demand for the product. In addition, the network contains some level of costs to move products between connected locations.

### 8.4.1 Transportation Algorithms

**Example 8.7:** Transportation Problem

Three suppliers A, B, and C each produce an item that has to be delivered to companies W, X, Y, and Z. The stock held at each supplier and the demand from each company is known in advance. The cost, in dollars, of transporting one load of the item from the supplier to the company is also known. This information needed to model this situation is provided in the Table 8.10.

As seen in this Cost/Matrix the amount of supply equals the amount of demand. If this is not the case, where supply does not equal demand, we simply introduce a dummy variable to absorb the excess supply with transportations cost all equal to zero. We also point out that since all the values in the table are integers, the solution to all unknowns also will be integers (see Theorem 7.3.1 from Albright and Fox).

Our solution goal is to minimize total cost while meeting supply and demand goals. We saw in Chapter 6 how to formulate a linear program.

**TABLE 8.10**

Cost/Matrix Table for Supply/Demand

| | Company W | Company X | Company Y | Company Z | Stock Available (loads) |
|---|---|---|---|---|---|
| Supplier A | 180 | 110 | 130 | 290 | 14 |
| Supplier B | 190 | 250 | 150 | 280 | 16 |
| Supplier C | 240 | 270 | 190 | 120 | 20 |
| Demand (loads) | 11 | 15 | 14 | 10 | 50 |

Variables:

$x_{ij}$ = *from suppler i to company demand j for i* = 1, 2, 3 *and j* = 1, 2, 3, 4.

The model formulation yields
Minimize

$$C = 180\ x_{11} + 110\ x_{12} + 130\ x_{13} + 290\ x_{14} + 190\ x_{21} + 250\ x_{22} + 150\ x_{23} + 280\ x_{24} + 240\ x_{31} + 270\ x_{32} + 190\ x_{33} + 120\ x_{3}$$

Subject to

$$x_{11} + x_{12} + x_{13} + x_{14} = 14$$
$$x_{21} + x_{22} + x_{23} + x_{24} = 16$$
$$x_{31} + x_{32} + x_{33} + x_{34} = 20$$
$$x_{11} + x_{21} + x_{31} + x_{41} = 11$$
$$x_{12} + x_{22} + x_{32} + x_{42} = 15$$
$$x_{13} + x_{23} + x_{33} + x_{43} = 14$$
$$x_{14} + x_{24} + x_{34} + x_{44} = 10$$
$$x_{ij} \geq 0 \text{ for all } i = 1, 2, 3 \text{ and } j = 1, 2, 3, 4$$

Solution:

Because of the theorem that we referenced earlier, we may solve this as a LP using any of other technologies (that we have previously illustrated) and we obtain our integer solution. We obtain a total cost of $7,560 when $x_{12} = 4$, $x_{21} = 11$, $x_{23} = 5$, $x_{32} = 1$, $x_{32} = 9$, $x_{34} = 10$. This allows supply to equal demand. We used Excel to obtain our results.

**Example 8.8:** We will now modify our previous problem just slightly so that supply does not equal demand. We will allocate more supply that then demand needed. We display the information in Table 8.11.

**TABLE 8.11**
Example 2 Cost/Matrix

| | Company W | Company X | Company Y | Company Z | Stock Available (loads) |
|---|---|---|---|---|---|
| Supplier A | 180 | 110 | 130 | 290 | 21 |
| Supplier B | 190 | 250 | 150 | 280 | 19 |
| Supplier C | 240 | 270 | 190 | 120 | 22 |
| Demand (loads) | 13 | 10 | 12 | 20 | 55/62 |

| Supplier/ Company | W | X | Y | Z | |
|---|---|---|---|---|---|
| A | 0 | 14 | 0 | 0 | 14 |
| B | 11 | 0 | 5 | 0 | 16 |
| C | 0 | 1 | 9 | 10 | 20 |
| | 11 | 15 | 14 | 10 | |
| TOTAL COST | 7560 | | | | |

| Supplier/ Company | W | X | Y | Z | |
|---|---|---|---|---|---|
| A | 0 | 0 | 4 | 10 | 14 |
| B | 0 | 15 | 1 | 0 | 16 |
| C | 11 | 0 | 9 | 0 | 20 |
| | 11 | 15 | 14 | 10 | |
| TOTAL COST | 11670 | | | | |

Our formulation is similarly done as in Example 8.7.
Minimize

$$C = 180\ x_{11} + 110\ x_{12} + 130\ x_{13} + 290\ x_{14} + 190\ x_{21} + 250\ x_{22} + 150\ x_{23} + 280\ x_{24} + 240\ x_{31} + 270\ x_{32} + 190\ x_{33} + 120\ x_{34} + 0\ x_{15} + 0x_{25} + 0x_{35}.$$

Subject to

$$x_{11} + x_{12} + x_{13} + x_{14} = 21$$
$$x_{21} + x_{22} + x_{23} + x_{24} = 19$$
$$x_{31} + x_{32} + x_{33} + x_{34} = 22$$
$$x_{11} + x_{21} + x_{31} + x_{41} = 13$$
$$x_{12} + x_{22} + x_{32} + x_{42} = 10$$
$$x_{13} + x_{23} + x_{33} + x_{43} = 12$$
$$x_{14} + x_{24} + x_{34} + x_{44} = 20$$

$x_{ij} > 0$ for all $i = 1, 2, 3$ and $j = 1, 2, 3, 4, 5$ *(dummy variable)*

Solution:

Our solution methodology requires the use of a dummy variable that has no transportation cost but has demand of the excess supply (in this case, 7 items). Again, since all our inputs are integers, the Simplex procedure will solve it and yields integer solutions.

| Supplier/Company | | | | | | |
|---|---|---|---|---|---|---|
| | W | X | Y | Z | dummy | |
| A | 0 | 10 | 11 | 0 | 0 | 21 |
| B | 13 | 0 | 1 | 0 | 5 | 19 |
| C | 0 | 0 | 0 | 20 | 2 | 22 |
| | 13 | 10 | 12 | 20 | 7 | |
| TOTAL COST | | | | | | |
| 7550 | | | | | | |

**FIGURE 8.18**
Excel solution to Example 8.8

Because of the same theorem that we referenced earlier, we may solve this as a LP using any of other technologies (that we have previously illustrated), and we obtain our integer solution. We obtain a total cost of $7,550 (Figure 8.18) when $x_{12} = 10, x_{13} = 11, x_{21} = 13, x_{23} = 1, x_{34} = 20$ and our dummy variables *are* $x_{25} = 5, x_{35} = 2$.

### 8.4.2 Assignment Algorithms

Typically, we have a group of $n$ "applicants" applying for $n$ "jobs," and the non-negative cost $C_{ij}$ of assigning the $i$th applicant to $j$th job is known. The objective is to assign one job to each applicant in such a way as to achieve the minimum possible total cost. Define binary variables $X_{ij}$ with value of either 0 or 1. When $X_{ij} = 1$, it indicates that we should assign applicant $i$ to job $j$. Otherwise ($X_{ij} = 0$), we should not assign applicant $i$ to job $j$.

A special case of the transportation problem is the assignment problem, which occurs when each supply is 1 and each demand is 1. In this case, the integrality implies that every supplier will be assigned one destination and every destination will have one supplier. The costs give the basis for assigning a supplier and destination to each other.

Suppose we want to impose the condition that either person $i$ should not perform job $j$ or person $k$ should not perform job $m$. That is, $X_{ij}.X_{km} = 0$.

This nonlinear condition is equivalent to the linear constraint $X_{ij} + X_{km} = 1$. This constraint should be added to the set of constraints as a side constraint. With this additional constraint, the AP becomes a binary ILP, which could be solved by many software packages such as EXCEL, LINDO, or QSB.

In Example 8.9, the goal is to assign people to particular tasks while minimizing total cost. The objective function takes into account the cost involved for each person to do a particular task. The constraints say that each person must be assigned to a task, and each task must be given to a person.

**Example 8.9:** Assignment Problem

A major company has training centers in five cities (let's call them {A, B, C, D, E}). As the manager, you must decide how to assign training teams from five offices to the training centers. (We will call the offices {I1, I2, I3, I4, I5}).

**TABLE 8.12**

Distances between Training Centers

| | A | B | C | D | E |
|---|---|---|---|---|---|
| I1 | 14 | 7 | 3 | 7 | 27 |
| I2 | 20 | 7 | 12 | 6 | 30 |
| I3 | 10 | 3 | 4 | 5 | 21 |
| I4 | 8 | 12 | 7 | 12 | 21 |
| I5 | 13 | 25 | 24 | 26 | 8 |

What we know is the distance between each in hundreds of miles. We put these in Table 8.12, as we did with transportation problems.

Let $x_{ij}$ be the assignment of training team $i$ to training center $j$. To save costs, we desire to minimize miles. We define $c_{ij}$ as the cost (usually given in dollars or miles).

The above definition can be developed into mathematical model as follows:

Determine $x_{ij} > 0$ $(i, j = 1, 2, 3 \ldots n)$ in order to
Minimize $\sum_{j=1}^{n}\sum_{i=1}^{n} c_{ij}x_{ij}$
Subject to the constraint

$\sum_{i=1}^{n} x_{ij}$ for j = 1, 2,…,n
$\sum_{j=1}^{n} x_{ij}$ for i = 1, 2,…,n
and $x_{ij}$ is either zero or one.
Minimize

$$C = 14x_{11} + 7x_{12} + 3x_{13} + 7x_{14} + 27x_{15} + 20x_{21} + 7x_{22} + 12x_{23} + 6x_{24} + 30x_{25} + 10x_{31} + 3x_{32} + 4x_{33} + 5x_{34} + 21x_{35} + 8x_{41} + 12x_{42} + 7x_{43} + 12x_{44} + 21x_{45} + 13x_{51} + 25x_{52} + 24x_{53} + 26x_{54} + 8x_{25}$$

Subject to

$$x_{11} + x_{12} + x_{13} + x_{14} + x_{15} = 1$$
$$x_{21} + x_{22} + x_{23} + x_{24} + x_{25} = 1$$
$$x_{31} + x_{32} + x_{33} + x_{34} + x_{35} = 1$$
$$x_{41} + x_{42} + x_{43} + x_{44} + x_{45} = 1$$
$$x_{51} + x_{52} + x_{53} + x_{54} + x_{55} = 1$$
$$x_{11} + x_{21} + x_{31} + x_{41} + x_{51} = 1$$
$$x_{12} + x_{22} + x_{32} + x_{42} + x_{52} = 1$$
$$x_{13} + x_{23} + x_{33} + x_{43} + x_{53} = 1$$
$$x_{14} + x_{24} + x_{34} + x_{44} + x_{54} = 1$$
$$x_{15} + x_{25} + x_{35} + x_{45} + x_{55} = 1$$

$x_{ij} > 0$ *for all* $i = 1, 2, 3, 4, 5$ *and* $j = 1, 2, 3, 4, 5$
*and binary* $\{0, 1\}$

We will state there are many methods to use such as the Hungarian method or Integer programming. Since our table is all integers, we may use a similar technique as we did in solving the transportation problems.

Assignment Problems

| | A | B | C | D | E | |
|---|---|---|---|---|---|---|
| I1 | 14 | 7 | 3 | 7 | 27 | 1 |
| I2 | 20 | 7 | 12 | 6 | 30 | 1 |
| I3 | 10 | 3 | 4 | 5 | 21 | 1 |
| I4 | 8 | 12 | 7 | 12 | 21 | 1 |
| I5 | 13 | 25 | 24 | 26 | 8 | 1 |
| | 1 | 1 | 1 | 1 | 1 | |

| | A | B | C | D | E | | Must = 1 |
|---|---|---|---|---|---|---|---|
| I1 | 0 | 0 | 1 | 0 | 0 | 1 | 1 |
| I2 | 0 | 0 | 0 | 1 | 0 | 1 | 1 |
| I3 | 0 | 1 | 0 | 0 | 0 | 1 | 1 |
| I4 | 1 | 0 | 0 | 0 | 0 | 1 | 1 |
| I5 | 0 | 0 | 0 | 0 | 1 | 1 | 1 |
| | 1 | 1 | 1 | 1 | 1 | | |
| Must =1 | 1 | 1 | 1 | 1 | 1 | | |
| | | | Obj Func | | | | |
| | | Min | 28 | | | | |

Our solution is that the minimum miles is 28 (in hundreds) when I4 is assigned to A, I3 assigned to B, I1 is assigned to C, I2 is assigned to D, and I5 is assigned to E.

## 8.5 Case Studies in Linear Programming

**Example 8.10:** Supply Chain Operations (Fox & Garcia, 2013)

In our case study, we present LP for supply chain design. We consider producing a new mixture of gasoline. We desire to minimize the total cost of manufacturing and distributing the new mixture. There is a supply chain involved with a product that must be modeled. The product is made up of components that are produced separately (Table 8.13).

Demand information is as follows (Table 8.14):

Let $i$ = crude type 1, 2, 3 (X10, X20, X30, respectively)

Let $j$ = gasoline type 1,2,3 (Premium, Super, Regular, respectively)

**TABLE 8.13**

Crude Oil Availability

| Crude Oil Type | Compound A (%) | Compound B (%) | Compound C (%) | Cost/ Barrel | Barrel Avail (000 of barrels) |
|---|---|---|---|---|---|
| X10 | 35 | 25 | 35 | $26 | 15,000 |
| X20 | 50 | 30 | 15 | $32 | 32,000 |
| X30 | 60 | 20 | 15 | $55 | 24,000 |

**TABLE 8.14**

Oil Demand

| Gasoline | Compound A (%) | Compound B (%) | Compound C (%) | Expected Demand (000 of barrels) |
|---|---|---|---|---|
| Premium | $\geq 55$ | $\leq 23$ | | 14,000 |
| Super | | $\geq 25$ | $\leq 35$ | 22,000 |
| Regular | $\geq 40$ | | $\leq 25$ | 25,000 |

We define the following decision variables:

$Gij$ = amount of crude i used to produce gasoline j

For example,

$G_{11}$ = *amount of crude X10 used to produce Premium gasoline.*
$G_{12}$ = *amount of crude type X20 used to produce Premium gasoline*
$G_{13}$ = *amount of crude type X30 used to produce Premium gasoline*
$G_{12}$ = *amount of crude type X10 used to produce Super gasoline*
$G_{22}$ = *amount of crude type X20 used to produce Super gasoline*
$G_{32}$ = *amount of crude type X30 used to produce Super gasoline*
$G_{13}$ = *amount of crude type X10 used to produce Regular gasoline*
$G_{23}$ = *amount of crude type X20 used to produce Regular gasoline*
$G_{33}$ = *amount of crude type X30 used to produce Regular gasoline*

LP formulation

Minimize Cost = \$86 (G11 + G21 + G31) + \$92(G12 + G22 + G32) + \$95(G13 + G23 + G33)

Subject to:

Demand

$G11 + G21 + G31 > 14000$ (Premium)

$G12 + G22 + G32 > 22000$ (Super)

$G13 + G23 + G33 > 25000$ (Regular)

Availability of products

$G11 + G12 + G13 < 15000$ (crude 1)

$G21 + G22 + G23 < 32000$ (crude 2)

$G31 + G32 + G33 < 24000$ (crude 3)

Product mix in mixture format

$(0.35\ G11 + 0.50\ G21 + 0.60\ G31)/(G11 + G21 + G31) > 0.55$ (X10 in Premium)

$(0.25\ G11 + 0.30\ G21 + 0.20\ G31)/(G11 + G21 + G31) < 0.23$ (X20 in Premium)

$(0.35\ G13 + 0.15\ G23 + 0.15\ G33)/(G13 + G23 + G33) > 0.25$ (X20 in Regular)

$(0.35\ G13 + 0.15\ G23 + 0.15\ G33)/(G13 + G23 + G33) < 0.35$ (X30 in Regular)

$(0.35\ G12 + 0.50\ G22 + 0.60\ G23)/(G12 + G22 + G32) < 0.40$ (Compound X10 in Super)

$(0.35\ G12 + 0.15\ G22 + 0.15\ G32)/\ (G12 + G22 + G32) < 0.25$ (Compound X30 in Super)

The solution was found using LINDO, and we noticed an alternate optimal solution:

Two solutions are found yielding a minimum cost of \$1,904,000 (Table 8.15).

Depending on whether we want to additionally minimize delivery (across different locations) or maximize sharing by having more distribution point involved, we have choices.

**Example 8.11:** Army Recruiting (modified from McGrath, 2007)

Although this is a simple model, it was adopted by the U.S. Army Recruiting Command for Operations. The model determines the optimal mix of prospecting strategies that a recruiter should use in a given week. The two prospecting strategies initially modeled and analyzed are phone and email prospecting. The data came from the Raleigh Recruiting

**TABLE 8.15**

Oil Distribution Solution for Example 8.10

| Decision Variable | Z = \$1,940,000 | Z = \$1,940,000 |
|---|---|---|
| $G_{11}$ | 0 | 1,400 |
| $G_{12}$ | 0 | 3,500 |
| $G_{13}$ | 14,000 | 9,100 |
| $G_{21}$ | 15,000 | 1,100 |
| $G_{22}$ | 7,000 | 20,900 |
| $G_{23}$ | 0 | 0 |
| $G_{31}$ | 0 | 12,500 |
| $G_{32}$ | 25,000 | 7,500 |
| $G_{33}$ | 0 | 4,900 |

Company U.S. Army Recruiting Command in 2006. On average, each phone lead yields 0.041 enlistments, and each email lead yields 0.142 enlistments. The 40 recruiters assigned to the Raleigh recruiting office prospected a combined 19,200 minutes of work per week via phone and email. The company's weekly budget is \$60,000 (Table 8.16).

The decision variables are:

$$x_1 = \text{Number of phone leads}$$
$$x_2 = \text{number of email leads}$$

Maximize $Z = 0.041x_1 + 0.142x_2$

Subject to

$$60x_1 + 1x_2 \leq 19200 (\textit{Prospecting minutes available})$$
$$10x_1 + 37x_2 \leq 60000 (\textit{Budget dollars available})$$
$$x_1, x_2 \geq 0 \ (\textit{non-negativity})$$

If we examine all the intersection points, we find a sub-optimal point, $x_1 =$ 294.29, $x_2 = 154.082$, achieving 231.04 recruitments.

We examine the sensitivity analysis report Figure 8.19,

**TABLE 8.16**

Recruiting Strategies

| | Phone ($x_1$) | Email ($x_2$) |
|---|---|---|
| Prospecting time (Minutes) | 60 minutes per lead | 1 minute per lead |
| Budget (Dollars) | \$10 per lead | \$37 per lead |

**Microsoft Excel 14.0 Sensitivity Report**
**Worksheet: [Book4]Sheet1**
**Report Created: 5/5/2015 2:21:25 PM**

Variable Cells

| Cell | Name | Final Value | Reduced Cost | Objective Coefficient | Allowable Increase | Allowable Decrease |
|---|---|---|---|---|---|---|
| $B$3 | x1 | 294.2986425 | 0 | 0.041 | 8.479 | 0.002621622 |
| $B$4 | x2 | 1542.081448 | 0 | 0.142 | 0.0097 | 0.141316667 |

Constraints

| Cell | Name | Final Value | Shadow Price | Constraint R.H. Side | Allowable Increase | Allowable Decrease |
|---|---|---|---|---|---|---|
| $C$10 | | 19200 | 4.38914E -05 | 19200 | 340579 | 17518.64865 |
| $C$11 | | 60000 | 0.003836652 | 60000 | 648190 | 56763.16667 |
| $C$12 | | 294.2986425 | 0 | 1 | 293.2986425 | 1E+30 |
| $C$13 | | 1542.081448 | 0 | 1 | 1541.081448 | 1E+30 |

**FIGURE 8.19**
Sensitivity report in Excel.

First, we see we maintain a mixed solution over a fairly large range of values for the coefficient of $x_1$ and $x_2$. Further, the shadow prices provide additional information. A one unit increase in prospecting minutes available yields an increase of approximately 0.00004389 in recruits, whereas an increase in budget of $1 yields an additional 0.003836652 recruits. At initial look, it appears as though we might be better off with an additional $1 in resource.

Let's assume that it cost only $0.01 for each additional prospecting minute. Thus, we could get 100∗0.00004389 or a 0.004389 increase in recruits for the same unit cost increase. In this case, we would be better off obtaining the additional prospecting minutes.

## 8.6 Sensitivity Analysis in Maple

We have looked at Maple to solve LP problems and to generate Tableaus for further analysis on problems. In this section, we explore the use of Maple to generate sensitivity analysis. This sensitivity analysis is essential for good decision making. We will present methods to obtain sensitivity analysis on parameters of (1) coefficient of non-basic variables (2) coefficient of basic variables, and (3) changes in the resources (the RHS values). We will examine these one at a time.

We must reduce the analysis to formula form using the following matrix notation for our analysis:

$CBV$: matrix of the original cost coefficients of the basic variables.
$B^{-1}$: the inverse of the matrix of the basic variables in the order in which they enter the basis.
$a_j$: the original column of the resources coefficient of the $j$th variable.
$cj$: the original column of the cost coefficient of the $j$th variable.
$b_j$: the resource limit of the $j$th constraint.

Let's start with the LP problem given by
Maximize $Z = 60d + 30t + 20c$
Subject to:

$$8d + 6t + c \le 48$$
$$4d + 2t + 1.5c \le 20$$
$$2d + 1.5t + 0.5c \le 8$$
$$d, t, c \ge 0$$

We solve this LP to obtain the following the solution and also to create the tableaus.

```
> with(Optimization) : with(LinearAlgebra) : with(simplex) : with(
    linalg) :

> obj := 60·d + 30·t + 20·c;

obj := 60 d + 30 t + 20 c

> constr := {8·d + 6·t + c ≤ 48, 4·d + 2·t + 1.5·c ≤ 20, 2·d
    + 1.5·t + 0.5·c ≤ 8};

constr := {8 d + 6 t + c≤48, 4 d + 2 t + 1.5 c≤20,
    2 d + 1.5 t + 0.5 c≤8}

> LPSolve(obj, {8·d + 6·t + c ≤ 48, 4·d + 2·t + 1.5·c ≤ 20,
    2·d + 1.5·t + 0.5·c ≤ 8}, maximize, assume = nonnegative)

[280., [t = 0., d = 2., c = 7.99999999999999912]]

> m := LPSolve(obj, {8·d + 6·t + c ≤ 48, 4·d + 2·t
    + 1.5·c ≤ 20, 2·d + 1.5·t + 0.5·c ≤ 8}, maximize,
    assume = nonnegative, output = solutionmodule);

m := module () export Results, Settings; end module

> m:-Results( );

["objectivevalue" = 280.,
  "solutionpoint" = ([t = 0., d = 2., c = 7.99999999999999912]),
  "iterations" = 2]
```

```
>
initTab := matrix(4, 8
  , [1,-60,-30,-20, 0, 0, 0, 0, 0, 8, 6, 1, 1, 0, 0, 48, 0, 4, 2, 1.5
  0, 1, 0, 20, 0, 2, 1.5, .5, 0, 0, 1, 8]);
>
```

$$initTab := \begin{bmatrix} 1 & -60 & -30 & -20 & 0 & 0 & 0 & 0 \\ 0 & 8 & 6 & 1 & 1 & 0 & 0 & 48 \\ 0 & 4 & 2 & 1.5 & 0 & 1 & 0 & 20 \\ 0 & 2 & 1.5 & 0.5 & 0 & 0 & 1 & 8 \end{bmatrix}$$

```
finalTabl := matrix(4, 8
  , [1, 0, 5, 0, 0, 10, 10, 280, 0, 0,-2, 0, 1, 2,-8, 24, 0, 0,-2, 1,
  0, 2,-4, 8, 0, 1, 1.25, 0, 0,-.5, 1.5, 2]);
>
```

$$finalTabl := \begin{bmatrix} 1 & 0 & 5 & 0 & 0 & 10 & 10 & 280 \\ 0 & 0 & -2 & 0 & 1 & 2 & -8 & 24 \\ 0 & 0 & -2 & 1 & 0 & 2 & -4 & 8 \\ 0 & 1 & 1.25 & 0 & 0 & -.5 & 1.5 & 2 \end{bmatrix}$$

The solution is $Z = 280$ when $s_1 = 24$, $d = 2$ and $c = 8$. The importance of the initial and final tableau is that they provide information about the variables needed in the sensitivity analysis.

### 8.6.1 Coefficient of Non-basic Variables

We will reduce the analysis to formula form using the following matrix notation. We will use

$$\bar{c}_j = c_{BV} B^{-1} a_j - c_j$$

For the current solution basis to remain the optimal basis, then all non-basic variable coefficients must remain "greater than or equal to" zero in the final tableau (for a maximization problem). Thus, $\bar{c}_j \geq 0$.

### 8.6.2 Change of Non-basic Variables Coefficient

```
> cbv := ⟨⟨0⟩|⟨−20⟩|⟨−60⟩⟩;
```

$cbv := \begin{bmatrix} 0 & -20 & -60 \end{bmatrix}$

```
> cbv1 := convert(cbv, Matrix);
```

$cbv1 := \begin{bmatrix} 0 & -20 & -60 \end{bmatrix}$

```
> NewB := ⟨⟨1, 0, 0⟩|⟨1, 1.5, .5⟩|⟨8, 4, 2⟩⟩;
```

$$NewB := \begin{bmatrix} 1 & 1 & 8 \\ 0 & 1.5 & 4 \\ 0 & 0.5 & 2 \end{bmatrix}$$

```
> NewBinv := MatrixInverse(NewB);
```

NewBinv := [1., 2., −8.], [0., 1.99999999999999956, 3.99999999999999912], [0., −.499999999999999888, 1.49999999999999978]

```
> CbvBinv := MatrixMatrixMultiply(−1·cbv1, NewBinv);
```

$CbvBinv := \begin{bmatrix} 0. & 10. & 10. \end{bmatrix}$

```
> a2 := ⟨⟨6, 2, 1.5⟩⟩;
```

$$a2 := \begin{bmatrix} 6 \\ 2 \\ 1.5 \end{bmatrix}$$

```
> MatrixMatrixMultiply(CbvBinv, a2);
```

$\begin{bmatrix} 35. \end{bmatrix}$

```
> c2 :=−⟨⟨30 + delta⟩⟩;
```

$c2 := \begin{bmatrix} -30-\delta \end{bmatrix}$

```
> solve(35 − 30 − delta ≥ 0, delta);
```

RealRange(−∞, 5)

This means that if the coefficient of table, *t*, that is currently at a value of 30, is less than 35, then tables will never become basic variables.

**Change in a Basic Variable Cost Coefficient**

**First, change the coefficient for *d* and then the coefficient for *c*.**

**Change in *d***

```
> cbv := <<0>|<-20>|<-60 - delta>>;
```

$$cbv := \begin{bmatrix} 0 & -20 & -60-\delta \end{bmatrix}$$

```
> CbvBinv1 := MatrixMatrixMultiply(-1·cbv, NewBinv);
```

$$CbvBinv1 := [0., 10.00000000 - 0.499999999999999888\,\delta, 10.00000000 + 1.49999999999999978\,\delta]$$

```
> a1 := <<8, 4, 2>>;
```

$$a1 := \begin{bmatrix} 8 \\ 4 \\ 2 \end{bmatrix}$$

```
> a3 := <<1, 1.5, .5>>;
```

$$a3 := \begin{bmatrix} 1 \\ 1.5 \\ 0.5 \end{bmatrix}$$

```
> t1 := MatrixMatrixMultiply(CbvBinv1, a1);
```

$$t1 := \begin{bmatrix} 60.00000000 + 1.000000000\,\delta \end{bmatrix}$$

```
> t2 := MatrixMatrixMultiply(CbvBinv1, a2);
```

$$t2 := \begin{bmatrix} 35.00000000 + 1.250000000\,\delta \end{bmatrix}$$

```
> t3 := MatrixMatrixMultiply(CbvBinv1, a3);
```

$$t3 := \begin{bmatrix} 20.00000000 \end{bmatrix}$$

```
> solve({10 - .5·delta ≥ 0, 10 + 1.5·delta ≥ 0, 5 + 1.25·delta ≥
    }, {delta});
```

$$\{-4. \leq \delta, \delta \leq 20.\}$$

Change in coefficient for *t:*

```
> cbv := ⟨⟨0⟩|⟨−20 − delta⟩|⟨−60⟩⟩;
```

$$cbv := \left[\begin{array}{ccc} 0 & -20-\delta & -60 \end{array}\right]$$

```
> CbvBinv1 := MatrixMatrixMultiply(−1·cbv, NewBinv);
```

$CbvBinv1 := [0., 10.00000000 + 1.99999999999999956\ \delta, 10.00000000 - 3.99999999999999912\ \delta]$

```
> t1 := MatrixMatrixMultiply(CbvBinv1, a1);
```

$$t1 := \left[\begin{array}{c} 60.00000000 \end{array}\right]$$

```
> t2 := MatrixMatrixMultiply(CbvBinv1, a2);
```

$$t2 := \left[\begin{array}{c} 35.00000000 - 2.000000000\ \delta \end{array}\right]$$

```
> t3 := MatrixMatrixMultiply(CbvBinv1, a3);
```

$$t3 := \left[\begin{array}{c} 20.00000000 + 1.000000000\ \delta \end{array}\right]$$

```
solve({5−2·delta ≥ 0, 10 + 2·delta ≥ 0, 10−4·delta ≥ 0}, {
  delta});
>
```

$$\left\{-5 \le \delta,\ \delta \le \frac{5}{2}\right\}$$

Let's interpret these results. First, for the coefficient of *d* we find that if the value is between 56 and 80, then *d* will remain a basic variable. If the coefficient of *c* is between 15 and 22.5, then *c* remains a basic variable.

**Change in RHS values**

**Test with a change in *b1*, originally 48.**

```
> b := ⟨⟨48 + delta, 20, 8⟩⟩;
```

$$b := \begin{bmatrix} 48 + \delta \\ 20 \\ 8 \end{bmatrix}$$

```
> MatrixMatrixMultiply(NewBinv, b);
```

$$\begin{bmatrix} 24. + 1.\,\delta \\ 8.00000000 \\ 2.00000000 \end{bmatrix}$$

```
> solve(24 + delta ≥ 0, delta);
```

RealRange(−24, ∞)

```
>
```

This is interpreted as if the resource of constraint 1, currently at 48 units, is at least 24; then, $d$ and $c$ remain as basic variables.

## 8.7 Stochastic Optimization

Stochastic programming is an optimization model that deals with optimizing with uncertainty. For example, imagine a company that provides energy to households. This company is responsible for delivering energy to households based on how much they demand. Typically, this problem could be solved as a simpler linear program with constraints based on demand from households. Though this is convenient, future demand of households is not always known and is likely dependent on factors such as the weather and time of year. Therefore, there is uncertainty, and our basic LP model will not suffice.

Stochastic programming offers a solution to this issue by eliminating uncertainty and characterizing it using probability distributions. Many different types of stochastic problems exist. We present only a few basic examples.

Suppose we have the following optimization problem:

$$\text{Max} \qquad Z = 5x_1 + 10x_2 - 4y_1 - 6y_2$$

Subject to $y_1 + y_2 = 10$
$x_1 + 3x_2 \leq 14$
$x_1, x_2 \geq 0$

This is a simple linear optimization problem with optimal solution: $x_1 = 14$, $y_1 = 10$, $x_2 = y_2 = 0$, $Z = 30$.

So, let's now assume that variables $x_1$ and $x_2$ are not constants but uncertain based upon probability distributions. Let's assume that there are three different scenarios, 1.5X, X, and 0.7X for the values of $x_1$ and $x_2$. Let's assume that each occur with a probability of 1/3. This new problem involves uncertainty and is thus considered a stochastic problem. We must now partition $x_1$ and $x_2$ into $x_{11}$, $x_{12}$, $x_{21}$, $x_{22}$, and $x_{23}$ respectively. Once turned into the discrete version, the problem is reformulated as shown below and can be solved once again using LP.

Max $(1/3)(7/5x_{1,1} + 15x_{2,1}) + (1/3)(5_{x1,2} + 10x_{2,2}) + (1/3)(3.5x_{1,3} + 7.5x_{2,3}) - 4y_1 - 6y_2$

Subject to $y_1 + y_2 = 10$
$1.5x_{1,1} + 4.5x_{2,1} \leq 14$
$x_{1,2} + 3x_{2,2} \leq 14$
$0.7x_{1,3} + 2.1x_{2,3} \leq 14$
*All variables non-negative*

The new optimal solution is:
Solution: Z = 30 when

| Stochastic Programming | | | |
|---|---|---|---|
| | | OBJF | 30 |
| x11 | 9.333333 | | |
| x21 | 0 | | |
| x12 | 14 | | |
| x22 | 0 | constraints | |
| x13 | 20 | 10 | 10 |
| x23 | 0 | 14 | 14 |
| y1 | 10 | 14 | 14 |
| y2 | 0 | 14 | 14 |

In Maple, this would be:

```
obj := 1/3*(7.5*x11 + 15*x21) + 1/3*(5*x12 + 10*x22) + 1/3*(3.5*x13 +
7*x23) – 4*y1 – 6*y2;
                        5    10
obj := 2.500000000 x11 + 5 x21 + – x12 + – x22 + 1.166666667 x13
                        3    3
 7
+ – x23 – 4 y1 - 6 y2
 3
const := {y1 + y2 = 10, 1.5*x11 + 4.5*x21 <= 14, x12 + 3*x22 <= 14,
0.7*x13 + 2.1*x23 <= 14};
   const := {y1 + y2 = 10, 1.5 x11 + 4.5 x21 <= 14,
     x12 + 3 x22 <= 14, 0.7 x13 + 2.1 x23 <= 14}
LPSolve(obj, const, assume = nonnegative, maximize);
[30.0000000066667, [x11 = 9.33333333333333, x12 = 14., x13 = 20.,
x21 = 0., x22 = 0., x23 = 0., y1 = 10., y2 = 0.]]
```

## References

Fox, W. P., & Garcia, F. (2013). Modeling and Linear Programming in Engineering Management. In: Márquez, Fausto Pedro Garcia, & Lev, Benjamin (Eds.). *Engineering Management*. InTech. ISBN 978-953-51-1037-8.

McGrath, G. (2007). Email marketing for the U.S. Army and Special Operations Forces Recruiting. Master's Thesis, Naval Postgraduate School. December 2007.

Winston, W. (2005). *Mathematical Programming*, Cengage, Belmont, CA.

# 9

# *Simulation of Queuing Models*

**OBJECTIVES**

1. Understand the power and limitation to simulations.
2. Understand random numbers.
3. Understand the concept of algorithms.
4. Build simple deterministic and stochastic simulations in Excel.
5. Understand the law of large numbers in simulations.

## 9.1 Introduction

Simulation is the process of modeling an event that is occurring over time. One common practice is to build a computer model of the simulated reality to understand how the events are expected to change over time. Simulation models are an invaluable tool to help decision-makers understand the impacts of their operations on queuing or how look entities, typically individuals, wait for service.

**Example 9.1:** Bank Queue

Most of us have seen a queue (waiting line) in our daily lives: the line at any store, fast-food chains, or bank. We will start our discussion with the simplest queuing model case, one line in front of a single cashier. For now, ignore the role of the second cashier (occasionally open) and the others. We will present later the wide applicability (not just cafes) of the insights we draw from this model.

### 9.1.1 Simulation Inputs

The first step in the simulation process is to identify the necessary input.

Customers arrive at the establishment and either are served directly or join the line. The time between any two customer arrivals is variable and uncertain, but we can say, for example, on average two customers join the

DOI: 10.1201/9781003298762-9

line every minute. The rate at which customers join the line is what we call arrival rate. This is typically represented with the symbol lambda $\lambda$.

In this case, arrival rate lambda $\lambda$ is 2 per minute. This also means that the average time between two arrivals is ½ minute; we call this interarrival time. Let's assume this is exponential.

The time cashier takes to serve one customer is called service time. This, too, changes from customer to customer and is, therefore, variable and uncertain. We will assume that, on average, service time for a customer is 20 seconds = 1/3 minutes. This also means that the one cashier can serve customers at the rate of 1/(1/3) = 3 customers per minute; this is what we call service rate (symbol mu $\mu$). We can assume that this distribution is also exponential.

Since there is only one cashier or teller, number of cashiers/tellers is 1. We use the symbol $m$ for the number of servers. In case the establishment opens another cashier and if a *single line* is used to feed both servers, we will say number of servers $m$ is 2. In case there are two different lines in front of two cashiers, then we will say that these are two different queues, each with $m = 1$. For now, let us go with our original scenario of single queue with single server.

To summarize, at least three inputs are needed to define a queue: arrival rate ($\lambda$), service rate ($\lambda$), and number of servers ($m$). Note that we are interested in expressing our inputs in terms of *rates* (per minute, for example) and not in time (minutes, for example).

Depending on the situation, we may need other inputs describing the extent of variability in arrivals and service. For this basic model, we assume certain type of variability (Poisson distribution for number of arrivals with interarrival times being exponential and Exponential distribution for service times) and not worry about it for now.

### 9.1.2 Simulation Outputs

Simulation modeling can provide the decision-makers with multiple metrics to get a sense of how well their system is operating. Some of the most common metrics include utilization, number of customers waiting at any given time, waiting times, number of customers in the system at any given time, and the percent of time the system is idle waiting for customers. We will review and demonstrate each of these metrics in a little more detail in the following pages.

#### *9.1.2.1 Utilization*

The first output we can get is the utilization of our resource, the cashier/teller. We use the symbol rho, $\rho$, for the utilization. This is equal to the ratio of the rate at which work arrives and the capacity of the server (cashier or teller for a bank). We know that the rate at which work arrives is arrival rate lambda $\lambda$. One server can service the work at the rate of service rate $\mu$. If there are more than one server (that is, if number of servers $m$ is more than 1), then *total* rate at which work can be served, station capacity, will be $m$ multiplied by $\mu$. Therefore, utilization rho $\rho$ can be calculated as lambda $\lambda$ divided by ($m$ multiplied by $\mu$), $\rho = \frac{\lambda}{m\mu}$. Make sure arrival rate and service rate are expressed in same unit of time; for example, both should be per minute or per hour. In the case of our establishment,

$$\rho = 2/3 = 0.6667.$$

Our cashier/teller is busy about 67% of the times. For our calculations to work, *utilization rho $\rho$ must be less than 1.*

#### 9.1.2.2 Number of Customers Waiting

The second output is number of customers waiting. We use the symbol $L_q$ for this. Note that this *excludes* the person who is getting serviced by the cashier. We have a simple table to find this value. To look at the table we need two things: first, arrival rate/service rate, that is lambda/mu $\lambda/\mu$, and second, number of servers, $m$. In our establishment case, $\lambda/\mu$ is 2/3 = 0.6667 and $m$ is 1. You will find one row in the table that corresponds to these two values. In this row, read the number in column titled $L_q$. You will see that the closest number is 1.207. This means that, on an average, the number of people waiting in line $L_q$ is 1.207.

Only in the special case when there is only one server, $m = 1$, we can also use a simple formula to compute the number of customers waiting $L_q$. (The table works for all values of m, including $m = 1$). For $m = 1$: $L_q = \frac{\rho^2}{1-\rho}$. For example, in this case $L_q = \frac{(2/3)^2}{\left(1-\left(\frac{2}{3}\right)\right)} = 1.3333$ which is not the same as we got from table because $\rho = 0.6667$ is not a value in the table.

#### 9.1.2.3 Waiting Times

The third output is the time an average customer waits in line to receive service. We use the symbol $W_q$ for this. We have a simple formula to convert number-of-customers-waiting $L_q$ into time-a-customer-waits $W_q$. To get $W_q$, divide $L_q$ by arrival rate lambda $\lambda$ (Little's law). In our establishment, the time-a-customer-waits $W_q = \frac{L_q}{\lambda} = \frac{1.33333}{2} = 0.6667$ min *or* about 40 secounds

Sometimes we want to think about the whole system, that is, not just waiting but both waiting and getting service. We would like to know the time-a-customer-spends-in-the-system (we use symbol $W_s$ for this), including both time for waiting and time for service. Clearly, this is equal to time-a-customer-waits $W_q$ plus service time. In our example case, $W_s$ is just equal to the sum of waiting time (40 seconds) and service time (20 seconds). $W_s = 60$ seconds = 1 minute.

#### 9.1.2.4 Number-of-Customers-in-System

There is also the question of the number-of-customers-in-system (symbol $L_s$), including both customers who are waiting and who are getting service. Another application of Little's law shows that to get number-of-customers-in-system symbol $L_s$, multiply time-in-system $W_s$ by arrival rate lambda $\lambda$. In our case, $L_s = \lambda\, W_s = 2 * 1 = 2$.

#### 9.1.2.5 Idle Percentage

Finally, to compute the chance that system is idle, that is, there is no customer in the system, we can read the closed column value titled $P_0$ from the table, just the way we read $L_q$. For this model, $P_0 = 0.350$, that is 35% chance that cashier is free (idle). We are often concerned with idle and busy times of the server.

### 9.1.3 Other Performance Measures

For single-server case, some other performance measures can be computed as following:

Probability that there are $n$ customers in system $P_n = (1 - \rho)\rho^n P_n = (1 - \rho)\rho^n$

Probability that the wait is greater than $t$ *is* $\rho e^{-\mu(1-\rho)t}$

Probability that time-in-system is greater than $t$ *is* $e^{-\mu(1-\rho)t}$

For more than one server, spreadsheets are available to compute these measures.

### 9.1.4 Determining Capacity

If we increase capacity (by increasing $m$ or by increasing $\mu$), we expect that the cost of providing that capacity will increase. We also expect, however, that the customers will wait less and that the cost of customer waiting will decrease. This suggests that we should look at the total cost = (cost of providing service + cost of customer waiting) in order to make decision about how much capacity to provide.

For example, if our establishment pays $15 per hour to a cashier, then adding one more cashier/teller increases the cost of providing capacity by $15 per hour. If we add another server, ($m$ goes from 1 to 2) then $\rho = 2/(2 * 3)$

$= 1/3, L_q = 0.16667, W_q = 0.16677/2 = 0.0833, W_s = 5$ *seconds* + *20* seconds = *25 sec* = ¼ minute and $L_s = 0.5$.

But it also reduces the number of customer in system from $L_s = 2$ (from above) to $L_s = 0.5$. If we assume that a customer's time is worth $20 per hour, then system saves (1–0.5) ∗ $20 per hour = $10.00, which is less than the $15 that we pay the second server. Therefore, in this example, from total system cost perspective, we should not add another server.

### 9.1.5 Other Extensions

Without making much fuss about it, we have made two significant assumptions about the pattern of variability in arrivals and service: Poisson distribution for number of arrivals and Exponential distribution for service times. These assumptions mean the following: coefficient of variation = (standard deviation / mean) for interarrival times $C_a = 1$ and coefficient of variation = (standard deviation / mean) for service time $C_s = 1$. But what if based on measurement of real data, they are not 1? We call this the case of general arrivals and service.

It is easy to compute $L_q$ in this more general case as follows:

$L_q$ in Case $Ca \neq 1$ and/or $Cs \neq 1$ is

$$L_q = \frac{C_a^2 + C_s^2}{2}$$

Starting from $L_q$, other performance measures can be computed in the same way as earlier.

### 9.1.6 Summary of Variables and Formulas for Queuing Analysis

Arrival rate lambda $\lambda = 1$ / (interarrival time, that is time between two arrivals)

Service rate mu $\mu = 1$ / service time

Number of servers $m$

Utilization rho, $\rho = \frac{\lambda}{m\mu}$

Assume arrivals Poisson distribution and service time are exponentially distributed.

Average number in waiting line $L_q$ can be obtained from table (given $\lambda/\mu$ and $m$)

In case number of servers $m = 1$, we can also use $L_q = \frac{\rho^2}{1-\rho}$

Average waiting time $W_q = L_q/\lambda$ from *Little's Law*

Average time-in-system (waiting time + service time) $W_s = W_q + (1/\mu)$

Average number-in-system (waiting + getting served) $L_s = \lambda W_s$

Probability that there is nobody in the system $P_0$ is available in table.

For single-server case, $m = 1$, we have following three formulas:

Probability that there are $n$ customers in system $P_n = (1 - \rho)\rho^n P_n = (1 - \rho)\rho^n$

Probability that the wait is greater than $t$ *is* $\rho e^{-\mu(1-\rho)t}$

Probability that time-in-system is greater than $t$ *is* $e^{-\mu(1-\rho)t}$

Determination of capacity is a trade-off between cost of service capacity and cost of customer waiting.

Coefficient of variation of interarrival times $C_a$ = (Standard deviation/mean of interarrival times)

Coefficient of variation of service times $C_s$ = (Standard deviation/mean of service times)

$L_q$ in Case $Ca \neq 1$ and/or $Cs \neq 1$ is (Figure 9.1)

$$L_q = \frac{C_a^2 + C_s^2}{2}$$

We will start by thinking about a queue or waiting line most of us have seen, the line at any store, fast-food chain, or bank. Let us consider the simplest case, one line in front of a single cashier or teller. For now, ignore the role of second cashier (occasionally open) and the others.

We will later discuss the wide applicability (not just cafes) of the insights we draw from this model.

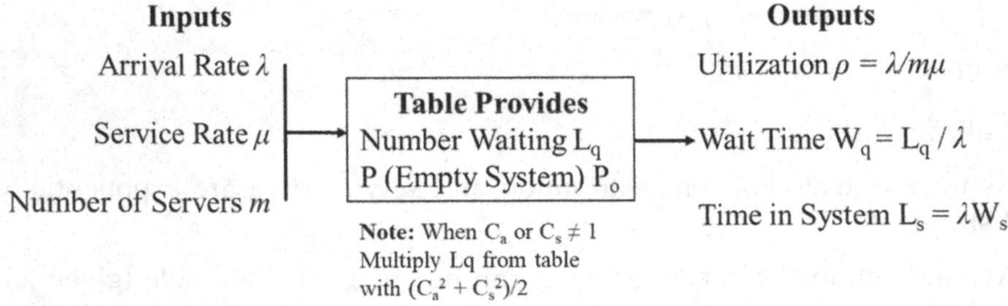


**FIGURE 9.1**
Queuing model logic flow.

Now that we have built our model, we have the ability to examine the impact of changes on the system.

### 9.1.7 Inputs

Let us say, on average, service time for a customer is 15 seconds = ¼ minute. This also means that the one cashier can serve customers at the rate of 1/(1/4) = 4 customers per minute; this is what we call service rate (symbol mu $\mu$). We will assume this distribution is also exponential.

Since there is only one cashier or teller, number of cashiers/tellers is 1. We use symbol $m$ for this. In case cafe opens another cashier and if a *single line* is used to feed both servers, we will say number of servers $m$ is 2. In case there are two different lines in front of two cashiers then we will say that these are two different queues, each with $m$ = 1. For now, let us go with our original scenario of single queue with single server.

To summarize, at least three inputs are needed to define a queue: arrival rate (lambda), service rate (mu) and number of servers ($m$). Note that we are interested in expressing our inputs in terms of *rates* (per minute, for example) and not in time (minutes, for example).

Depending on the situation, we may need other inputs describing the extent of variability in arrivals and service. For this basic model, we assume certain type of variability (Poisson distribution for number of arrivals with inter-arrival times being exponential and Exponential distribution for service times) and not worry about it for now.

### 9.1.8 Outputs

The first output we can get is the utilization of our resource, the cashier/teller. We use the symbol rho $\rho$ for the utilization. This is equal to the ratio of the rate at which work arrives and the capacity of the station. We know that the rate at which work arrives is arrival rate lambda $\lambda$. One cashier can service the work at the rate of service rate = μ. If there are more than one server (that is, if number of servers $m$ is more than 1), then *total* rate at which work can be served, station capacity, will be $m$ multiplied by μ. Therefore, utilization rho $\rho$ can be calculated as lambda λ divided by ($m$ multiplied by $\mu$). $\rho = \frac{\lambda}{m\mu}$. Make sure arrival rate and service rate are expressed in same unit of time; for example, both should be per minute or per hour. In the case of Orin Cafe, ρ = 2/(2 ∗ 4) = 0.5. Our cashier is busy 50% of the times. For our calculations to work, *utilization rho ρmust be less than 1.*

The second output is number of customers waiting. We use the symbol $L_q$ for this. Note that this *excludes* the person who is actually getting serviced by the cashier. We have a simple table to find this value. The table is attached. To look at the table we need two things: first, (arrival rate/service rate, that is λ/μ) and second, number of servers $m$. In Orin Cafe's case λ/μ is

2/4 = 0.5 and m is 1. You will find one row in the table that corresponds to these two values. In this row, read the number in column titled $L_q$. You will see that the number is 0.5. This means that, on an average, the number of people waiting in line $L_q$ is 0.5.

Only in the special case when there is only one server, $m = 1$, can we also use a simple formula to compute the number of customers waiting $L_q$. (The table works for all values of m, including $m = 1$). For $m = 1$: $L_q = \frac{\rho^2}{1-\rho}$. For example, in this case

$L_q = 0.5$, same as that we got from the table.

The third output is the time an average customer waits in line to receive service. We use the symbol $W_q$ for this. We have a simple formula to convert number-of-customers-waiting $L_q$ into time-a-customer-waits $W_q$. To get $W_q$, divide $L_q$ by arrival rate lambda $\lambda$ (Little's law) In Orin Cafe, the time-a-customer-waits $W_q = L_q/\lambda = 0.5/2 = 0.25$ min or 15 seconds.

Sometimes we want to think about the whole system, that is, not just waiting but both waiting and getting service. We would like to know the time-a-customer-spends-in-the-system (we use symbol $W_s$ for this) including both time for waiting and time for service. Clearly, this is equal to time-a-customer-waits $W_q$ plus service time. In Orin Cafe's case, $W_s$ is just equal to the sum of waiting time (15 seconds) and service time (15 seconds). $W_s$ = 30 seconds = 0.5 minute.

There is also the question of the number-of-customers-in-system (symbol $L_s$), including both, customers who are waiting and who are getting service. Another application of Little's law shows that to get number-of-customers-in-system symbol $L_s$, multiply time-in-system $W_s$ by arrival rate lambda $\lambda$. In Orin Cafe's case, $L_s = \lambda\ W_s = 2 * 0.5 = 1$.

Finally, to compute the chance that system is idle, that is, there is no customer in the system, we can read the column titled $P_0$ from the table, just the way we read $L_q$. For this model, $P_0 = 0.5$, that is 50% chance that cashier is free.

### 9.1.9 Determining Capacity

If we increase capacity (by increasing $m$ or by increasing $\mu$), we expect that the cost of providing that capacity will increase. We also expect, however, that the customers will wait less and that the cost of customer waiting will decrease. This suggests that we should look at the total cost = (cost of providing service + cost of customer waiting) in order to make decision about how much capacity to provide.

For example, if Orin Cafe pays $15 per hour to a cashier, then adding one more cashier increases the cost of providing capacity by $15 per hour. But it also reduces the number of customer in system from $L_s = 1$ (see above) to

$L_s = 0.533$ ($L_q = 0.33$ from tables with $m = 2$ and then repeat to get $Ls$). If we assume that a customer's time is worth \$20 per hour then system saves $(1 - 0.533) * \$20$ per hour = \$9.34. Therefore, in this example, from total system cost perspective, we should not add another server.

Starting from $L_q$, other performance measures can be computed in the same way as earlier.

## 9.2 Queueing Model Practice Problems: Solutions

The best way to get a fill and better understanding is to tackle a set of examples. We will cover many of the basic aspects of simulation modeling in a series of examples, with the necessary corresponding technology code.

**Example 9.2:** Ambulance Service

A small town with one hospital has two ambulances to supply ambulance service. Requests for ambulances during non-holiday weekend averages 0.8 per hour and tend to be Poisson distributed. Travel and assistance time averages one hour per call and follows an exponential distribution. What is the utilization of ambulances? On an average, how many requests are waiting for ambulances? How long will a request have to wait for ambulances? What is the probability that both ambulances are sitting idle at a given point in time?

Ambulances are resources or servers servicing the requests that are coming in (customers). Two ambulances mean $m = 2$. Ambulance request arrival rate is $\lambda = 0.8$ per hour. Service time = one hour. Service rate $\mu$ = (1/ service time) = 1 per hour.

$$\text{Utilization } \rho = \frac{\lambda}{m\mu} = \frac{0.8}{2 * 1} = 0.4$$

$$\lambda/\mu = 0.8 \text{ and m} = 2.$$

From table $L_q = 0.152$, $P_0 = 0.429$

On an average, how many requests are waiting for ambulances $L_q = 0.152$

How long will a request have to wait for ambulances $W_q = \frac{L_q}{\lambda} = \frac{0.152}{0.8} = 0.19\ hr$

What is the probability that both ambulances are sitting idle at a given point in time $P_0 = 0.429$

Other Performance measures include $W_s$, $L_s$

Time spent by a call in system $W_s = W_q + service \;\; time = 0.19 + 1 = 1.19 \; hr$

Number of requests in system $L_s = \lambda W_s = 0.8 * 1.19 = 0.952$

**Example 9.3:** Bank Service

At a bank's ATM location with a single machine, customers arrive at the rate of one every other minute. This can be modeled using a Poisson distribution. Each customer spends an average of 90 seconds completing their transactions. Transaction time is exponentially distributed. Determine (1) the average time customers spend from arriving to leaving, (2) the chance that the customer will not have to wait, and (3) the average number waiting to use the machine.

**Solution:**

One every other minute means arrival rate $\lambda = 0.5/\text{min}$

Service rate $\mu = 1/90 \text{ sec} = 0.667/\text{min} \; m = 1$

$$\rho = \frac{\lambda}{m\mu} = \frac{0.5}{1 * 0.667} = 0.75$$

$L_q = \frac{\rho^2}{(1-\rho)} = \frac{0.75^2}{1-0.75} = 2.25$ (we an use the formula because $m = 1$; table should give the same result)

$$W_q = \frac{L_q}{\lambda} = \frac{2.25}{0.5} = 4.5 \text{ min}$$

1. The average time customers spend from arriving to leaving $W_s$ = waiting + service time = $W_q$ + 90 sec = 4.5 minutes + 1.5 minutes = 6 minutes
2. The chance that the customer will not have to wait is the chance that there are 0 customers in the system $P_0 = 1 - \rho = 0.25$ (the formula is for m = 1 case).
3. The average number waiting to use the machine $L_q = 2.25$

**Example 9.4:** Tire Installation

The last two things that are done before a car is completed are engine marriage (station 1) and tire installation (station 2). On average 54 cars per hour arrive at the beginning of these two stations. Three servers are available for engine marriage. Engine marriage requires 3 minutes. The next stage is a single server tire installation. Tire installation requires 1 minute. Arrivals are Poisson and service times are exponentially distributed.

We can analyze these two stations as independent queues. One queue at station 1 followed by another queue at station 2.

1. What is the queue length at each station?
2. How long does a car spend waiting at the final two stations?

**Station 1: Engine Marriage**

$M = 3$

$\lambda = 54$ per hour

$\mu = 20$ per hour

$\lambda/\mu = 54/20 = 2.7,$

(1) From the table we know that $L_q = 7.354$

Waiting time at station 1 ⇒ $W_q = \frac{L_q}{\lambda} = 7.354/54 = 0.1362$ hours.

**Station 2: Tire Assembly**

The arrival rate at station 2 is the rate of departure from station 1. If 54 cars per hour arrive at station 1 then, just for the system to be stable, on an average 54 cars must depart from station 1. Thus, arrival rate at station 2:

$\lambda = 54$ per hour

$\mu = 60$ per hour

$\lambda/\mu = 54/(60) = 0.90$

(1) From table, $L_q = 8.1$.

Waiting time at station 2 ⇒ $W_q = \frac{L_q}{\lambda} = 8.1/54 = 0.15$ hour

b. Total wait time is therefore 0.1362 hour + 0.15 hour = 0.2862 hour

**Example 9.5:** Machine Shop

A machine shop leases grinders for sharpening their machine cutting tools. A decision must be made as to how many grinders to lease. The cost to lease a grinder is $50 per day. The grinding time required by a machine operator to sharpen his cutting tool has an exponential distribution, with an average of 1 minute. The machine operators arrive to sharpen their tools according to a Poisson process at a mean rate of one every 20 seconds. The estimated cost of an operator being away from his machine to the grinder is 10¢ per minute. The machine shop is open 8 hours per day. How many grinders should the machine shop lease?

**Solution:** The data for this problem are as follows:

$\lambda$ = 1 customer per 20 seconds = 3 per minute
$\mu$ = 1 per minute
Lease cost = \$50 per grinder per day = \$6.25 per grinder per hour
Waiting cost = 10¢ per minute – \$6.00 per operator hour
$M$ = the number of leased grinders

The trade-off is between costs of providing service (lease cost) and the cost of customer waiting (waiting cost). Note that customers are our own employees and therefore it is possible to have a good estimate of waiting cost as above. We will focus on minimizing the sum of these two costs so that the trade-off can be resolved optimally. Let us call the sum *total cost per hour.* Let us try three options for leasing: 4 grinders, 5 grinders and 6 grinders.

$M = 4$:
$Lq = 1.528$ from Table, $\lambda/\mu = 3$, $M = 4$)
$L_s = \lambda\, W_s = \lambda(W_q + (1/\mu)) = 1.528 + 3 - 4.528$
Total Cost/Hour = (\$6.25)(4) + (\$6)(4.528) = \$52.17

Note that 4 grinders result in a cost of providing service equal to \$6.25 ∗ 4 per hour. Since, on average, there are 4.528 operators away from their machines, the cost of customer waiting (being away from machine) is \$6 ∗ 4.528 per hour. The operator being services is also away from his machine

$M = 5$:
$Lq = 0.354$ from Table, $\lambda/\mu = 3$, $M = 5$)
$L_s = \lambda\, W_s = \lambda(W_q + (1/\mu)) = 0.354 + 3 = 3.354$
Total Cost/Hour = (\$6.25)(5) + (\$6)(3.354) = \$51.37

M = 6:
$L_q = 0.099$(from Table, $\lambda/\mu = 3$, M = 6),
$L_S = \lambda W_S = \lambda(W_q + (1/\mu)) = 0.099 + 3 = 3.099$,
Total Cost/Hour = (\$6.25)(6) + (\$6)(3.099) = \$56.09.

We should look at renting five (5) grinders since this yields the lowest total cost per hour (\$51.37).

a. Consider a queue with a single server, arrival rate of 5 per hour and service rate of 10 per hour. Assuming Poisson arrivals and exponential service time, what is the waiting time in queue?

b. Actual measurements show that interarrival time standard deviation is 24 minutes and service time standard deviation is 3 minutes. What is the waiting time in queue?

**Solution:**

a. Arrivals follow Poisson distribution. This means that interarrival time standard deviation is equal to interarrival time mean and that coefficient of variation (=standard deviation/mean) of interarrival time $C_a = 1$. Service time follows exponential distribution. This means that service time standard deviation is equal to service time mean and that coefficient of variation (=standard deviation/mean) of service time $C_s = 1$. This means that for part (a) we can use standard table or formulas to compute performance.

$\lambda = 5/hr;\ \mu = 10/hr;\ \lambda/\mu = 0.5;\ \text{m} = 1$
$\rho = \lambda/m\mu = 0.5$
$L_q = \rho^2/(1 - \rho) = 0.25/(1 - 0.5) = 0.5;\ \ W_q = L_q/\lambda = 0.5/5 = 0.1\ hr = 6\ \text{min}$

b. Now that standard deviations are not equal to mean, we have neither $C_a = 1$, nor $C_s = 1$. We need to modify above results to incorporate new values of *Ca* and $C_s$.

Since arrival rate is 5 per hour, average interarrival time is (1/5) hour = 12 minutes.
Standard deviation of interarrival time is 24 minutes.
Coefficient of variation of interarrival time $C_a$ = standard deviation/mean = 24/12 = 2.
Since service rate is 10 per hour, average service time is (1/10) hour = 6 minutes.
Standard deviation of service time is 3 minutes.
Coefficient of variation of service time $C_s$ = standard deviation/mean = 3/6 = 0.5.

$$L_q \text{ in case } C_a \neq 1 \text{ and/or } C_s \neq 1 = (L_q \text{ as computed fromtable})\left(\frac{C_a^2 + C_s^2}{2}\right)$$

$$L_q = 0.5_*\left(\frac{2^2 + 0.5^2}{2}\right) = 1.0625$$

$$W_q = L_q/\lambda = 1.0625/5 = 0.2125\ hr = 12.75\ \text{min}$$

### 9.2.1 Simulation

In this section, we present algorithms and Excel output for the following simulations.

**Example 9.6:** Missile Attack

1. An aircraft missile attack
2. The amount of gas that a series of gas stations will need
3. A simple single barber in a barbershop queue

An analyst plans a missile strike using F-15 aircraft. The F-15 must fly through air-defense sites that hold a maximum of eight missiles. It is vital to ensure success early in the attack. Each aircraft has a probability of 0.5 of destroying the target, assuming it can get to the target through the air-defense systems and then acquire and attack its target. The probability that a single F-15 will acquire a target is approximately 0.9. The target is protected by air-defense equipment with a 0.30 probability of stopping the F-15 from either arriving at or acquiring the target. How many F-15s are needed to have a successful mission assuming we need a 99% success rate?

#### *9.2.1.1 Algorithm: Missiles*

Inputs: $N$ = number of F-15s

$M$ = number of missiles fired

$P$ = probability that one F-15 can destroy the target

$Q$ = probability that air defense can disable an F-15

Output: $S$ = probability of mission success

Step 1. Initialize $S = 0$

Step 2. For $I = 0$ to $M$ do

Step 3. $P(i) = [1 - (1 - P)^{N-I}]$

Step 4. $B(i)$ = binomial distribution for $(m, i, q)$

Step 5. Compute $S = S + P(i) \ast B(i)$

Step 6. Output $S$.

Step 7. Stop

| | | | | | | | | | | |
|---|---|---|---|---|---|---|---|---|---|---|
| | | | | | p | 0.5 | T | 0.9 | P*T | 0 |
| | Initial S | | Bombers | N | q | 0.3 | | | | |
| S | 0 | | | 15 | Quess | | | S > 99 | good | |
| | | | | | | | S_Final | 0.99313666 | | |
| i | B | P | P*B | New S | | | | | | |
| 0 | 0.004747562 | 0.9999 | 0.004747 | 0.004747 | | | | | | |
| 1 | 0.030520038 | 0.9998 | 0.030513 | 0.03526 | | | | | | |
| 2 | 0.091560115 | 0.9996 | 0.091522 | 0.126781 | | | | | | |
| 3 | 0.170040213 | 0.9992 | 0.16991 | 0.296691 | | | | | | |
| 4 | 0.218623131 | 0.9986 | 0.218319 | 0.51501 | | | | | | |
| 5 | 0.206130381 | 0.9975 | 0.205608 | 0.720618 | | | | | | |
| 6 | 0.147235986 | 0.9954 | 0.146558 | 0.867176 | | | | | | |
| 7 | 0.081130033 | 0.9916 | 0.080451 | 0.947627 | | | | | | |
| 8 | 0.034770014 | 0.9848 | 0.034241 | 0.981867 | | | | | | |
| 9 | 0.011590005 | 0.9723 | 0.011269 | 0.993137 | | | | | | |
| 10 | 0.002980287 | 0.9497 | 0.00283 | 0.995967 | | | | | | |
| 11 | 0.000580575 | 0.9085 | 0.000527 | 0.996494 | | | | | | |
| 12 | 8.29393E-05 | 0.8336 | 6.91E-05 | 0.996564 | | | | | | |
| 13 | 8.20279E-06 | 0.6975 | 5.72E-06 | 0.996569 | | | | | | |
| 14 | 5.02212E-07 | 0.45 | 2.26E-07 | 0.996569 | | | | | | |
| 15 | 1.43489E-08 | 0 | 0 | 0.996569 | | | | | | |

**FIGURE 9.2**
Excel Screenshot of missile attack.

We run the simulation letting the number of F-15s vary and calculate the probability of success. We guess $N = 15$ and find that we a probably of success greater than 0.99 when we send 9 planes. Thus, any number greater than 9 works (Figure 9.2).

We find that nine F-15s gives us $P(s) = 0.99313$.

Actually, any number of F-15s greater than nine provides a result with the probability of success we desire. Fifteen F-15s yielding a $P(s) = 0.996569$. Any more would be overkill.

**Example 9.7:** Gasoline-Inventory Simulation

You are a consultant to an owner of a chain of gasoline stations along a freeway. The owner wants to maximize profits and meet consumer demand for gasoline. You decide to look at the following problem.

#### *9.2.1.2 Problem Identification Statement*

Minimize the average daily cost of delivering and storing sufficient gasoline at each station to meet consumer demand.

#### *9.2.1.3 Assumptions*

For an initial model, consider that, in the short run, the average daily cost is a function of demand rate, storage costs, and delivery costs. You also assume that you need a model for the demand rate. You decide that historical date will assist you. This is displayed in Table 9.1 (Table 9.2).

**TABLE 9.1**
Gasoline Demand

| Demand: Number of Gallons | Number of Occurrences (days) |
|---|---|
| 1,000–1,099 | 10 |
| 1,100–1,199 | 20 |
| 1,200–1,299 | 50 |
| 1,300–1,399 | 120 |
| 1,400–1,499 | 200 |
| 1,50–1,599 | 270 |
| 1,600–1,699 | 180 |
| 1,700–1,799 | 80 |
| 1,800–1,899 | 40 |
| 1,900–1,999 | 30 |
| **Total number of days** | **1,000** |

**TABLE 9.2**
Gasoline Demand Probabilities

| Demand: Number of Gallons | Probabilities |
|---|---|
| 1,000 | 0.01 |
| 1,150 | 0.02 |
| 1,250 | 0.05 |
| 1,350 | 0.12 |
| 1,450 | 0.20 |
| 1,550 | 0.27 |
| 1,650 | 0.18 |
| 1,750 | 0.04 |
| 1,850 | 0.03 |
| 2,000 | 1.00 |

#### 9.2.1.4 Model Formulation

We convert the number of days into probabilities by dividing by the total and we use the midpoint of the interval of demand for simplification.

Because cumulative probabilities will be more useful we convert to a cumulative distribution function (CDF) (Table 9.3).

We might use cubic splines to model the function for demand as we discussed in Chapter 2.

#### 9.2.1.5 Algorithm: Inventory

Inputs: $Q$ = delivery quantity in gallons

$T$ = time between deliveries in days

**TABLE 9.3**

CDF of Demand

| Demand: Number of Gallons | Cumulative Probabilities |
|---|---|
| 1,000 | 0.01 |
| 1,150 | 0.03 |
| 1,250 | 0.08 |
| 1,350 | 0.20 |
| 1,450 | 0.40 |
| 1,550 | 0.67 |
| 1,650 | 0.85 |
| 1,750 | 0.93 |
| 1,850 | 0.97 |
| 2,000 | 1.00 |

$D$ = delivery cost in dollars per delivery

$S$ = storage costs in dollars per gallons

$N$ = number of days in the simulation

Output: C = average daily cost

Step 1. Initialize: Inventory → $I = 0$ and C = 0.

Step 2. Begin the next cycle with a delivery:

$$I = I + Q$$

$$C = C + D$$

Step 3. Simulate each day of the cycle.
For $i = 1, 2, \ldots, T$, do Steps 4–6.

Step 4. Generate a demand, $q_i$. Use cubic splines to generate a demand based on a random CDF value, $x_i$.

Step 5. Update the inventory: $I = I - q^i$.

Step 6. Calculate the updated cost: $C = C + s * I$ if the inventory is positive.
If the inventory is ≤ 0, then set $I = 0$ and go to Step 7.

Step 7. Return to Step 2 until the simulation cycle is completed.

Step 8. Compute the average daily cost: $C = C/n$.

Step 9. Output C.
Stop.

We run the simulation and find that the average cost is about $5,753.04, and the inventory on hand is about 199,862.45 gallons.

**Example 9.8:** Queuing Model

A queue is a waiting line. An example would be people in line to purchase a movie ticket or in a drive through line to order fast food. There are two important entities in a queue: customers and servers. There are some important parameters to describe a queue:

1. The number of servers available.
2. Customer arrival rate: average number of customers arriving to be serviced in a time unit.
3. Server rate: average number of customers processed in a time unit.
4. Time.

In many simple queuing simulations, as well as theoretical approaches, assume that arrivals and service times are exponentially distributed with a mean arrival rate of $\lambda_1$ and a mean service time of $\lambda_2$.

Theorem 9.3 If the arrival rate is exponential and the service rate is given by any distribution, then the expected number of customers waiting in line, $L_q$, and the expected waiting time, $W_q$, are given by

$$L_q = \frac{\lambda^2\sigma^2 + \rho^2}{2(1-\rho)} \text{ and } W_q = \frac{L_q}{\lambda}$$

where $\lambda$ is the mean number of arrival per time period; $\mu$ is the mean number of customers serviced per time unit, $\rho = \lambda/\mu$ and $\sigma$ is the standard deviation of the service time.

Here, we have a barber shop where we have two customers arrive every 30 minutes. The service rate of the barber is three customers every 60 minutes. This implies the time between arrivals is 15 minutes, and the mean service time is one customer every 20 minutes. How many customers will be in the queue, and what is their average waiting time?

#### *9.2.1.6 Possible Solution with Simulation*

We provide an algorithm for use.

#### *9.2.1.7 Algorithm*

For each customer 1…$N$

Step 1. Generate an inter-arrival time, an arrival time, start time based on finish time of the previous customer, service time, completion time, amount of time waiting in a line, cumulative wait time, average wait time, number in queue, average queue length.

Step 2. Repeat $N$ times.

Step 3. Output average wait time and queue length.

Stop

You will be asked to calculate the theoretical solution in the exercise set. We illustrate the simulation.

We will use the following to generate exponential random numbers,

$$x = -1/\lambda \ln(1 - \text{rand}())$$

We generate a sample of 5,000 runs and plot customers versus average weight time (Figures 9.3 and 9.4).

| D | E | F | G | H |
|---|---|---|---|---|
| Customer number | Time between arrivals | Arrival time | Start time | Service time |
| 1 | =-(1/$B$1)*LN(1-RAND()) | =E2 | =F2 | =-(1/$B$2)*LN(1-RAND()) |
| =D2+1 | =-(1/$B$1)*LN(1-RAND()) | =F2+F3 | =MAX(I2,F3) | =-(1/$B$2)*LN(1-RAND()) |
| =D3+1 | =-(1/$B$1)*LN(1-RAND()) | =F3+E4 | =MAX(I3,F4) | =-(1/$B$2)*LN(1-RAND()) |
| =D4+1 | =-(1/$B$1)*LN(1-RAND()) | =F4+E5 | =MAX(I4,F5) | =-(1/$B$2)*LN(1-RAND()) |
| =D5+1 | =-(1/$B$1)*LN(1-RAND()) | =F5+E6 | =MAX(I5,F6) | =-(1/$B$2)*LN(1-RAND()) |
| =D6+1 | =-(1/$B$1)*LN(1-RAND()) | =F6+E7 | =MAX(I6,F7) | =-(1/$B$2)*LN(1-RAND()) |
| =D7+1 | =-(1/$B$1)*LN(1-RAND()) | =F7+E8 | =MAX(I7,F8) | =-(1/$B$2)*LN(1-RAND()) |
| =D8+1 | =-(1/$B$1)*LN(1-RAND()) | =F8+E9 | =MAX(I8,F9) | =-(1/$B$2)*LN(1-RAND()) |
| =D9+1 | =-(1/$B$1)*LN(1-RAND()) | =F9+E10 | =MAX(I9,F10) | =-(1/$B$2)*LN(1-RAND()) |

| B | C | D | E | F | G | H | I | J | K | L |
|---|---|---|---|---|---|---|---|---|---|---|
| 2 | | Customer number | Time between arrivals | Arrival time | Start time | Service time | Completion time | wait time | Cumulative wait time | average wait |
| 3 | | 1 | 1.934408754 | 1.934408754 | 1.9344088 | 0.071524668 | 2.005933422 | 0 | 0 | 0 |
| | | 2 | 0.116601281 | 2.051010035 | 2.05101 | 0.714947959 | 2.765957994 | 0 | 0 | 0 |
| | | 3 | 0.055768834 | 2.106778869 | 2.765958 | 0.36811946 | 3.134077454 | 0.659179 | 0.659179125 | 0.219726375 |
| | | 4 | 0.879801355 | 2.986580224 | 3.1340775 | 0.206478939 | 3.340556393 | 0.147497 | 0.806676355 | 0.201669089 |
| | | 5 | 0.095844504 | 3.082424728 | 3.3405564 | 1.055590069 | 4.396146462 | 0.258132 | 1.06480802 | 0.212961604 |
| | | 6 | 1.043432803 | 4.125857531 | 4.3961465 | 0.63308224 | 5.029228702 | 0.270289 | 1.335096951 | 0.222516159 |
| | | 7 | 0.223185659 | 4.34904319 | 5.0292287 | 0.818579146 | 5.847807847 | 0.680186 | 2.015282463 | 0.287897495 |
| | | 8 | 2.251848324 | 6.600891514 | 6.6008915 | 0.393204228 | 6.994095741 | 0 | 2.015282463 | 0.251910308 |
| | | 9 | 0.384299775 | 6.985191288 | 6.9940957 | 0.320344496 | 7.314440237 | 0.008904 | 2.024186916 | 0.224909657 |
| | | 10 | 0.163595249 | 7.148786537 | 7.3144402 | 0.066657268 | 7.381097506 | 0.165654 | 2.189840616 | 0.218984062 |
| | | 11 | 0.000502847 | 7.149289384 | 7.3810975 | 0.792646337 | 8.173743842 | 0.231808 | 2.421648738 | 0.220149885 |
| | | 12 | 0.102456472 | 7.251745856 | 8.1737438 | 1.062486891 | 9.236230734 | 0.921998 | 3.343646724 | 0.278637227 |
| | | 13 | 0.384817067 | 7.636562923 | 9.2362307 | 0.2167925 | 9.453023234 | 1.599668 | 4.943314534 | 0.380254964 |
| | | 14 | 0.625581112 | 8.262144036 | 9.4530232 | 0.342525761 | 9.795548994 | 1.190879 | 6.134193732 | 0.438156695 |
| | | 15 | 0.5489886 | 8.811132636 | 9.795549 | 0.392117518 | 10.18766651 | 0.984416 | 7.118610091 | 0.474574006 |
| | | 16 | 0.540099845 | 9.351232481 | 10.187667 | 0.500628779 | 10.68829529 | 0.836434 | 7.955044122 | 0.497190258 |
| | | 17 | 0.025796165 | 9.377028647 | 10.688295 | 0.051349505 | 10.7396448 | 1.311267 | 9.266310766 | 0.545077104 |
| | | 18 | 0.199860228 | 9.576888875 | 10.739645 | 0.497924558 | 11.23756935 | 1.162756 | 10.42906669 | 0.579392594 |
| | | 19 | 0.422003799 | 9.998892674 | 11.237569 | 0.610221593 | 11.84779095 | 1.238677 | 11.66774337 | 0.614091756 |
| | | 20 | 1.086641979 | 11.08553465 | 11.847791 | 0.139853034 | 11.98764398 | 0.762256 | 12.42999966 | 0.621499983 |
| | | 21 | 0.085067941 | 11.17060259 | 11.987644 | 0.304856673 | 12.29250065 | 0.817041 | 13.24704105 | 0.630811478 |
| | | 22 | 0.688452558 | 11.85905515 | 12.292501 | 0.13728232 | 12.42978297 | 0.433446 | 13.68048655 | 0.621840298 |

| I | J | K | |
|---|---|---|---|
| Completion time | wait time | Cumulative wait time | a |
| =G2+H2 | =G2-F2 | =J2 | =K2/D2 |
| =G3+H3 | =G3-F3 | =K2+J3 | =K3/D3 |
| =G4+H4 | =G4-F4 | =K3+J4 | =K4/D4 |
| =G5+H5 | =G5-F5 | =K4+J5 | =K5/D5 |
| =G6+H6 | =G6-F6 | =K5+J6 | =K6/D6 |
| =G7+H7 | =G7-F7 | =K6+J7 | =K7/D7 |

**FIGURE 9.3**
Excel screenshot of customers average wait time.

We note that the plot appears to be converging at values slightly higher than 0.66. Thus, we will run 100 more trials of 5,000 and compute the average.

We obtain the descriptive statistics from Excel. We note the mean is 0.6601 that is very close to our theoretical mean.

| *Column1* | |
|---|---|
| Mean | 0.660147135 |
| Standard Error | 0.006315375 |
| Median | 0.658168429 |
| Mode | #N/A |
| Standard Deviation | 0.063153753 |
| Sample Variance | 0.003988397 |
| Kurtosis | -0.319393469 |
| Skewness | 0.155656707 |
| Range | 0.318586462 |
| Minimum | 0.500642393 |
| Maximum | 0.819228855 |
| Sum | 66.01471348 |
| Count | 100 |

### 9.2.2 Maple Applied Simulation

**Example 9.9:** Revisiting Missile Attacks (Example 9.6)

We will update the problem we presented in Example 9.6.

Inputs: $N$ = number of F-15s

$M$ = number of missiles fired

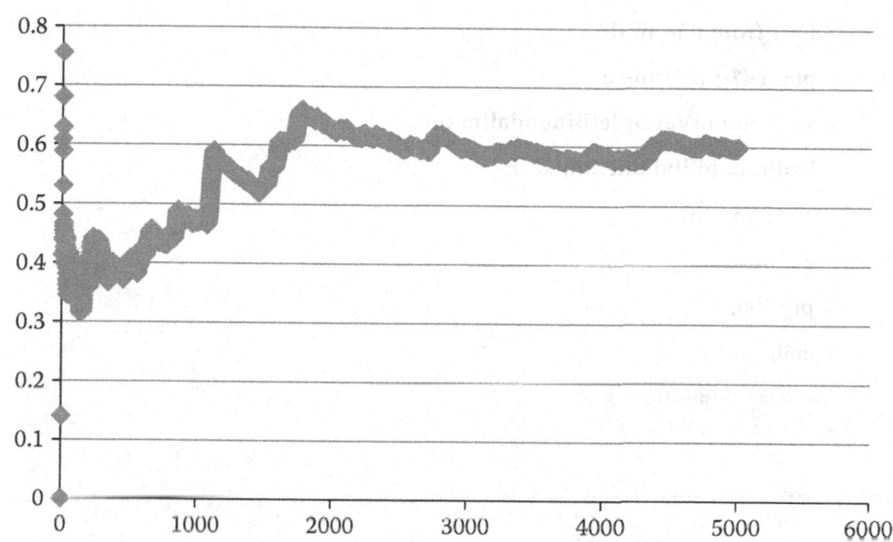


**FIGURE 9.4**
Customers average wait time.

$P$ = probability the one F-15 can destroy the target

$Q$ = probability the air-defense can disable the F-15

Output: $S$ = probability of mission success

Step 1. Initialize $S = 0$.
Step 2. For $I = 0$ to $M$ do.
Step 3. $P(i) = (1\text{-}(1\text{-P})^{\text{N-I}}$
Step 4. $B(i)$ = Binomial Distribution for $(m,i,q)$.
Step 5. Compute $S = S + P(i) \,*\, B(i)$.
Step 6. Output $S$.
Step 7. Stop.

MAPLE Program to run the Missile Attacks Algorithm

```
> restart;
> with(Statistics):

> bombsaway:=proc(n,m,p,q)
> s:=0:qn:=q:nn:=n:pn:=p;
```

```
> for i from 0 to m do
> pn:=1-(1-pn)^(nn-i);
> x:=RandomVariable(Binomial(m,qn));
> b:=ProbabilityFunction(x, i);
> s:=s+pn * b;
> end do;
> print(s);
> end;
```

```
bombsaway := proc(n, m, p, q)
local s, qn, nn, pn, i, x, b;
    s := 0;
    qn := q;
    nn := n;
    pn := p;
    for i from 0 to m do
        pn := 1 - (1 - pn)^(nn - i);
        x := Statistics :- RandomVariable ( Binomial (m, qn));
        b := Statistics :- ProbabilityFunction (x, i);
        s := s + pn×b
    end do;
    print(s)
end proc
```

We run the simulation letting the number of F-15 vary and calculate the probability of success.

```
> for
    i from 1 to 10 do
    > bombsaway(i,8,.45,.4);
    > end do;
```

0.0075582720

0.07419703680

0.3043347039

0.5924805227

0.8254837008

0.9497277101

0.9912246093

0.9992039991

0.9999226475

0.9999574561

We find that the number of F-15s equaling 7 gives us $P(s) = 0.99122$.

Actually, any number of F-15s greater than 7 works to provide a result with the probability of success we desire. We would think the 10 F-15s yielding a $P(s) = 0.999957$ would suffice. Any more would be overkill.

**Example 9.10:** Revisiting Gasoline Inventory Simulation (Example 9.7)

Background: We presented this problem in Example 9.7 and will modify here to demonstrate he program process.

The Inventory Algorithm is designed to help decision-makers adjust for and anticipate peaks in demands and sales.

Inputs: $Q$ = delivery quantity in gallons

$T$ = time between deliveries in days

$D$ = delivery cost in dollars per delivery

$S$ = storage costs in dollars per gallons

$N$ = number of days in the simulation

Output: $LC$ = average daily cost

Step 1. Initialize: Inventory→ $I = 0$, and $C = 0$.

Step 2. Begin the next cycle with a delivery.

$$I = I + Q$$

$$C = C + D$$

Step 3. Simulate each day of the cycle.
For i = 1,2, ....T do Steps 4–6.

Step 4. Generate a demand, $q_i$. Use cubic splines to generate a demand based on a random CDF value $x_i$.

Step 5. Update the inventory: $I = I - q^i$

Step 6. Calculate the updated cost, $C = C + s * I$ if the Inventory is positive. If the inventory is $\leq 0$ then set $I = 0$ go to Step 7.

Step 7. Return to Step 2 until the simulation cycle is completed.

Step 8. Compute the average daily cost: $C = C/n$.

Step 9. Output C.

Step 10 Stop.

**MAPLE** program for running the inventory algorithm simulation

```
> inventorygas:=proc(q,t,d,s,n,xdat,qdat)
> #print(xdat,qdat);
> k:=n:i:=0:c:=0:Flag:=0:nq:=q: nt:=t:nd:=d:ns:=s;
> label_2:i:=i+nq;#print(i,nt);
> c:=c+nd:#print(c);
> if(nt>=k) then
> nt:=k: Flag:=1;
> readlib(spline):
> spline(xdat,qdat,x,cubic);
> nfunc:=unapply(%,x):
> end if;
> for j from 1 to nt do
> readlib(spline):
> spline(xdat,qdat,x,cubic);
> nfunc:=unapply(%,x):
> nx:=evalf(rand()/(1.0 * 10^12));
> newq:=evalf(nfunc(nx));
> i:=i-newq:#print(i);
> if (i<=0) then
> i:=0;
> goto(label_9);
> else
> c:=c+i * ns:#print(c);
> end if;
> label_9;
> k:=k-1: #print(k);
> if k>0 then
> goto(label_2);
> else
> newc:=c/n:print(newc);
```

```
> goto(label_3);
>
>
> end if;
> print(newc);
> end do;
> label_3;
> print(i,newc);
> end;
>
> xdat:=[0,.01,.03,.08,.2,.4,.67,.85,.93,.97,1];
            xdat := [0, 0.01, 0.03, 0.08, 0.2, 0.4, 0.67, 0.85, 0.93, 0.97, 1]
> qdat:=[1000,1050,1150,1250,1350,1450,1550,1650,1750,1850,2000];

qdat := [1000, 1050, 1150, 1250, 1350, 1450, 1550, 1650, 1750, 1850, 2000]
> inventorygas(11500,7,500,.05,20,xdat,qdat);
5753.039330
199862.4518, 5753.039330
```

The average cost is $5,753.04, and the inventory on hand is 199,862.4518 gallons.

### 9.2.3 R Applied Simulation

**Example 9.11:** Revisiting Missile Attacks (Example 9.6) in R

We presented the problem in Example 9.6 and will now use it to demonstrate how to program it in R.

Inputs: $N$ = number of F-15s

$M$ = number of missiles fired, $M$ = 10

$P$ = probability the one F-15 can destroy the target, P==.9 ∗ .5=.45

$Q$ = probability that one air-defense missile can disable the F-15 Q=.6

$X$ = number of F-15s disabled in the attack

$Pi$ = probability of mission success given X=i → $1\text{-}(1\text{-}P)^{N\text{-}i}$

$B$ = binomial probability given

$$X = i\binom{m}{i}q^i(1-q)^{m-i},\ i = 0, 1, 2, \ldots, m$$

$S = \sum P_i B_i$

Output: $S$ = probability of mission success

Step 1. Initialize $S = 0$.
Step 2. For i = 0 to M do.
Step 3. $P(i) = (1\text{-}(1\text{-}P)^{N\text{-}I}$
Step 4. $B(i)$ = Binomial Distribution for $(m,i,q)$.
Step 5. Compute $S = S + P(i) * B(i)$.
Step 6. Output $S$.
Step 7. Stop.

**R Code** for running the xxx Algorithm simulation

```
S = 0
> n=c(10,11,12,13,14,15,16,17,18,19,20)
> for (i in 0:10) {
+ pn= 1-(1-.45)^(n-i)
+ x=rbinom(i,10,.45)
+ xs=dbinom(i,10,.6)
+ s=s+pn * xs
+ print (s)}
[1] 0.0001045920 0.0001047115 0.0001047773 0.0001048134 0.0001048333
[6] 0.0001048442 0.0001048502 0.0001048536 0.0001048554 0.0001048564
[11] 0.0001048569
[1] 0.001670212 0.001673592 0.001675450 0.001676472 0.001677034
[6] 0.001677344 0.001677514 0.001677607 0.001677659 0.001677687
[11] 0.001677703
[1] 0.01219815 0.01224153 0.01226539 0.01227851 0.01228573 0.01228970
[7] 0.01229188 0.01229309 0.01229375 0.01229411 0.01229431
[1] 0.05401894 0.05435326 0.05453714 0.05463827 0.05469390 0.05472449
[7] 0.05474132 0.05475057 0.05475566 0.05475846 0.05476000
[1] 0.1624099 0.1641328 0.1650804 0.1656016 0.1658883 0.1660459
[7] 0.1661326 0.1661803 0.1662066 0.1662210 0.1662289
[1] 0.3529692 0.3592366 0.3626837 0.3645796 0.3656223 0.3661958
[7] 0.3665112 0.3666847 0.3667801 0.3668326 0.3668615
[1] 0.5808401 0.5974358 0.6065634 0.6115836 0.6143447 0.6158633
[7] 0.6166986 0.6171579 0.6174106 0.6175496 0.6176260
[1] 0.7600618 0.7927536 0.8107341 0.8206234 0.8260625 0.8290540
[7] 0.8306993 0.8316042 0.8321019 0.8323757 0.8325262
```

```
[1] 0.8444121 0.8935658 0.9206004 0.9354694 0.9436473 0.9481452
[7] 0.9506190 0.9519796 0.9527280 0.9531396 0.9533659
[1] 0.8625520 0.9216826 0.9542045 0.9720915 0.9819293 0.9873402
[7] 0.9903161 0.9919529 0.9928531 0.9933482 0.9936205
[1] 0.8625520 0.9244036 0.9584220 0.9771321 0.9874226 0.9930825
[7] 0.9961954 0.9979074 0.9988491 0.9993670 0.9996519

> print (s)
[1] 0.8625520 0.9244036 0.9584220 0.9771321 0.9874226 0.9930825
[7] 0.9961954 0.9979074 0.9988491 0.9993670 0.9996519
>
```

$N = 15$ is the first time our probability of success is greater than 99% at 99.30825%.

We find that the number of F-15s equaling 15 gives us $P(s) = 0.9931$.

Actually, any number of F-15s greater than 15 works to provide a result with the probability of success we desire. We would think the 15 F-15s yielding a$P(s) = 0.9931$ would suffice. Any more would be overkill.

## 9.3 Exercises

1. A small town with one hospital has two ambulances to supply ambulance service. Requests for ambulances during non-holiday weekend averages 0.8 per hour and tend to be Poisson distributed. Travel and assistance time averages one hour per call and follows an exponential distribution. What is the utilization of ambulances? On an average, how many requests are waiting for ambulances? How long will a request have to wait for ambulances? What is the probability that both ambulances are sitting idle at a given point in time?
2. At a bank's ATM location with a single machine, customers arrive at the rate of one every other minute. This can be modeled using a Poisson distribution. Each customer spends an average of 90 seconds completing their transactions. Transaction time is exponentially distributed. Determine (1) the average time customers spend from arriving to leaving, (2) the chance that the customer will not have to wait, and (3) the average number waiting to use the machine.
3. The last two things that are done before a car is completed are engine marriage (station 1) and tire installation (station 2). On average, 54 cars per hour arrive at the beginning of these two stations. Three servers are available for engine marriage. Engine

marriage requires 3 minutes. The next stage is a single server tire installation. Tire installation requires 1 minute. Arrivals are Poisson and service times are exponentially distributed.

a. What is the queue length at each station?

b. How long does a car spend waiting at the final two stations?

4. A machine shop leases grinders for sharpening their machine cutting tools. A decision must be made as to how many grinders to lease. The cost to lease a grinder is $50 per day. The grinding time required by a machine operator to sharpen his cutting tool has an exponential distribution, with an average of one minute. The machine operators arrive to sharpen their tools according to a Poisson process at a mean rate of one every 20 seconds. The estimated cost of an operator being away from his machine to the grinder is 10¢ per minute. The machine shop is open 8 hours per day. How many grinders should the machine shop lease?

   a. Consider a queue with a single server, arrival rate of 5 per hour and service rate of 10 per hour. Assuming Poisson arrivals and exponential service time, what is the waiting time in queue?

   b. Actual measurements show that interarrival time standard deviation is 24 minutes and service time standard deviation is 3 minutes. What is the waiting time in queue?

# 10

# *Modeling of Financial Analysis*

**OBJECTIVES**

1. Understand and apply the different multi-attribute decision-making algorithms.
2. Understand the strengths and weakness of each approach.
3. Know DEA, SAW, and TOPSIS methods.
4. Know weighting methods.

## 10.1 Introduction

This chapter discusses mathematical methods and formulas that are used in many businesses and financial organizations. The discrete models in the chapter on discrete dynamical systems (DDS) are used to derive many of these formulas. We will begin our discussions with a quick review of some sequences from DDS as it gives essential understanding to the basic principles that we will develop for these formulas. We also present some advanced modeling in finance using previous algorithms from past chapters.

### 10.1.1 Conjecturing Solutions

Recall the drug dosage problem. Suppose we have the DDS that described the amount of drug in the bloodstream after $n$ hours, where no additional drug is added to the system, $a(n+1) = 0.75\, a(n)$. Let's see if we can **conjecture** (make an educated guess) the analytical solution. We will iterate the system to see if there is a pattern to how we get the answers. Being able to develop an analytical solution for these type problems will help us develop the formulas needed in many financial mathematics applications.

DOI: 10.1201/9781003298762-10

$$a(1) = 0.75a(0)$$
$$a(2) = 0.75a(1)$$
$$a(3) = 0.75a(2)$$
$$a(4) = 0.75a(3)$$

This is not very revealing, but what if we substitute into each difference equation to get them all in terms of $a(0)$:

$$a(1) = 0.75a(0)$$
$$a(2) = 0.75a(1) = 0.75[0.75a(0)]$$
$$a(3) = 0.75a(2) = 0.75[0.75[0.75a(0)]]$$
$$a(4) = 0.75a(3) = 0.75[0.75[0.75[0.75a(0)]]]$$

Now, let's look for a pattern:

$$a(1) = 0.75a(0)$$
$$a(2) = 0.75[0.75a(0)] = (0.75)^2a(0)$$
$$a(3) = 0.75[0.75[0.75a(0)]] = (0.75)^3a(0)$$
$$a(4) = 0.75[0.75[0.75[0.75a(0)]]] = (0.75)^4a(0)$$

We can see that $a(1)$ is just $(0.75)^1\ a(0)$. This results in the following pattern:

$$a(1) = (0.75)^1a(0)$$
$$a(2) = (0.75)^2a(0)$$
$$a(3) = (0.75)^3a(0)$$
$$a(4) = (0.75)^4a(0)$$

We might now conjecture that the solution to the DDS is $a(k) = (0.75)^k a(0)$ for any value $k$.

Suppose we have an initial value $a(0) = 100$. By substitution, the solution is $a(k) = (0.75)^k\ 100$. If we want to know how much drug is in the bloodstream after 10 hours, we just substitute 10 for $k$ and solve the equation

$$a(10) = (0.75)^{10}\ 100 = 5.6314$$

Once we have the solution to the DDS, we can get an answer without having to iterate the system.

Our solution, $a(k) = (0.75)^k a(0)$, is a general solution because it will work for any initial value. If we have $a(0) = c$, the solution is $a(k) = (0.75)^k c$, where $c$ is a constant. This solution will satisfy the DDS $a(n+1) = 0.75\ a(n)$.

To verify our conjecture, we write the solution in terms of $n$ and $n + 1$, and substitute them into the DDS.

$$a(k) = (0.75)^k c$$
$$a(n) = (0.75)^n c$$
$$a(n+1) = (0.75)^{n+1} c$$

Substituting these into $a(n+1) = 0.75\, a(n)$, we have

$$(0.75)^{n+1}c \stackrel{?}{=} 0.75[(0.75)^n c]$$

Since we are trying to establish equality (we do not know if the expressions are equal yet), we use the symbol $\stackrel{?}{=}$.

$$(0.75)^n(0.75)^1 c \stackrel{?}{=} 0.75(0.75)^n c$$
$$(0.75)(0.75)^n c = 0.75(0.75)^n c$$

This verifies that our conjecture is the solution that satisfies the DDS. We call this the **general solution,** because it satisfies the DDS and has an undetermined constant, $c$, which will work for any given initial value.

Once we have an initial value, we can check to see if the solution satisfies the DDS with the initial condition. For the example $a(0) = 100$, we had $a(k) = (0.75)^k\, 100$. To verify that this satisfies the initial condition, we substitute 0 for k and the initial condition ($a(0) = 100$) into our solution and solve

$$a(0) \stackrel{?}{=} (0.75)^0\, 100$$
$$100 \stackrel{?}{=} (0.75)^0\, 100$$
$$100 \stackrel{?}{=} (1)100$$
$$100 = 100$$

We say that a solution that satisfies the DDS, with a given initial value, is a **particular solution**. Notice there are no constants left to be determined.

Suppose that we have a general DDS, $a(n+1) = ra(n)$, where $r$ is a undetermined coefficient. In the example above, $r$ was 0.75. At this point, we can conjecture a solution:

$$a(1) = ra(0) = r^1 a(0)$$
$$a(2) = ra(1) = r[ra(0)] = r^2 a(0)$$
$$a(3) = ra(2) = r[r[ra(0)]] = r^{3a}(0)$$
$$a(4) = ra(3) = r[r[r[ra(0)]]] = r^4 a(0)$$
$$\vdots$$
$$a(k) = r^k a(0)$$

When $a(0) = c$, we have $a(k) = r^k c$, or $a(k) = cr^k$. This is the solution to this DDS of the form $a(n+1) = r\,a(n)$.

### 10.1.2 Developing a Financial Model Formula

Now, consider \$1,000 being deposited in a money market that earns 2.75% per year. Using the DDS modeling paradigm, Future = Present + Change, we have the following:

Let $A(n)$ = amount of money in the money market account after time period $n$, $n$ defined in years.

$$A(n + 1) = A(n) + .0275\,A(n), \quad A(0) = 1000$$

Or

$$A(n + 1) = (1.0275)A(n), \quad A(0) = 1{,}000$$

By the method above, we know the solution is

$$A(k) = (1.0275^k)A(0) = 1{,}000(1.0275^k).$$

After 3 years, we have 1,000 $(1.0275^3)$ = \$1,084.78 in the account.

So, what exactly have we accomplished. We have derived a formula for "compound interest". Let's remind ourselves that we can always iterate a solution. However, at times we only need one value and it might be easier to know and use a formula if one is available.

## 10.2 Simple and Compound Interest

Simple interest can be expressed in terms of three variables:

$P$ = principal (amount borrowed)
$r$ = interest rate (in decimals per time period)
$t$ = time (period over which to repay the principal)

By definition, simple interest is computed by the formula.

$$Prt = principal \times interest\ rate \times time$$

Thus, if money is borrowed at a simple interest, the amount paid back $A$ that must be repaid after $t$ years is:

$$A = P + Prt \text{ or } P(1 + rt)$$

We can now apply the formula to an example (Example 10.1).

**Example 10.1:** A principal of $1,500 is borrowed at 4.5% per year simple interest. Find the amount owed (or the future value) after:

a. two years
b. four months
c. 180 days

**Solution**

a. $P = 1500$; $r = 0.045$; and $t = 2$, so

$$A = 1500[1 + .045(2)] = 1500(1.\ 09) = 1635$$

b. Since 4 months is equivalent to 4/12 or 1/3 of a year$t = 1/3$; $r = 0.045$; and $P = 1500$ so that we obtain from the simple interest formula

$$A = 1500\ [1 + .045(1/3)] = 1500(1.015) = \$1522.50$$

c. Since 180 days is 0.4931 years (180/365), we have $t = 0.4931$. From our simple interest formula,

$$A = 1500\{1 + .045(.4931)] = 1500(1.0222) = \$1533.287$$

If you are paying this, round down to $1,533.28. (If requesting the amount be paid, then round up to $1,533.29.)

### 10.2.1 Compound Interest

A more interesting and more widely accepted method uses compound interest. This is a method whereby the interest previous earned also earns interest.

Compound Interest Formula. If $P$ dollars are invested at an annual interest rate $r$, compounded $k$ times a year, then after $n$ conversion periods (where a conversion period is the number of periods × time) the investment has grown to an amount $A$ given by

$$A = P(1 + 4/k)^n$$

Again, this is a familiar form. It is the general solution to DDS of the form: $A(n+1) = A(n)\ (1+\Delta)$ where $\Delta$ represents change. In previous chapters, we learned that DDS are always solvable using iteration and graphical methods. Now, we are finding analytical solutions to certain forms of DDS.

We can take a look at some examples to reinforce this point.

**Example 10.2:** Find the future value after 2 years of $1,500 invested at 4.5% annual interest compounded.

a. Quarterly
b. Semi-annually
c. Annually
d. Monthly
e. Daily

In each case, our general formula is $A = P(1 + rt)^k$. What we notice is changing in each part of our example is the compounding period. We can see the effect of the compounding period on the money earned using the analytical formula.

a. $k = 4$; $r/k = 0.045/4$; $P = \$1{,}500$; $n = (4)(2)$ = Eight conversion periods $n$ 2 years
   $r/k = (0.045/4) = 0.01125$ or 1.125% each compounding period.
   $A = P(1+r/k)^n = 1{,}500(1+0.01125)^8 = 1{,}500(1.01125)^8 = \$1{,}640.43$
b. $k = 2$; $r/k = 0.045/2 = 0.0225$; $P = \$1{,}500$; $n = (2)(2)$ = Four conversion periods in 2 years
   $A = P(1+r/k)^n = 1{,}500(1+0.0225)^4 = 1{,}500(1.0225)^4 = \$1{,}639.62$
c. $k = 1$; $r/k = 0.045/1 = 0.045$; $P = \$1{,}500$; $n = (2)(1)$ = Two conversion periods in 2 years
   $A = P(1+r/k)^n = 1{,}500(1 + 0.045)^2 = 1{,}500(1.045)^2 = \$1{,}638.03$

**TABLE 10.1**

Compounding Growth of Money

| Compounding Period | Future Money (based on $1,500 deposited) |
|---|---|
| Daily | $1,641.14 |
| Monthly | $1,640.98 |
| Quarterly | $1,640.43 |
| Semi-annually | $1,639.62 |
| Annual | $1,638.03 |

d. $k = 12$; $r/k = 0.045/12 = 0.00375$; $P = \$1{,}500$; $n = (2)(12) = 24$ conversion periods in 2 years.
$A = P(1+r/k)^n = 1{,}500(1 + 0.00375)^{24} = 1{,}500(1.00375)^{24} = \$1{,}640.98$

e. $k = 365$; $r/k = 0.045/365 = 0.0001232$; $P = \$1{,}500$; $n = (2)(365) = 730$ conversion periods in 2 years.

$A = P(1+r/k)^n = 1{,}500(1+0.0001232)^{730} = 1{,}500(1.0001232)^{730} = \$1{,}641.14$ (Table 10.1)

Do you see a trend in the amount earned based on the compounding period?

### 10.2.2 Continuous Compounding

If we imagine a situation where the investment grows while the number of conversions increase indefinitely and the investment grows in proportion to its current value, then we have continuous compounding.

Of the 4,935 institutions Bankrate surveys nationally that sell one-year CDs, 3,639 of them offer daily compounding, 697 offer monthly compounding and 222 offer quarterly compounding. For consumers, more-frequent compounding produces a higher total yield. None offer continuous compounding. Finding continuous compounding might not be worth the effort in finding, but we will introduce the concept nevertheless.

Continuous Compounding Interest Formula

$$A = Pe^{rt}$$

**Example 10.3:** Suppose $10,000 was invested at 5% per year compounded continuously. What is the value after 2 years? Compare the result to daily compounding after 2 years.

**Solution**

By substitution into $A = Pe^{rt}$, where $P = 10{,}000$; $r = 0.05$; and $t = 2$ years, we have

$$A = 10{,}000e^{(.05)(2)} = \$11051.70$$

If we compounded daily (assuming 365 days in year),

$$A = 10{,}000(1 + 0.\ 05/365)^{720} = \$11036.50$$

## 10.3 Rates of Interest, Discounting, and Depreciation

As we saw in the previous section, compound interest is affected by both the annual interest rate and the frequency of the compounding. Often it is difficult to compare options when the rates and compounding periods are different. For example, which is better to invest at, 4% compounded monthly or 4.5% compounded semi-annually? To make these comparisons, it is common to use effective interest rates.

### 10.3.1 Annual Percentage Rate (APR)

Just watch any TV advertisement about buying a car and you will see in fine print the term APR. Let's examine this more closely. The APR is a standard measure to compare interest rates when the compounding periods are different.

If interest is compounded once a year at APR rate, the investment would yield exactly the same return, $P_t$ at the end of $t$ years as it would if interest were compounded $m$ times a year at a nominal rate.

$P_t = P(1 + (r/m))^{mt}$ return after $t$ years on a nominal rate $r$ compounded $m$ times per year.

$P_t = P(1 + APR)^t$ return after $t$ years on APR rate, compounded once a year.

$$\begin{aligned}
&P(1 + (r/m))^{mt} = P(1 + APR)^t \\
&ln(P) + mt\ ln(1 + r/m) = ln(P) + t\ ln(1 + APR) \\
&mt\ ln(1 + r/m) = t\ ln(1 + APR) \\
&m\ ln(1 + r/m) = ln(1 + APR) \\
&ln(1 + r/m)^m = ln(1 + APR) \\
&e^{ln(1+r/m)m} = e^{ln(1+APR)} \\
&(1 + r/m)^m = (1 + APR) \\
&APR = (1 + r/m)^m - 1
\end{aligned}$$

where

$$APR = \textit{effective interest rate}$$
$$r = \textit{annual rate}$$
$$m = \textit{number of conversion period per year}$$

**Example 10.4:** Calculate the APR for a 6.5% nominal interest rate, which is compounded,

a. quarterly
b. monthly

**Solution**

a. $APR = (1 + 0.065/4)^4 - 1 = 6.66\%$
b. $APR = (1 + 0.065/12)^{12} = 6.697\%$

**Example 10.5:** Which is a better option for investment, 2.5% per year compounded monthly or 2.6% simple interest?

**Solution**

We will compute the APR for the first investment. We have $r = 0.025$; $m = 12$

$$APR = (1 + (0.025/12))^{12} - 1 = 2.52\%$$

Since this value is less than 2.6%, the simple interest rate is better.

**Example 10.6:** You are offered two credit plans. One has an interest rate of 9.8% annual interest compounded monthly. The other offers 10% compounded quarterly.

**Solution**

$$APR_1 = (1 + 0.098/12)^{12} - 1 = 10.25\%$$
$$APR_2 = (1 + 0.1/4)^4 - 1 = 10.38\%$$

Since we are paying back the money on this credit plan, we would select the smaller APR, option 1 the 9.8% annual interest compounded monthly.

#### *10.3.1.1 APR for Continuous Compounding*

For the situation, where we have a continuous compounding interest, we will need to modify our model:

$$APR = er - 1$$

**Example 10.7:** Find the APR if money is invested at 7.5% per year with continuous compounding.

**Solution**

$$APR = e^{0.075} - 1 = 0.0778 \text{ or } 7.78\%$$

### 10.3.2 Discounts

There are some situations in which the lender may deduct the interest due in advance. This situation is known as discounting, and the money deducted in advance is called the discount. The money received by the borrower is called the proceeds.

Let's define

$P$ = proceeds (amount received by the borrower)
$d$ = discount rate
$t$ = time in years
$S$ = amount paid back by the borrower

The simple discount is $S \times d \times t$ so that the proceeds are given by $P = S - Sdt$. We should note that this is also the form for Future Value = Present Value + Change.

The Simple Discount Formula is one such discounting model

$$P = S(1 - dt)$$

**Example 10.8:** Simple Discounting

Bill borrows $600 for 2 years at 6% simple discount. In this case,

$$P = S(1 - dt) = 600(1 - 0.06(2)) = \$528$$

Therefore, Paul receives $528 and pays back $600 in two years

**Example 10.9:** Simple Discounting APR

What is the APR on the discounting transaction in Example 10.8?

$$600 - 528 = \$72 \textit{ paid in interest over two years.}$$
$$72 = 528(r\ (2) = 1{,}056\ r$$
$$r = 72/1{,}056 = 0.06818 \textit{ or } 6.18\%\ APR$$

### *10.3.3 Depreciation*

Reducing balance depreciation is the converse of compound interest with larger amounts being subtracted from the original asset value each year. The formula for depreciation is:

$$At = A_0(1 - r)t$$

where
$At$ = value of the asset after $t$ years accounting for depreciation
$A0$ = original value of the asset
$r$ = depreciation rate
$t$ = number of years

**Example 10.10:** Depreciation

We purchase some retooling machinery for $100,000. We know that this type of machinery historically depreciates at 10% per year. How much will the machinery be worth in 10 years?

**Solution**

$$A_{10} = 100{,}000(1 - 0.10)^{10} = \$34{,}867.84$$

## 10.4 Present Value

Another common financial analysis effort is to determine the principal $P$ that must be invested today to obtain a desired future amount.

### 10.4.1 Net Present Value and Internal Rate of Return

**Example 10.11:** Net Present Value

Consider the cash flow of an investment project given in Table 10.2. Is the investment worthwhile?

**TABLE 10.2**
Cash Flow for Example 10.11

| Year | 0 | 1 | 2 | 3 | 4 | 5 |
|---|---|---|---|---|---|---|
| Cash Flow (in $1,000) | –500 | 120 | 130 | 140 | 150 | 160 |

In order to answer this question, the costs and returns must be brought back to the present value.

$$NPV = \textit{present value of cash inflows–present value of cash outflows}$$

Decision Rule:

*If NPV > 0 then the investment project is worthwhile*
*If NPV < 0 then do not invest in the project*

For our example, we are assuming an 8% discounting rate ($r = 0.08$) and use the formula:

$$P_0 = P_t/(1 = r)^t$$

$$\begin{aligned} &Year\ 1{:}\ 120{,}000/(1 + 0.08)^1 = 111{,}111 \\ &Year\ 2{:}\ 130{,}000/(1 + 0.08)^2 = 111{,}454 \\ &Year\ 3{:}\ 140{,}000/(1 + 0.08)^3 = 111{,}137 \\ &Year\ 4{:}\ 150{,}000/(1 + 0.08)^4 = 110{,}254 \\ &Year\ 5{:}\ 160{,}000/(1 + 0.08)^5 = 108{,}893 \end{aligned}$$

Total Present Value is 552,849.

$$NPV = 500{,}000 - 552{,}849 = 52{,}849.$$

Since 52,849 > 0, then we should go with the investment project.

#### *10.4.1.1 Internal Rate of Return*

The internal rate of return (IRR) is the interest rate for which the NPV is zero. A project is viable if the interest rate is less than the IRR but not profitable if the interest rate is larger than the IRR.

**Example 10.12:** Internal Rate of Return

Let's return to our investment project Example 10.11. We want to explore at what interest rate would it no longer be advisable to invest in the project.

We computed the NPV for various interest rates from 8% to 12% (Table 10.3) and graphed the results for better visibility (Figure 10.1).

The IRR is the point where the curve crosses the horizontal axis. This is between 11% and 12%.

We can interpolate this point as follows:

$$IRR = \frac{(R_1 \cdot NPV_2) - (R_2 \cdot NPV_1)}{(NPV_2 - NPV_1)} = \frac{(0.11 \cdot -3456.7 - (0.12 \cdot 9747.67)}{(-3456.7 - 9747.67)} = 11.73\%$$

**TABLE 10.3**
NPV Calculations for Various Interest Rates

| Rate (%) | NPV |
|---|---|
| 8 | 52,849.46 |
| 9 | 37,868.63 |
| 10 | 23,512.43 |
| 11 | 9,747.68 |
| 12 | –3,456.70 |

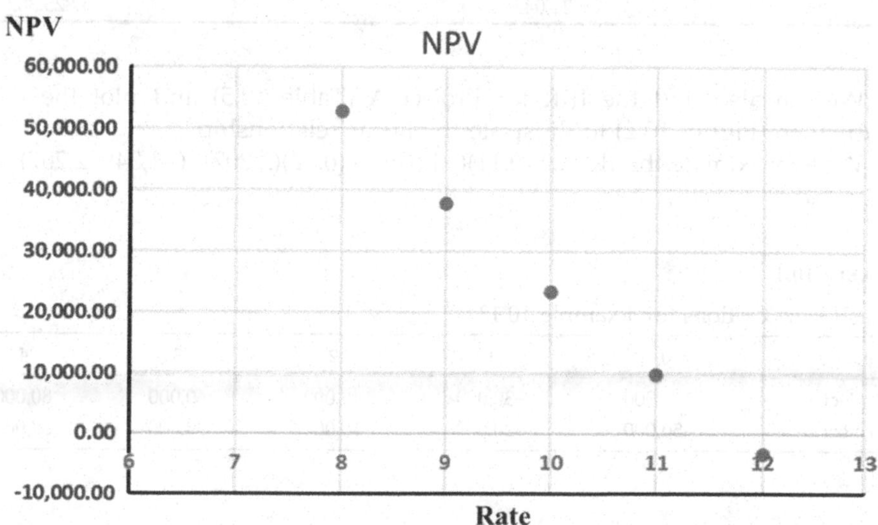


**FIGURE 10.1**
Net present value of interest rate changes for Example 10.12.

The power of determining the NPV is in comparing multiple options that are available to you.

**Example 10.13:** Net Present Value: Multiple Options

Consider for a moment a situation where you have two viable alternatives (Project A and Project B) from which to choose for possible investments. The predicted cash flow for these options are included in Table 10.4.

We will assume that we have a discount rate of 6% and that we want to maximize profits.

*Project A*: *NPV* = −100,000–28,301 + 35,599 + 50,377 + 63,367 = $21,042

*Project B*: *NPV* = −50,000–18,867 + 8,899 + 25,188 + 39,604 = $4,825

Project A is more profitable than Project B.

| Project A | | | Project B | | |
|---|---|---|---|---|---|
| Investment | (1+r)^t | I/(1+r)^t | Investment | (1+r)^t | P/(1+r)^t |
| −100,000 | 1 | −100,000 | −50,000 | 1 | −50,000 |
| −30,000 | 1.06 | −28,301.9 | −20,000 | 1.06 | −18,867.9 |
| 40,000 | 1.1236 | 35,599.86 | 10,000 | 1.1236 | 8,899.964 |
| 60,000 | 1.191016 | 50,377.16 | 30,000 | 1.191016 | 25,188.58 |
| 80,000 | 1.262477 | 63,367.49 | 50,000 | 1.262477 | 39,604.68 |
| | | 21,042.62 | | | 4,825.302 |

We can also find the IRR for Project A (Table 10.5) and plot the information (Figure 10.2) to illustrate the linear relationship.

We approximate the IRR as (0.11)(−1,349) – (0.12)(2,207)/(−1,349–2,207) = 11.62%

**TABLE 10.4**

Cash Flow Options for Example 10.13

| Year | 0 | 1 | 2 | 3 | 4 |
|---|---|---|---|---|---|
| Project A | −100,000 | −30,000 | 40,000 | 60,000 | 80,000 |
| Project B | −50,000 | −20,000 | 10,000 | 30,000 | 50,000 |

**TABLE 10.5**
NPV for Various Interest Rates for Example 10.13

| Rate (%) | NPV |
|---|---|
| 6 | 21,042 |
| 7 | 16,909 |
| 8 | 12,918 |
| 9 | 9,149 |
| 10 | 5,505 |
| 11 | 2,007 |
| 12 | −1,349 |

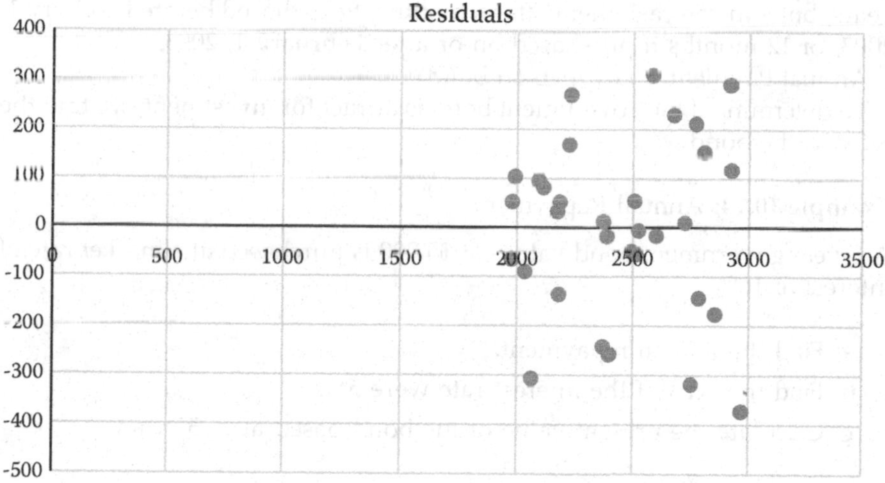


**FIGURE 10.2**
Plot of NPV vs rate.

## 10.5 Bond, Annuities, and Shrinking Funds

### 10.5.1 Government Bonds

A bond is a cash investment made to the government for an agreed number of years. In return, the government pays the investor a fixed sum at the end of each year; in addition, the government repays the original value of the bond with the final payment.

Treasury notes and bonds are securities that pay a fixed rate of interest every 6 months until your security matures, which is when we pay you their par value. The only difference between them is their length until maturity. Treasury notes mature in more than a year, but not more than 10 years from their issue date. Bonds, on the other hand, mature in more than 10 years from their issue date. You usually can buy notes and bonds for a price close to their par value.

Treasury sells two kinds of notes, fixed-principal and inflation-indexed. Both pay interest twice a year, but the principal value of inflation-indexed securities is adjusted to reflect inflation as measured by the Consumer-Priced Index – the Bureau of Labor Statistics' Consumer Price Index for All Urban Consumers (CPI-U). With inflation-indexed notes and bonds, we calculate your semiannual interest payments and maturity payment based on the inflation-adjusted principal value of your security.

Saving bonds are treasury securities that are payable only to the person to whom they are registered. Savings bonds can earn interest for up to 30 years, but you can cash them after 6 months if purchased before February 1, 2003, or 12 months if purchased on or after February 1, 2003.

Annual Payment = $rx$ (price on bond)

To determine if the government bond is attract for investment, we find the NPV of the bond.

**Example 10.14:** Annual Repayment

A 5-year government bond values at $5,000 is purchased at a market rate of interest of 10%

a. Find the annual repayment.
b. Find the NPV if the interest rate were 5%.
c. Calculate the present value of the bond based after 5 years.

**Solution**

a. 0.20 + 5,000 = $1,000 a year
b. NPV = $1{,}000/1.05 + 1{,}000/1.05^2 + 1{,}000/1.05^3 + 1{,}000/1.05^4 + 1{,}000/1.05^5$ = $4,329.47
c. $5{,}000/(1.1)^5$ = $3,104.60 so

Total NPV = 3,104.60 + 4,329.47 = $7,434.07
The investment made $2,434.07 over the 5 years.

### 10.5.2 Annuities

An annuity is a series of regular periodic payments.

Examples of annuities include:

- Interest-bearing saving account
- Mortgage payments
- Insurance premium payments
- Retirement benefits
- Social Security benefits

### *10.5.2.1 Ordinary Annuities*

A series of equal payments, all made at the end of a compounding period. The term is the period from the beginning of the first payment to the end of the last payment.

**Example 10.15:** Ordinary Annuitiy

At the end of each month, $100 is invested into an annuity account paying 3.5% per year compounded monthly. What is the value of this account after the 5th payment?

**Solution**

First payment earns interest for 4 months.
Second payment earns interest for 3 months.
Third payment earns interest for 2 months.
Fourth payment earns interest for 1 month.
Fifth payment earns interest for 0 months.

Using the compound interest formula,

$$\text{Value} = 100(1 + .035/12)^4 + 100(1 + .035/12)^3 + 100(1 + .035/12)^2 + 100(1 + .035/12)^1 + 100(1 + .035/12)^0 = \$503.21$$

To derive a general formula let,

$A_0$ = value of regular payment
$i$ = interest rate per compounding period
$n$ = number of regular payments (number of periods in the terms)
$V_n$ = value of the annuity at the end of $n$ compounding periods

Using DDS, find

$$V_n = A_o((1 + i)^{n-1} + (1 + i)^{n-2} + \ldots + (1 + i) + 1)$$

$$A_0\left[\frac{(1+i)^n-1}{i}\right]$$

This is also called the Uniform Series Amount.

**Example 10.16:** A small company invests $3,000 at the end of 6 months in a fund paying 5% compounded semi-annually. Find the value of the investment after 8 years.

**Solution**

$A_0 = 3000,\ r = .08\ so\ i = .08/2 = .04,\ n = (8)(2) = 16\ periods\ in\ 8\ years$
$V_{18} = 3000[1 + 0.4]^{18} - 1]/0.04 = \$76,936.23$

### 10.5.3 Sinking Funds

A series of regular deposits used to accumulate a sum of money at some future time is called a sinking fund.

**Example 10.17:** Sinking Funds

A small firm anticipates capital expenditures of $120,000 to buy new equipment in 6 years. How much should the firm be depositing in a funding earning 8% per year compounded quarterly in order to buy the new equipment?

**Solution**

$A_0 = ?\ i = .08/4 = .02,\ n = (6)(4) = 24\ periods,\ V_{24} = 120{,}000$
$V_{24} = 120000 = A_o = [1 + .02]^{24} - 1]/0.02$

We solve for $A_0$, $A_0$ = $3,944.53. Thus, the firm must invest $3,499.53 each period.

### 10.5.4 Present Value of an Annuity

The present value of an annuity is the amount of money today that is equivalent to a series of equal payments at some time in the future.

Let's start with an example.

**Example 10.18:** Present Value of an Annuity

Retirement plans are on most Americans minds. Consider a couple who wishes to make a lump sum investment paying 8% per year compounded annually in order to receive payments of \$20,000 per year for 5 years. How much should they invest?

**Solution**

$$A_0 = \$20{,}000\ i = 0.08,\ n = 5$$

The present value of the annuity is

$$20000/(1.08) + 20000/(1.08^2) + 20000/(1.08^3) + 20000/(1.08)^4 + 20000 /(1.08^5) = \$79{,}854$$

To derive a general formula, let's again look at the geometric series:

Let $1 + i = r$ and let the term be $n$ years. Then $A_0[1/r+1/r^2+...+1/r^n]$ represents the present value.

A geometric series $1+r+r^2+...+r^n$ can be represented by its sum that makes the sum of the first $n$ terms $S_n = a_1\ (1-r^n)/(1-r)$.

The sum $[1/(1+i)+1/(1+i)^2+...+1/(1+i)^n]$ can be represented as $((1+i)^n-1)/i$. So,

$$A_0\frac{(1 + i)^n - 1}{i}$$

is the value of the series of payments.

If $V_0$ is sufficient to provide for these payments, then the future value of $V_0$ at the end of the term $n$ is $V_0(1+i)^n$, and this value should be equal to the value of the payments. Therefore,

$$V_0(1 + i)^n = A_0\frac{(1 + i)^n - 1}{i}$$

Solving for $V_0$, we find

$$V_o = A_o\left[\frac{1 - (1 + i)^{-n}}{i}\right]$$

**Example 10.19:** Annuity Payments

How much should you pay for annuity of quarterly payments of $15,000 for 5 years, assuming an interest rate of 5% per year.

**Solution**

$$V_0 = ?\ A_0 = 15{,}000\ n = (5)(4) = 20\ periods$$
$$i = .05/4 = 0.0125$$
$$V_o = 1500\left[\frac{1-(1+0.0125)^{-20}}{0.0125}\right] = \$262989.74$$

**Example 10.20:** A loan of $50,000 is to be repaid in equally monthly payments over 5 years, assuming fixed rate of 4.75% per year. Find the amount of each payment and the total interest paid.

**Solution**

$$V_0 = 50000\quad A_0 = ?\quad n = (5)(12) = 60\ periods$$
$$i = .0475/12 = 0.\ 003958$$

$$V_o = A_o\left[\frac{1-(1+i)^{-n}}{i}\right]$$

$$5000 = A_o\left[\frac{1-(1+0.003958)^{-60}}{0.003958}\right]$$

$$A_0 = \$2605.\ 19$$

The total payments are $2,605.19 (60) = $156,311.82
The interest paid is $156,311.82 – 50,000 = $106,311.82

## 10.6 Mortgages and Amortization

An interest-bearing debt is said to be amortized if the principal and interest are paid by a sequence of equal payments made over a prescribed period. What items do we buy over time? We buy many items such as cars, homes, and boats but also many convenience items using a money card.

Recall the DDS mortgage problem from Chapter 2 where a home was purchased at $80,000 at a monthly interest rate of 1%.

Suppose we have a linear, first order, nonhomogeneous DDS, where the nonhomogeneous part is a constant, $a(n+1) = ra(n) + b$. The following is a model example with a 1% interest rate:

$$a(n + 1) = a(n) + 0.01a(n) - b$$

This DDS is composed of a homogeneous part and a nonhomogeneous part.

$$a(n) = 1.01a(n) - b$$

We already know that the solution to the homogeneous part for $a(n+1) = 1.01\, a(n)$ is $a(k) = c(1.01)^k$. Let's conjecture that for the system above, the solution to the homogeneous part is not directly affected by the solution to the non-homogeneous part. For example, the amount of drug in the bloodstream after $n$ hours, $0.75\, a(n)$, is not affected by the 100 mg added after $n + 1$ hours. So, let's say that the solution to the DDS is a linear combination of the solution of the homogeneous part and the solution for the nonhomogeneous part:

$$a(k) = c(1.01)^k + \text{solution for nonhomogeneous part}$$

Since the nonhomogeneous part is a constant, we conjecture that the solution might be a constant, say $d$.

$$a(k) = c(1.01)^k + d$$

We need to verify the conjectured solution to see if it satisfies the DDS, for some value of $d$.

$$\begin{aligned} a(k) &= c(1.01)^k + d \\ a(n) &= c(1.01)^n + d \\ a(n+1) &= c(1.01)^{n+1} + d \end{aligned}$$

Substituting these into the DDS, we perform the following algebraic manipulation to find $d$:

$$\begin{aligned} &c(1.01)^{n+1} + d = 1.01[c(1.01)^n + d] - \mathrm{b} \\ &c(1.01)^n(1.01)^1 + d = c(1.01)(1.01)^n + 1.01d - \mathrm{b} \\ &c(1.01)^n(1.01) + d = c(1.01)^n(1.01) + 1.01d - \mathrm{b} \\ &d = 1.01d - \mathrm{b} \\ &\quad - 0.01d = -\mathrm{b} \\ &d = b/.01 = 100\mathrm{b} \end{aligned}$$

Therefore, $d$ must equal $100b$ for our conjecture to satisfy the DDS and $a(k) = c(1.01)^k + 100b$ must be the general solution. A linear, 1st order, nonhomogeneous DDS, where the nonhomogeneous term is a constant, may have an equilibrium value.

For our problem, $a(n + 1) = .1.01\ a(n) - b$, we can find the equilibrium value, $ev = 100b$. Thus, the $ev$ is the missing part of the general solution.

$$A(k) = C(1.01^k) + 100b$$

But we know that initially $A(0) = 80{,}000$ and finally after 240 months or payments $A(240) = 0$. We can substitute these into our general solution to obtain two equations and two unknowns.

$$\begin{aligned} 80000 &= C(1.01^0) + 100\,b \\ 0 &= C(1.01^{240}) + 100\,b \\ 80000 &= C + 100\,b \\ 0 &= 10.89255\,C + 100\,b \\ C &= -8086.893\ b = 880.8693 \end{aligned}$$

Therefore, the amount owed is

$$A(k) = -8086.893(1.01)^k + 8800.8693$$

**Example 10.21:** You charge $1,500 on your Discover card that charges 19.99% annual interest compounded monthly on the unpaid balance. You can afford $35 a month. How long until you pay off your bill, assuming you do not charge anything until it is paid off?

$$A(n + 1) = A(n) + (.1999/12)A(n) - 35,\ A(0) = 1500$$

It will take you over 75 months to pay this off. You bought $1,500 worth of merchandise and paid $35 a month for 75 months. You paid (75)(35) = $2, 625 for the items.

Periodic Payment on an Amortized Loan

$$A_o = V_o\left[\frac{i}{1 - (1 + i)^{-n}}\right]$$

You want to invest in a certain stock portfolio. The minimum investment value is $25,000. You currently have $10,000 cash and can set aside $500 per month in other investments until the total reaches $25,000. How long will it take?

An interest-bearing company is said to be amortized if the principle and the interest are paid by a sequence of equal payments over equal time periods. Mortgages and loans (car loans, for example) are applications that use the formula,

$$A_o = V_o\left[\frac{i}{1-(1+i)^{-n}}\right]$$

Recall our mortgage problem from before. If we have a mortgage of $80,000 for 20 years at 12% per year, find the monthly payments.

$A_0 = ?\ V_0 = 80{,}000$

$N = 20(12) = 240\ \textit{periods}\ i = .12/12 = .01$

$80{,}000 = A_0\left[\frac{1-(1+.01)^{-240}}{.01}\right]$

$A_0 = \$880.87$

Remark, this now gives us several methods to find these payments.

**Example 10.22:** Car Purchasing Decision

You go out car shopping, and you budget car payments at most $275 per month. You found several cars that you are interested in buying a Ford Focus, which is now selling for $15,500 at 3.75% over 5 years, and a Toyota Corolla, selling for $16,500 at 4.25% over 6 years. Which of these cars, if any, can you afford to buy? Assume you can put no money down.

**Solution**

Car 1 (Ford Focus)

$A_0 = ?\ V_0 = 15{,}500$

$N = 5(12) = 60\ \textit{periods}\ i = .0.0375/12 = .003125$

$15{,}500 = A_0\left[\frac{1-(1+.003125)^{-60}}{.003125}\right]$

$A_0 = \$283.71$

Car 2 (Toyota Corolla)

$A_0 = ?\ V_0 = 16{,}500$

$N = 6(12) = 72\ \textit{periods}\ i = .0.0425/12 = .003542$

$16{,}500 = A_0\left[\frac{1-(1+.003542)^{-72}}{.003542}\right]$

$A_0 = \$260.03$

Decision. You can afford to buy the Toyota Corolla but not the Ford Focus.

**Example 10.23:** Investment Opportunity Decision

A parent must decide between investment alternatives for their child's college fund. Your child is currently in the 5th grade. You have heard about the South Carolina Tuition Prepayment Program (SCTPP) but you wonder if it is a good deal. It allows you to prepay your child's tuition and guarantee payment to attend any South Carolina public 4-year institution for 4 years. The options are as follows.

a. You can pay a lump sum right now,
b. You can pay 48 fixed monthly payments starting now, or
c. You can pay fixed monthly installments starting now until you child attends college.

Below is a table of in-state tuition and fees for some SC public colleges:

| Public School | 93–94 | 94–95 | 96–97 | 97–98 | 98–99 | 99–00 | 00–01 | 01–02 | 02–03 | 03–04 |
|---|---|---|---|---|---|---|---|---|---|---|
| Citadel | 3,080 | 3,176 | 3275 | 3,297 | 3,631 | 3,396 | 3,404 | 3,727 | 4,067 | 4,999 |
| Clemson | 2,954 | 3,036 | 3,112 | 3,112 | 3,252 | 3,344 | 3,470 | 3,590 | 5,090 | 5,834 |
| Coastal | 2,470 | 2,710 | 2,800 | 2,910 | 3,100 | 3,220 | 3,340 | 3,500 | 3,770 | 4,350 |
| College of Charleston | 2,950 | 3,060 | 3,090 | 3,190 | 3,290 | 3,390 | 3,520 | 3,630 | 3,760 | 4,858 |
| Francis Marion | 2,800 | 2,920 | 3,010 | 3,010 | 3,270 | 3,350 | 3,350 | 3,600 | 3,790 | 4,340 |
| Winthrop | 3,470 | 3,620 | 3,716 | 3,818 | 3,918 | 4,032 | 4,126 | 4,262 | 4,668 | 5,600 |
| Medical Univ. SC | 2,560 | 2,819 | 2,910 | 3,202 | 3,648 | 4,034 | 4,626 | 5,180 | 5,824 | 6,230 |
| USC Beaufort | | | | | | | | | | 4,208 |

http://www.che.sc.gov/Finance/Abstract/Abstract2003.pdf

Situation: Current Lump Sum Payment

| | | | | | | |
|---|---|---|---|---|---|---|
| **10th Grade** | **3** | **2006–2007** | **25,566** | **NA** | **954.00** | **29** |
| **9th Grade** | **4** | **2007–2008** | **25,386** | **NA** | **694.00** | **41** |
| **8th Grade** | **5** | **2008–2009** | **25,207** | **601.00** | **552.00** | **53** |
| **7th Grade** | **6** | **2009–2010** | **25,028** | **597.00** | **464.00** | **65** |
| **6th Grade** | **7** | **2010–2011** | **24,849** | **592.00** | **403.00** | **77** |
| **5th Grade** | **8** | **2011–2012** | **24,671** | **588.00** | **358.00** | **89** |
| **4th Grade** | **9** | **2012–2013** | **24,494** | **584.00** | **325.00** | **101** |

Your child is currently in the 5th grade, and your lump sum payment is \$24,671.

Suppose you have the money but decide to put the money into a money market account currently paying 5.84% annual interest compounded monthly. How much would be in the account when your child enters college. Which colleges, if any of those listed, will you be able to afford for your child?

Monthly payments

Another option is 48 monthly payments of $588 or 89 payments of $358. What is the NPV of each option?

If you could invest that amount in a money market account paying 5.85% interest compounded monthly, how much money would you have available when your child needs to enter college?

## 10.7 Advanced Financial Models

In this section, we address a few advanced financial models. We examine estimating growth rates and accuracy of forecasts, multiple regression to forecast sales, portfolio optimization, and MADM for project selection. Many of these topics use material previous covered.

### 10.7.1 Estimating Growth Rates

Assume we want to model sales, $y$, as a function of years, $x$. We have data shown in Table 10.6.

Using the steps from our regression chapter, we find the trend of the actual data is increasing, and perhaps exponential. The exponential model might be expected from experience in sales growth models.

We start with an approximating model by taking the natural logarithm of both sides of $y = ae^{bx}$. This gives $ln\ (y) = ln\ (a) + bx$. We plot this and see it appears linear, so we build a regression of model and find our model is $ln(y) = 4.154 + 0.417\ x$. Since this is a transformed model, we make each side of the exponent the exponential to transform the model back into the real $x$, $y$ space. This yields $y = 63.68\ e^{0.417x}$. This result is an approximation, and we should use these coefficients ($a = 63.68$, $b = 0.417$) as starting points in any nonlinear numerical solution method. We have previously shown that all our technologies can find our solution.

The exponential model using nonlinear regression techniques from Chapter 5 is

$$sales = 77.33212\ e^{(0.389978\ years)}.$$

The approximate $R^2$ is 0.99, which implies the percent of $y$ explained by our model is 99%. The mean average percent error, $MAPE = 12.63$. If we plot the

**TABLE 10.6**

Data for Sales

| year | 1 | 2 | 3 | 4 | 5 | 6 | 7 | 8 | 9 | 10 | 11 | 12 |
|---|---|---|---|---|---|---|---|---|---|---|---|---|
| sales | 98 | 140 | 198 | 346 | 591 | 804 | 1,183 | 1,843 | 2,759 | 3,753 | 4,710 | 5,725 |

residuals, we see no pattern. Our model is deemed adequate. If we attempt to forecast *sales(12)*, we get that sales will be 8,331.94 units. Our results appear reasonable.

### 10.7.2 Multiple Regression for Sales

Suppose we want to forecast sales. We examine variables that influence sales such as the quarter (season), interest rate available, unemployment rate in the region, and GDP.

Our data used in this example is:

| sales | year | qtr | GDP | Unempl | Int |
|---|---|---|---|---|---|
| 2739 | 1 | 1 | 3811.5 | 2.95 | 6.266667 |
| 2910 | 1 | 2 | 3960 | 2.85 | 6.266667 |
| 2562 | 1 | 3 | 3892.5 | 2.95 | 6.466667 |
| 2385 | 1 | 4 | 4051.5 | 3 | 7.933333 |
| 2520 | 2 | 1 | 4177.5 | 3.1 | 8.933333 |
| 2142 | 2 | 2 | 3763.5 | 3.65 | 6.4 |
| 2130 | 2 | 3 | 3855 | 3.85 | 6.133333 |
| 2190 | 2 | 4 | 4000.5 | 3.7 | 9.066667 |
| 2370 | 3 | 1 | 4317 | 3.7 | 9.6 |
| 2208 | 3 | 2 | 4252.5 | 3.7 | 10.2 |
| 2196 | 3 | 3 | 4345.5 | 3.7 | 10.06667 |
| 1758 | 3 | 4 | 4116 | 4.15 | 7.866667 |
| 1944 | 4 | 1 | 3873 | 4.4 | 8.533333 |
| 2094 | 4 | 2 | 3919.5 | 4.7 | 8.266667 |
| 1911 | 4 | 3 | 3793.5 | 5 | 6.2 |
| 2031 | 4 | 4 | 3816 | 5.35 | 5.2 |
| 2046 | 5 | 1 | 3949.5 | 5.2 | 5.2 |
| 2502 | 5 | 2 | 4317 | 5.05 | 5.6 |
| 2238 | 5 | 3 | 4576.5 | 4.7 | 6.066667 |
| 2394 | 5 | 4 | 4911 | 4.25 | 5.866667 |
| 2586 | 6 | 1 | 5391 | 3.95 | 6.133333 |
| 2898 | 6 | 2 | 5661 | 3.75 | 6.533333 |
| 2448 | 6 | 3 | 5791.5 | 3.6 | 6.866667 |
| 2460 | 6 | 4 | 5878.5 | 3.75 | 5.866667 |
| 2646 | 7 | 1 | 6060 | 3.7 | 5.466667 |
| 2988 | 7 | 2 | 6199.5 | 3.65 | 5 |
| 2967 | 7 | 3 | 6454.5 | 3.55 | 4.733333 |
| 2439 | 7 | 4 | 6589.5 | 3.5 | 4.8 |
| 2598 | 8 | 1 | 6840 | 3.55 | 5.933333 |
| 3045 | 8 | 2 | 6880.5 | 3.55 | 5.133333 |
| 3213 | 8 | 3 | 7074 | 3.45 | 4.933333 |
| 2685 | 8 | 4 | 7194 | 3.4 | 4.933333 |

We desire to examine a multiple regression model for *sales* = $b_0 + b_1$ *year* + $b_2$ *quarter* + $b_3$ *GDP* + $b_4$ *unemployment* + $b_5$ *interest rate*.

Our multiple regression model is

| SUMMARY OUTPUT | | | | | | | | | |
|---|---|---|---|---|---|---|---|---|---|
| | | | | | | | | | |
| *Regression Statistics* | | | | | | | | | |
| Multiple R | 0.868606 | | | | | | | | |
| R Square | 0.754477 | | | | | | | | |
| Adjusted R Square | 0.707261 | | | | | | | | |
| Standard Error | 196.2857 | | | | | | | | |
| Observations | 32 | | | | | | | | |
| | | | | | | | | | |
| ANOVA | | | | | | | | | |
| | *df* | *SS* | *MS* | *F* | *Significance F* | | | | |
| Regression | 5 | 3078249 | 615649.8 | 15.97925 | 3.23321E-07 | | | | |
| Residual | 26 | 1001730 | 38528.07 | | | | | | |
| Total | 31 | 4079979 | | | | | | | |
| | | | | | | | | | |
| | *Coefficients* | *Standard Err* | *t Stat* | *P-value* | *Lower 95%* | *Upper 95%* | *Lower 95.0%* | *Upper 95.0%* | |
| Intercept | 882.666 | 2463.355 | 0.358319 | 0.722998 | -4180.832795 | 5946.165 | -4180.83 | 5946.165 | |
| year | -244.778 | 215.0698 | -1.13813 | 0.265444 | -686.8605416 | 197.304 | -686.861 | 197.304 | |
| qtr | -128.002 | 55.5992 | -2.30223 | 0.029583 | -242.2879729 | -13.7164 | -242.288 | -13.7164 | |
| GDP | 0.613797 | 0.440256 | 1.394184 | 0.175063 | -0.29116117 | 1.518756 | -0.29116 | 1.518756 | |
| Unempl | 118.8209 | 348.035 | 0.341405 | 0.735542 | -596.5751792 | 834.217 | -596.575 | 834.217 | |
| Int | -75.1856 | 27.51302 | -2.73273 | 0.011143 | -131.7393793 | -18.6317 | -131.739 | -18.6317 | |

Our residuals and percent relative error are

| *Observation* | *Predicted sales* | *Residuals* | *Standard Residuals* | *% Relative Error* |
|---|---|---|---|---|
| 1 | 2728.732388 | 10.26761 | 0.057118268 | 0.374867192 |
| 2 | 2679.997001 | 230.003 | 1.279496383 | 7.90388313 |
| 3 | 2507.408495 | 54.5915 | 0.303690097 | 2.130815961 |
| 4 | 2372.668958 | 12.33104 | 0.068597035 | 0.51702483 |
| 5 | 2525.932212 | -5.93221 | -0.033000628 | 0.235405228 |
| 6 | 2399.639606 | -257.64 | -1.43323759 | 12.02799282 |
| 7 | 2371.613542 | -241.614 | -1.344085312 | 11.34335877 |
| 8 | 2094.551397 | 95.4486 | 0.530976304 | 4.3583837 |
| 9 | 2387.947518 | -17.9475 | -0.09984124 | 0.757279249 |
| 10 | 2175.244086 | 32.75591 | 0.182219684 | 1.483510609 |
| 11 | 2114.34979 | 81.65021 | 0.454216462 | 3.718133431 |
| 12 | 2064.358818 | -306.359 | -1.704260381 | 17.4265539 |
| 13 | 2034.015904 | -90.0159 | -0.500754443 | 4.63044774 |
| 14 | 1990.251059 | 103.7489 | 0.577150712 | 4.954581707 |
| 15 | 1975.940214 | -64.9402 | -0.361259504 | 3.398232046 |
| 16 | 1978.521363 | 52.47864 | 0.291936305 | 2.583881667 |
| 17 | 2181.868428 | -135.868 | -0.75582998 | 6.640685635 |
| 18 | 2231.539353 | 270.4606 | 1.504560466 | 10.80977806 |
| 19 | 2186.143626 | 51.85637 | 0.288474689 | 2.317085543 |
| 20 | 2225.024303 | 168.9757 | 0.940004232 | 7.058299774 |
| 21 | 2603.17947 | -17.1795 | -0.095568621 | 0.664325978 |
| 22 | 2587.064123 | 310.9359 | 1.729722359 | 10.72932633 |
| 23 | 2496.277486 | -48.2775 | -0.268565492 | 1.972119522 |
| 24 | 2514.684368 | -54.6844 | -0.304206689 | 2.222941781 |
| 25 | 2789.450016 | -143.45 | -0.798006012 | 5.421391368 |
| 26 | 2776.218097 | 211.7819 | 1.178133241 | 7.087747758 |
| 27 | 2812.901594 | 154.0984 | 0.857242531 | 5.193744723 |
| 28 | 2756.808621 | -317.809 | -1.767955124 | 13.03028376 |
| 29 | 2970.523836 | -372.524 | -2.072333416 | 14.33886975 |
| 30 | 2927.528897 | 117.4711 | 0.653486484 | 3.857835892 |
| 31 | 2921.451496 | 291.5485 | 1.621871268 | 9.074027499 |
| 32 | 2861.163936 | -176.164 | -0.979992088 | 6.56104046 |

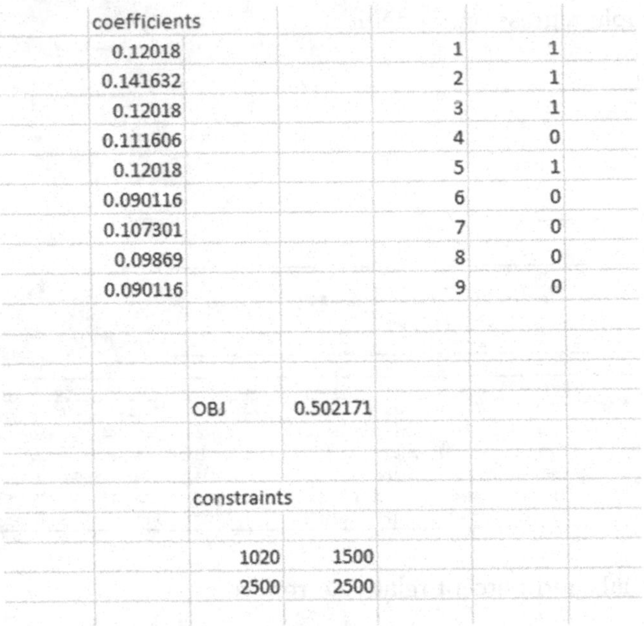

| coefficients | | | | |
|---|---|---|---|---|
| 0.12018 | | | 1 | 1 |
| 0.141632 | | | 2 | 1 |
| 0.12018 | | | 3 | 1 |
| 0.111606 | | | 4 | 0 |
| 0.12018 | | | 5 | 1 |
| 0.090116 | | | 6 | 0 |
| 0.107301 | | | 7 | 0 |
| 0.09869 | | | 8 | 0 |
| 0.090116 | | | 9 | 0 |
| | OBJ | 0.502171 | | |
| | constraints | | | |
| | 1020 | 1500 | | |
| | 2500 | 2500 | | |

**FIGURE 10.3**
Residual plot.

Our residual plot in Figure 10.3 shows no pattern.

### 10.7.3 Portfolio Optimization

**Minimum Variance of Expected Investment Returns:**

**Example 10.24:** Expected Investment Returns

A new company has $5,000 to invest, but the company needs to earn about 12% interest. A stock expert has suggested three mutual funds {A, B, and C} in which the company could invest. Based upon the previous year's returns, these funds appear relatively stable. The expected return, variance on the return, and covariance between funds are shown in Table 10.7.

**TABLE 10.7**
Expected Return Statistics

| Expected Value | A | B | C |
|---|---|---|---|
| | 0.14 | 0.11 | 0.10 |
| Variance | **A** | **B** | **C** |
| | 0.02 | 0.08 | 0.18 |
| Covariance | **A** | **B** | **C** |
| | 0.05 | 0.02 | 0.03 |

**Model Formulation:**

We use laws of expected value, variance, and covariance in our model. Let $x_j$ be the number of dollars invested in funds $j$ ($j$ = 1, 2, 3).

$$Minimize\ V_I = var(Ax_1 + Bx_2 + Cx_3) = x_1^2 Var(A) + x_2^2 Var(B) + x_3^2 Var(C)$$
$$+ 2x_1x_2 Cov(AB) + 2x_1x_3 Cov(AC) + 2x_2x_3 Cov(BC)$$
$$= 0.2x_1^2 + 0.08x_2^2 + 0.18x_3^2 + 0.10x_1x_2 + 0.04x_1x_3 + 0.06x_2x_3$$

Our constraints include (1) the expectation to achieve at least the expected return of 12% from the sum of all the expected returns:

$$0.14x_1 + 0.11x_2 + 0.10x_3 \geq 0.12(5{,}000)\ or\ 0.14x_1 + 0.11x_2 + 0.10x_3 \geq 600$$

(2) the sum of all investments must not exceed the $5,000 capital.

$$x_1 + x_2 + x_3 \leq \$5000$$

**Solution**

We set up the Lagrangian function, $L$.

$$Min\ L = 0.2x_1^2 + 0.08x_2^2 + 0.18x_3^2 + 0.10x_1x_2 + 0.04x_1x_3 + 0.06x_2x_3$$
$$+ x_1[0.14x_1 + 0.11x_2 + 0.10x_3 - U_1^2 - 600]$$
$$+ x_2[x_1 + x_2 + x_3 + U_2^2 - \$5000]$$

**Necessary Conditions:**

(32) $Lx_1 = 0.4x_1 + 0.10x_2 + 0.04x_3 + 0.14\ U_1 + U_2 = 0$
(33) $Lx_2 = 0.16x_2 + 0.10x_1 + 0.06x_3 + 0.11\ U_1 + U_2 = 0$
(34) $L_{x3} = 0.36x_3 + 0.04x_1 + 0.06x_2 + 0.10\ U_1 + U_2 = 0$
(35) $Lx_1 = 0.14x_1 + 0.11x_2 + 0.10x_3 - U_1^2 - 600 = 0$
(36) $Lx_2 = x_1 + x_2 + x_3 + U_2^2 - \$5{,}000 = 0$

There are only two constraints, so we need to consider four cases. The solution is found in the case where $\lambda_1$ and $\lambda_2$ both do not equal zero. This is the case where the solution lies at the intersection of the constraints. Both constraints are binding constraints/.

```
obj7 := 0.2 * x1^2 + .08 * x2^2 + .18 * x3^2 + .1 * x1·x2
    + .04 * x1 * x3 + .06 * x2 * x3;
>
```

$$obj7 := 0.2\,x1^2 + 0.08\,x2^2 + 0.18\,x3^2 + 0.1\,x1\,x2 + 0.04\,x1\,x3 + 0.06\,x2\,x3$$

```
> cons17 := .14·x1 + .11·x2 + .1·x3 - U1^2 - 600;
```

$$cons17 := 0.14\,x1 + 0.11\,x2 + 0.1\,x3 - U1^2 - 600$$

```
> cons27 := x1 + x2 + x3 + U2^2 - 5000;
```

$$cons27 := x1 + x2 + x3 + U2^2 - 5000$$

```
h := Hessian(obj7, [x1, x2, x3]);
```

$$h := \begin{bmatrix} 0.4 & 0.1 & 0.04 \\ 0.1 & 0.16 & 0.06 \\ 0.04 & 0.06 & 0.36 \end{bmatrix}$$

The Hessian is positive definite, so our solution is a minimum.
We solve the following system of equations:

| $x_1$ | $x_2$ | $x_3$ | $\lambda_1$ | $\lambda_2$ |
|---|---|---|---|---|
| .4 | .10 | .04 | .14 | 1 |
| .1 | .16 | .06 | .11 | 1 |
| .04 | .06 | .36 | .10 | 1 |
| .14 | .11 | .10 | 0 | 0 |
| 1 | 1 | 1 | 0 | 0 |

equal to the right-hand side matrix [0 0 0 600 5,000]. The solution is

$$x_1 = 1{,}904.80$$
$$x_2 = 2{,}381.00$$
$$x_3 = 714.20$$
$$\lambda_1 = -13{,}809.50$$
$$\lambda_2 = 904.80$$

z = $1,880,942.29 or a standard deviation of *$1,371.50.*

The expected return is 12% found by 0.14(1,904.8) + 0.11(2,381) + 0.1(714.2)/5,000.

This solution is optimal. The Hessian matrix, *H*, has all positive leading principal minors. Therefore, since *H* is always positive definite, then our solution is the optimal minimum.

$$H = \begin{bmatrix} .4 & .1 & .04 \\ .1 & .16 & .06 \\ .04 & .06 & .36 \end{bmatrix}$$

### 10.7.4 MADM for Project Selection

Let us assume we have nine projects to choose from for a specific project. The data is provided in Table 10.8.

Examining our criteria, we would prefer to have lower costs and smaller man-hours while larger NPV, market growth, success, and chance for approval by the FDA. Since not all our criteria will be maximized, we decide to use TOPSIS. Since our experts are not certain about the priority of the criteria, we decide to use entropy for obtaining the weights.

We find that Project 2 ranks first.

Now, we can add some real constraints to modify the problem. First, we will assume we can do more than one project, but we are limited by 1,500 in capital and 2,500 in manpower. How might we proceed.

We will use only NPV, Market Growth, Success, and FDA approval in our MADM method to obtain values for the projects and then optimize (using integer programming) while adding the constraints. We might also use SAW as our MADM method.

**TABLE 10.8**

Project Data

| Project | Cost | Man-hours | NPV | Market Growth | Success | FDA Approval |
|---|---|---|---|---|---|---|
| 1 | 300 | 500 | 4 | 1 | 3 | 2 |
| 2 | 250 | 600 | 3 | 2 | 4 | 3 |
| 3 | 350 | 550 | 3 | 3 | 2 | 2 |
| 4 | 380 | 750 | 2 | 3 | 3 | 1 |
| 5 | 120 | 850 | 2 | 4 | 2 | 2 |
| 6 | 420 | 950 | 2 | 2 | 1 | 3 |
| 7 | 360 | 400 | 3 | 1 | 3 | 2 |
| 8 | 260 | 1,100 | 1 | 2 | 2 | 4 |
| 9 | 180 | 1,200 | 1 | 3 | 1 | 3 |

| | | | | Criterion | | | | | TOPSIS | Final | |
|---|---|---|---|---|---|---|---|---|---|---|---|
| Alternatives | Cost | Manhours | NPV | Market Growth | Success | FDA Approval | NA | NA | Value | Rank | |
| 1 | 300 | 500 | 4 | 1 | 3 | 2 | 0 | 0 | 0.529 | 3 | 1 |
| 2 | 250 | 600 | 3 | 2 | 4 | 3 | 0 | 0 | 0.634 | 1 | 2 |
| 3 | 350 | 550 | 3 | 3 | 2 | 2 | 0 | 0 | 0.525 | 4 | 3 |
| 4 | 380 | 750 | 2 | 3 | 3 | 1 | 0 | 0 | 0.470 | 6 | 4 |
| 5 | 120 | 850 | 2 | 4 | 2 | 2 | 0 | 0 | 0.538 | 2 | 5 |
| 6 | 420 | 950 | 2 | 2 | 1 | 3 | 0 | 0 | 0.305 | 9 | 6 |
| 7 | 360 | 400 | 3 | 1 | 3 | 2 | 0 | 0 | 0.471 | 5 | 7 |
| 8 | 260 | 1100 | 1 | 2 | 2 | 4 | 0 | 0 | 0.372 | 7 | 8 |
| 9 | 180 | 1200 | 1 | 3 | 1 | 3 | 0 | 0 | 0.366 | 8 | 9 |

**FIGURE 10.4**
Excel NPV project screenshot.

First, we use TOPSIS to get values for our projects (Figure 10.4).

We find that Project 1, 2, 3, and 5 are selected. We also see that only man-hours is a binding constraint so if we get more man-hours, we will improve the solution without increasing capital (if man-hours were free)

We suggest sensitivity analysis on this problem as an exercise.

## 10.8 Exercises

### 10.8.1 Compounding Interests

1. A principal of $3,500 is borrowed at 6% per year simple interest. Find the value after
   a. one year
   b. three years
   c. three months
   d. three days
2. If $500 were deposited in a bank account earning 5% per year, find the value after
   a. 5 years
   b. 10 years
   c. 25 years
3. You are saving for a new set of golf clubs that costs $2,500. You find a money market account that pays 6.5% per year simple interest. How much should be invested so that this purchase can made within 2 years?

4. Find the interest on $5,500 for 1 year at
   a. 6.5% per year simple interest
   b. 6.5 % per year compounded monthly.
   c. 6.5% per year compounded quarterly.
   d. 6.5% per year compounded semi-annually.
5. Find the interest on $5,000 for
   a. 2 years at 4.7% compounded monthly.
   b. 3 years at 4.7% compounded quarterly.
   c. 10 years at 4.7% compounded semi-annually.
   d. 60 months at 4.7% per year compounded daily.
6. Find the future value of $12,000 invested at a rate of 9% compounded.
   a. monthly
   b. quarterly
   c. semi-annually
   d. daily
7. Suppose $5,500 is invested and compounded continuously at 5% per year. Find the value after
   a. one year
   b. two years
   c. six months
8. Assume that $5,000 is deposited in an account that pays simple interest. If the account grows to $5,500 in two years, find the interest rate.

### 10.8.2 Rate of Interests

1. Determine the APR on each of the following investments.
   a. 5% per year compounded monthly.
   b. 5% per year compounded quarterly.
   c. 5% per year compounded semi-annually.
2. Determine the APR on each of the following investments.
   a. 5% per year compounded monthly.
   b. 7% per year compounded quarterly.
   c. 9% per year compounded semi-annually.
3. Determine which is a better investment: an investment paying 6.5% per year compounded monthly or 6.75% compounded semi-annually.

4. Which is a better option for investment, 3.5% per year compounded monthly or 3.6% simple interest?
5. Which is a better option for investment, 9% per year compounded monthly or 9.1% simple interest?
6. You are offered two credit plans. One has an interest rate of 19.8% annual interest compounded monthly. The other offers 19% compounded quarterly. Which plan should you choose?
7. Find the APR if money is invested at 15% per year with continuous compounding.
8. Find the APR if money is invested at 5.5% per year with continuous compounding.
9. Sam borrows $6,450 for 4 years at 6% simple discount. Determine the amount received and the APR for this transaction.
10. Sally borrows $450 for 4 years at 7% simple discount. Determine the amount received and the APR for this transaction.
11. We buy new machinery for our printing business for $12,500. The depreciation schedule is 5.5% per year. What will this machine be worth in 5 years?
12. We buy new computers for our company and spend $20,000. The depreciation is 25% per year. What is the value of the computers after 3 years?

### 10.8.3 Present Value

1. Given the cash flow for investments below, determine the NPV and determine if the investment is worthwhile.

   a. Assume an 8% discount rate

| Year | 0 | 1 | 2 | 3 | 4 |
|---|---|---|---|---|---|
| Cash Flow | –200 | 120 | 130 | 140 | 150 |

   b. Assume a 5% discount rate

| Year | 1 | 2 | 3 | 4 | 5 | 6 | 7 | 8 |
|---|---|---|---|---|---|---|---|---|
| Cash Flow | –2,000 | 1,000 | 700 | 550 | 400 | 400 | 350 | 250 |

2. A 10-year government bond valued at $1,000 is purchased at a market rate of 8% interest.
   a. Find the annual repayment.
   b. Find the NPV if the interest rate were 4%.
   c. Calculate the present value of the bond based after 5 years.
3. A $500 saving bonds is bought in 2000. It matures in 2005 (meaning it is worth $500 in 2005). It will continue to earn 2.5% each year after 2005. How much will it be worth in 2010? 2020?
4. A 15-year government bond valued at $50,000 is purchased at a market rate of 4.5% interest.
   a. Find the annual repayment.
   b. Find the NPV if the interest rate were 2%.
   c. Calculate the present value of the bond based after 20 years.
5. A small company invests $5,000 at the end of six months in a fund paying 6% compounded semi-annually. Find the value of the investment after 10 years.
6. Microsoft invests $3 million at the end of nine months in a fund paying 5% compounded monthly. Find the value of the investment after 2 years.
7. A university anticipates capital expenditures of $220,000 to buy new equipment within 10 years. How much should the firm be depositing in a funding earning 4% per year compounded quarterly in order to buy the new equipment?
   a. A loan of $150,000 is to be repaid in equal monthly payments over 15 years assuming a fixed rate of 5.75% per year. Find the amount of each payment and the total interest paid.
   b. A loan of $5,000 is to be repaid in equal monthly payments over 4 years assuming fixed rate of 4.25% per year. Find the amount of each payment and the total interest paid.

### 10.8.4 Mortgages and Amortization

1. You owe $2,000 on your card that charges 1.5% interest per month. Determine your monthly payment if you want to pay off the card in
   a. 60 months
   b. 1 year
   c. 3 years
2. You are considering a 30-year $150,000 mortgage that charges 0.5% each month. Determine the monthly payment schedule and total amount paid.

3. You charge $2,500 on your Mastercard that charges 9.9% annual interest compounded monthly on the unpaid balance. You can afford $50 a month. How long until you pay off your bill assuming you do not charge anything until it is paid off?
4. You charge $5,000 on your Mastercard that charges 12% annual interest compounded monthly on the unpaid balance. You can afford at most $100 a month. How long until you pay off your bill assuming you do not charge anything until it is paid off?
5. You go out car shopping, and you budget car payments at most $275 per month. You found several cars that you are interested in buying: a Ford Focus, which is now selling for $15,500 at 3.75% over 5 years, and a Toyota Corolla, selling for $16,500 at 4.25% over 6 years. Which of these cars, if any, can you afford to buy? Assume you can put no money down.
6. You go out car shopping, and you budget car payments at most $475 per month. You found several cars that you are interested in buying: a pre-owned SUV, which is now selling for $19,500 at 7.75% over 5 years, and a Toyota Highlander, selling for $24,500 at 2.25% over 6 years. Which of these cars, if any, can you afford to buy? Assume you can put no money down.

## 10.9 Projects

1. In 1945, Noah Sentz died in a car accident, and his estate was handled by the local courts. The state law stated that 1/3 of all assets and property go to the wife and 2/3 of all assets go to the children. There were four children. Over the next four years, three of the four children sold their shares of the assets back to the mother for a sum of $1,300 each. The original total assets were mainly 75.43 acres of land. This week, the fourth child has sued the estate for his rightful inheritance from the original probate ruling. The judge has ruled in favor of the fourth son and has determined that he is rightfully due monetary compensation. The judge has picked your group as the jury to determine the amount of compensation. Use the principles of financial mathematical modeling to build a model that enables you to determine the compensation. Additionally, prepare a short one-page summary letter to the court that explains your results. Assume the date is November 10, 2003.

2. *Buying a new car:*
   Part 1 – You wish to buy a new car soon. You initially narrow your choices to a Saturn, a Cavalier, and a Toyota Tercel. Each dealership offers you their prime deal:

   | | | | |
   |---|---|---|---|
   | Saturn | $11,900 | $1,000 down | 3.5% interest for up to 60 months |
   | Cavalier | $11,500 | $1,500 down | 4.5% interest for up to 60 months |
   | Tercel | $10,900 | $ 500 down | 6.5% interest for up to 48 months |

   You have allocated at most $475 a month on a car payment.

   Use a dynamical system to compare the alternatives, choose a car, and establish your exact monthly payment.

   Part 2 – Nissan is offering a first-time buyer special. Nissan is offering 6.9% for 24 months or $500 "cash back". If the regular interest rate is 9.00%, determine the amount financed at 6.9% so that these two options are equivalent for a new car buyer.

   What kind of car could you get from Nissan with the equivalent option?

3. On November 24, 1971, a cold, rainy Thanksgiving evening, a middle-aged man giving the name Dan B. Cooper purchased a plane ticket on Northwest Airlines Flight 305 from Portland, Oregon, to Seattle, Washington. He boarded the plane and flew into history.

   After Cooper was seated, he demanded the flight attendant bring him a drink and a $200,000 ransom. His note read: "Miss, I have a bomb in my suitcase, and I want you to sit beside me." When she hesitated, Cooper pulled her into the seat next to him and opened his case, revealing several sticks of dynamite connected to a battery. He then ordered her to have the captain relay his ransom demand for the money and four parachutes: Two front-packs and two backpacks. The FBI delivered the parachutes and ransom to Cooper when the 727 landed in Seattle to be refueled. Cooper, in turn, allowed the 32 other passengers he had held captive to go free.

   When the 727 was once again airborne, Cooper instructed the pilot to fly at an altitude of 10,000 feet on a course destined for Reno, Nevada. He then forced the flight attendant to enter the cockpit with the remaining three crewmembers and told her to stay there until landing. Alone in the rear cabin of the plane, he lowered the aft stairs beneath the tail. Then, in the dark of the night, somewhere over Oregon, D. B.

Cooper stepped with the money into history. When the plane landed in Reno, the only thing the FBI found in the cabin was one of the backpack chutes. Despite the massive efforts of manhunts conducted by the FBI and scores of local and state police groups, no evidence was ever found of D. B. Cooper, the first person to hijack a plane for ransom and parachute from it. One package of marked bills from the ransom was found along the Columbia River near Portland in 1980. Some surmise that Cooper might have been killed in his jump to fame and part of the ransom washed downstream.

Consider the following items dealing with Flight 305 and its famous passenger. Using Excel to assist you, complete each one as directed.

a. Suppose Cooper invested the ransom over a period of 10 years, one $20,000 unit per year, in a small local bank in rural Utah, at a fixed rate of 6%, compounded annually, starting on January 1, 1972. Further, if he left the money to accumulate, what would be the value of the account December 31, 1993?
b. Suppose that instead of playing it safe, Cooper gambled half of the money away in Reno the night after the hijacking and then invested the other half with Nightwings Federal in Reno on January 1, 1972. If the account paid 6% interest compounded quarterly, what would this money be worth at the end of this year?
c. Another possibility for Cooper would have been to invest 75% of the total with a "loco" investor, Smiling Pete's Federal Credit Union, at a rate of 3% compounded monthly. What would this investment be worth at the end of this year?
d. Compare these three methods of investing. Contrast their different patterns of growth over time.

# 11

# *Reliability Models*

**OBJECTIVES**

1. Know and understand the concept of reliability.
2. Know and understand the concept of mean time to failure.
3. Know and apply series and parallel components.
4. Know redundant and standby redundant applications.
5. Be able to work with large systems.

Reliability modeling is the mathematical process of examining the ability of equipment or systems to function without failure. Reliability describes the ability of the equipment to function under prescribed conditions and the probability of its failure. At its heart, reliability looks at the prediction, prevention, and management of uncertainty and risks. The concepts of reliability models are seen in fields such as quality and safety engineering.

## 11.1 Introduction to Total Conflict (Zero-Sum) Games

You are a New York City police detective on a stakeout and must occupy a position for at least the next 24 hours. You are providing hourly situation reports back to your supervisor. The stakeout is ineffective unless it can communicate with you in a timely manner. Therefore, radio communications must be reliable. The radio has several components that affect its reliability, an essential one being the battery.

Batteries have a useful life that is not deterministic; in other words, we do not know exactly how long the battery will last when we install it. Its lifetime is a variable that may depend on previous use, manufacturing defects, weather, etc. The battery that is installed in the radio prior to leaving for the stakeout could last only a few minutes or for the entire 24 hours.

DOI: 10.1201/9781003298762-11

Since communications are so important to this mission, we are interested in modeling and analyzing the reliability of the battery.

For this analysis effort, we will use the following definition for reliability:

If $T$ is the time to failure of a component of a system, and $f(t)$ is the probability distribution function of $T$, then the components' *reliability* at time $t$ is

$$R(t) = P(T > t) = 1 - F(t).$$

$R(t)$ is called the reliability function, and $F(t)$ is the cumulative distribution function of $f(t)$.

A measure of this reliability is the probability that a given battery will last more than 24 hours. If we know the probability distribution for the battery life, we can use our knowledge of probability theory to determine the reliability. If the battery reliability is below acceptable standards, one solution is to have the police persons carry spares. Clearly, the more spares they carry, the less likely there is to be a failure in communications due to batteries. Of course, the battery is only one component of the radio. Others include the antenna, handset, etc. Failure of any one of the essential components causes the system to fail. This is a relatively simple example of one of many military applications of reliability.

This chapter will show we can use elementary probability to generate models that can be used to determine the reliability of equipment.

## 11.2 Modeling Component Reliability

In this section, we will discuss how to model component reliability. The reliability function, $R(t)$ is defined as:

$$R(t) = P(T > t) = P(\text{component fails after time } t).$$

This can also be stated, using $T$ as the component failure time, as $t$

$$R(t) = P(T > t) = 1 - P(T \leq t) = 1 - \int_{-\infty}^{t} f(x)f(x)dx = 1 - F(t).$$

Thus, if we know the probability density function $f(t)$ of the time to failure $T$, we can use probability theory to determine the reliability function $R(t)$. We normally think of these functions as being time dependent; however, this is not always the case. The function might be discrete such as the lifetime of a cannon tube. It is dependent on the number of rounds fired through it (a discrete random variable).

A useful probability distribution in reliability is the exponential distribution. Recall that its density function is given by

$$f(t) = \begin{cases} \lambda e^{-\lambda t} & t \geq 0 \\ 0 & otherwise \end{cases}$$

where the parameter λ.

We know λ is such that its reciprocal, $\frac{1}{\lambda}$ equals the mean of the random variable *T*. If *T* denotes the time to failure of a piece of equipment or a system, then $\frac{1}{\lambda}$ is the mean time to failure, which is expressed in units of time. For applications of reliability, we will use the parameter λ. Since, $\frac{1}{\lambda}$ is the mean time to failure, λ is the average number of failures per unit time or the failure rate. For example, if a light bulb has a time to failure that follows an exponential distribution with a mean time to failure of 50 hours, then its failure rate is 1 light bulb per 50 hours or 1/50 per hour, so in this case, λ = 0.02 per hour. Note that the mean of *T*, the mean time to failure of the component, is 1/50.

**Example 11.1:** Radio Reliability

We can take a more formal look at our police officer's situation that we presented in the introduction. Let the random variable *T* be defined as follows:

*T* = time until a randomly selected battery fails.

Suppose radio batteries have a time to failure that is exponentially distributed with a mean of 30 hours. In this case, we could write

$$T \sim \exp\left(\lambda = \frac{1}{30}\right)$$

Therefore λ = 1/30 so that we know the density function is *F(t)* = 1/30 $e^{-t/30}$ for $t \geq 0$ and *F(t)* is as follows:

$$F(t) = \int_0^t \frac{1}{30} e^{-x/30} dx$$

*F(t)*, the CDF of the exponential distribution, can be integrated to obtain

$$1 - e^{(-t/30)} \text{for t} \geq 0$$

$$F(t) = 1 - e^{-t}/30, t > 0$$

Now, we can compute the reliability function for a battery:

Recall that in the earlier example, the police must occupy the stakeout for 24 hours. The reliability of the battery for 24 hours is

$$R(24) = 1 - (1 - e^{(-24/30)}) = 1 - 0.55067 = 0.44933$$

so the probability that the battery lasts more than 24 hours is 0.4493.

**Example 11.2:** Nickel Cadmium Battery

We have the option to purchase a new nickel cadmium battery for our stakeout. Testing has shown that the distribution of the time to failure can be modeled using a parabolic function:

$$f(x) = \begin{cases} \frac{x}{384}\left(1 - \frac{x}{48}\right) & 0 \leq x \leq 48 \\ 0 & \textit{otherwise} \end{cases}$$

Let the random variable $T$ be defined as follows:

$$T = \text{time until a randomly selected battery fails.}$$

In this case, we could write

$$f(t) = (t/384)(1 - t/48), 0 \leq t \leq 48$$

and

$$F(t) = \int_0^{24} (x/384)(1 - x/48)dx.$$

Recall that in the earlier example the soldiers must man the stakeout for 24 hours. The reliability of the battery for 24 hours is therefore

$$R(24) = 1 - F(24) = \int_0^{24} (x/384)(1 - x/48)dx = 0.500.$$

which is an improvement over the batteries from Example 11.1. Therefore, we should use the new battery.

## 11.3 Modeling Series and Parallel Components

### 11.3.1 Modeling Series Systems

Now, we consider a system with $n$ components $C_1, C_2 ..., Cn$ where each of the individual components must work in order for the system to function. A model of this type of system is shown in Figure 11.1.

If we assume these components are mutually independent, the reliability of this type of system is easy to compute. We denote the reliability of component $i$ at time $t$ by $R_i(t)$. In other words, $R_i(t)$ is simply the probability that component $I$ will function continuously from time 0 through until time $t$. We are interested in the reliability of the entire system of $n$ components, but since these components are mutually independent, the system reliability is

$$R(t) = R_1(t) * R_2(t) * \dots * R_n(t)$$

**Example 11.3:** Multiple Components

Our radio has several components. Let us assume that there are four major components – they are (in order) the handset, the battery, the receiver-transmitter, and the antenna. Since they all must function properly for the radio to operate, we can model the radio with the diagram shown in Figure 11.2.

Suppose we know that the probability that the handset will work for at least 24 hours is 0.6703, and the reliabilities for the other components are 0.4493, 0.7261, and 0.9531, respectively. If we assume that the components work *independently* of each other, then the probability that the entire system works for 24 hours is:

$$R(24) = R_1(24) \times R_2(24) \times R_3(24) \times R_4(24) = (0.6703)(0.4493)(0.7261)(0.9531)$$
$$= (0.2084).$$

Recall that two events $A$ and $B$ are *independent* if $P(A \mid B) = P(A)$.

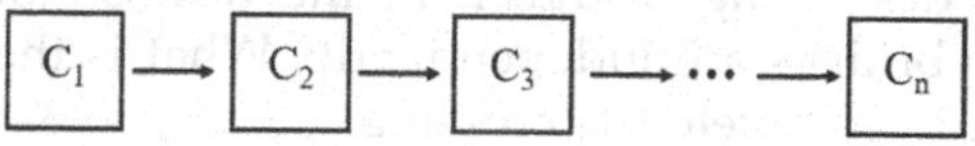

$C_1$ → $C_2$ → $C_3$ → ... → $C_n$

**FIGURE 11.1**
Series system.

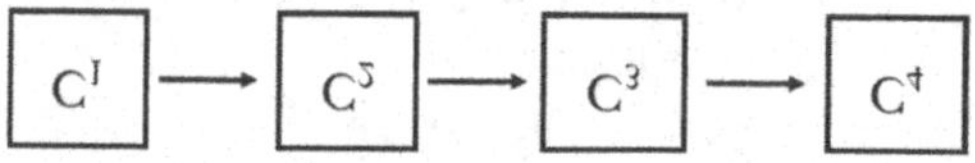

$C_1$ → $C_2$ → $C_3$ → $C_4$

**FIGURE 11.2**
Radio system.

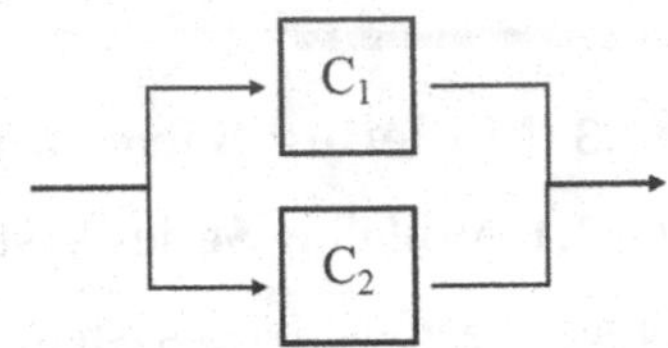


**FIGURE 11.3**
Parallel system of two components.

### 11.3.2 Modeling Parallel Systems (Two Components)

Now, we consider a system with two components where only one of the components must work for the system to function. A system of this type is depicted in Figure 11.3.

Notice that in this situation the two components are *both* put in operation at time 0; they are both subject to failure throughout the period of interest. Only when *both* components fail before time $t$ does the system fail. Again, we also assume that the components are independent. The reliability of this type of system can be found using the following well-known addition model:

$$P(A \cup B) = P(A) + P(B) - P(A \cap B).$$

In this case, $A$ is the event that the first component functions for longer than some time, $t$, and $B$ is the event that the second component functions longer than the same time, $t$. Since reliabilities *are* probabilities, we can translate the above formula into the following:

$$R(t) = R_1(t) + R_2(t) - R_1(t)R_2(t).$$

**Example 11.4:** Bridge Reliability

Suppose we have two bridges in the area for a Boy Scout hike. It will take 3 hours to complete the crossing for all the hikers. The crossing will be successful as long as at least one bridge remains operational during the entire crossing period. You estimate that the bridges are in bad shape and that bridge number 1 has a one-third chance of being destroyed and a one-fourth chance of destroying bridge 2 in the next 3 hours. Assume the destruction of the bridges are independently. What is the probability that your Boy Scouts can complete the crossing?

Solution: First, we compute the individual reliabilities:

$$R_1(3) = 1 - 1/3 = 2/3$$

and

$$R_2(3) = 1 - 1/4 = 3/4$$

Now, it is easy to compute the system reliability:

$$R(3) = R_1(3) + R_2(3) - R_1(3)R_2(3)$$
$$= 2/3 + 3/4 - (2/3)(3/4) = 0.9167.$$

## 11.4 Modeling Active Redundant Systems

Consider the situation in which a system has $n$ components, all of which begin operating (are active) at time $t = 0$. The system continues to function properly as long as at least $k$ of the components do not fail. In other words, if $n - k + 1$ components fail, the system fails. This type of component system is called an active redundant system. The active redundant system can be modeled as a parallel system of components as shown in Figure 11.4.

We assume that all $n$ components are identical and will fail independently. If we let $T_i$ be the time to failure of the $i$th component, then the $T_i$ terms are independent and identically distributed for $i = 1, 2, 3, \ldots, n$. Thus $R_i(t)$ the reliability at time $t$ for component $i$ is identical for all components.

Recall that our system operates if at least $k$ components function properly. Now, we define the random variables X and $T$ as follows:

X = number of components functioning at timet, and
T = time to failure of the entire system.

Then, we have

$$R(t) = P(T > t) = P(X \geq k)$$

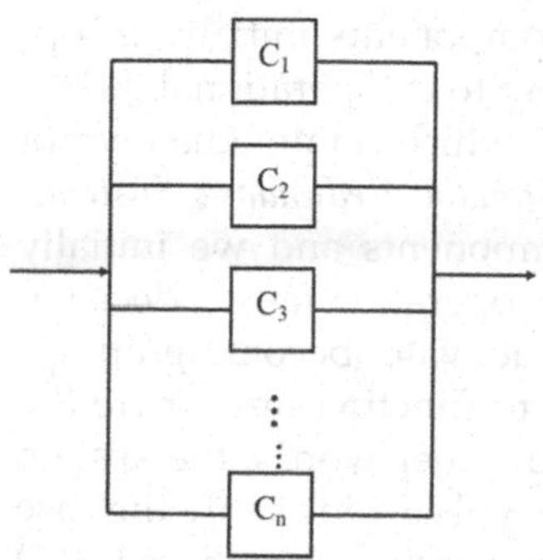


**FIGURE 11.4**
Active redundant system.

It is easy to see that we now have $n$ identical and independent components with the same probability of failure by time $t$. This situation corresponds to a binomial experiment, and we can solve for the system reliability using the binomial distribution with parameters $n$ and $p = R_i(t)$.

**Example 11.5:** Manufacturing Reliability

A manufacturing company has 15 different machines to make item A. They estimate if at least 12 are operating that they will be able to make all the items necessary to meet demand. Machines are assumed to be in parallel (active redundant), i.e. they fail independently. If we know that each machine has a 0.6065 probability of operating properly for at least 24 hours, we can compute the reliability of the entire machine system for 24 hours.

Define the random variable: X = number of machines working after 24 hours.

Clearly, the random variable X is binomially distributed with $n = 15$ and $p = 0.6065$. In the language of mathematics, we write this sentence as

$$X \sim b(15, 0.6065) \text{ or } X \sim \text{BINOMIAL}(15, 0.6065)$$

We know that the reliability of the machine system for 24 hours is

$$R(24) = P(x \geq 2) = P(12 \leq x \leq 15) = 0.0950$$

Thus, the reliability of the system for 24 hours is only 0.0990.

## 11.5 Modeling Standby Redundant Systems

Active redundant systems can sometimes be inefficient. These systems require only $k$ of the $n$ components to be operational, but all $n$ components are initially in operation and thus subject to failure. An alternative is the use of spare components. Such systems have only $k$ components initially in operation; exactly what we need for the whole system to be operational. When a component fails, we have a spare "standing by", which is immediately put into operation. For this reason, we call these *Standby Redundant Systems*. Suppose our system requires $k$ operational components and we initially have $n - k$ spares available. When a component in operation fails, a decision switch causes a spare or standby component to activate (becoming an operational component). The system will continue to function until there are less than $k$ operational components remaining. In other words, the system works until $n - k + 1$ components have failed. We will consider only the case where one operational component is required (the special case where $k = 1$)

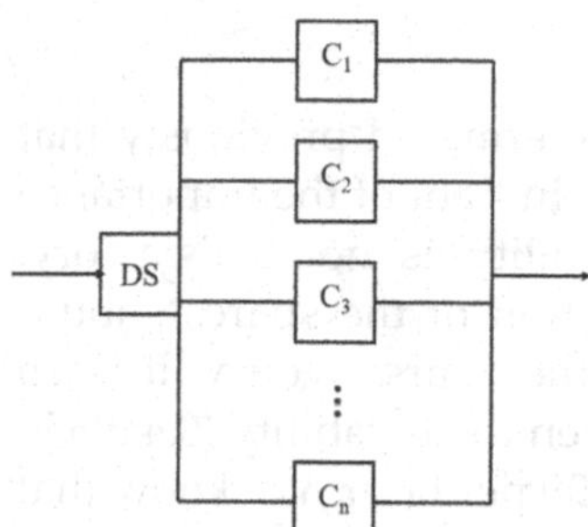


**FIGURE 11.5**
Standby redundant system.

and there are $n - 1$ standby (spare) components available. We will assume that a decision switch (DS) controls the activation of the standby components instantaneously and 100% reliably. We use the model shown in Figure 11.5 to represent this situation.

If we let $T_i$ be the time to failure of the $i$th component, then the $T_i$'s are independent and identically distributed for $i = 1, 2, 3, \ldots, n$. Thus, $R_i(t)$ is identical for all components.

Let, $T$ = time to failure of the entire system. Since the system fails only when all $n$ components have failed, and component $i + 1$ is put into operation only when component $i$ fails, it is easy to see that

$$T = T_1 + T_2 + \ldots + T_n.$$

In other words, we can compute the system failure time easily if we know the failure times of the individual components.

Finally, we can define a random variable X = number of components that *fail* before time $t$ in a standby redundant system. Now, the reliability of the system is simply equal to the probability that less than $n$ components fail during the time interval (0, $t$). In other words,

$$R(t) = P(x < n).$$

It can be shown that X follows a Poisson distribution with parameter *11* = $\alpha\, t$ where $\alpha$ is the failure rate, so we write

$$X \sim POISSON(\lambda).$$

For example, if time is measured in seconds, then $\alpha$ is the number of failures per second. The reliability for some specific time $t$ then becomes:

$$R(t) = P(x < n) = P(0 \le x \le n - 1).$$

**Example 11.6:** Revisiting Battery Reliability

Consider the reliability of a radio battery. We determined previously that one battery has a reliability for 24 hours of 0.4493. In light of the importance of communications, you decide that this reliability is not satisfactory. Suppose we carry two spare batteries. The addition of the spares should increase the battery system reliability. Later in the course, you will learn how to calculate the failure rate $\alpha$ for a battery given the reliability (0.4493 in this case). For now, we will give this to you: $\alpha = 130$ per hour. We know that $n = 3$ total batteries.

Therefore:

$$X \sim Poisson\,(\lambda = \alpha t = 24/30 = 0.8) \text{ and}'$$

and

$$R\,(24) = P\,(x < 3) = P\,(0 \leq x \leq 2) = 0.9526$$

The reliability of the system with two spare batteries for 24 hours is now 0.9526.

**Example 11.7:** Extended Stakeout

If the police stakeout must stay out for 48 hours without resupply, how many spare batteries must be taken to maintain a reliability of 0.95? We can use trial and error to solve this problem. We start by trying our current load of two spares. We have

$$X \sim Poisson\left(\lambda = \alpha t = \frac{48}{30} = 1.6\right)$$

and we can now compute the system reliability

$$R\,(48) = P\,(X < 3) = P\,(0 \leq X \leq 2) = 0.7834 \leq 0.95.$$

which is not good enough. Therefore, we try another spare so $n = 4$ (three spares) and we compute:

$$R\,(48) = P\,(X < 4) = P\,(0 \leq X \leq 3) = 0.9212 < 0.95.$$

which is still not quite good enough, but we are getting close! Finally, we try $n = 5$, which turns out to be sufficient:

$$R(48) = P(X < 5) = P(0 \le X \le 4) = 0.9763 > 0.95.$$

Therefore, we conclude that the stakeout should take out at least four spare batteries for a 48-hour mission.

## 11.6 Models of Large-Scale Systems

In our discussion of reliability up to this point, we have discussed series systems, active redundant systems, and standby redundant systems. Unfortunately, things are not always this simple. The types of systems listed above often appear as subsystems in larger arrangements of components that we shall call "large-scale systems". Fortunately, if you know how to deal with series systems, active redundant systems, and standby redundant systems, finding system reliabilities for large-scale systems is easy.

**Example 11.8:** Large-Scale System

The first and most important step in developing a model to analyze a large-scale system is to draw a picture. Consider the network that appears as Figure 11.6 below. Subsystem *A* is the standby redundant system of three components (each with failure rate 5 per year) with the decision switch on the left of the figure. Subsystem $B_1$ is the active redundant system of three components (each with failure rate 3 per year), where at least two of the three components must be working for the subsystem to work. Subsystem $B_1$ appears in the upper right portion of the figure. Subsystem $B_2$ is the two

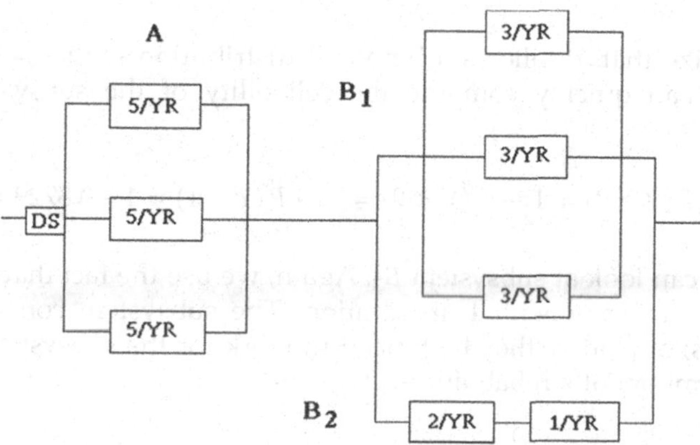


**FIGURE 11.6**
Network example.

component parallel system in the lower right portion of the figure. We define subsystem $B$ as being subsystems $B_1$ and $B_2$ together. We assume all components have exponentially distributed times to failure with failure rates as shown in the Figure 11.6.

Suppose we want to know the reliability of the whole system for 6 months. Observe that you already know how to compute reliabilities for the subsystems $A$, $B_1$, and $B_2$. Let's review these computations and then see how we can use them to simplify our problem.

Subsystem $A$ is a standby redundant system, so we will use the Poisson model. We let

$$X = \text{the number of components which fail in one year.}$$

Since 6 months is 0.5 years, we seek $R_A(0.5) = P(X < 3)$ where $X$ follows a Poisson distribution with parameter $11 = \alpha t = (5)(0.5) = 2.5$. Then,

$$R_A(0.5) = P(X < 3) = P(0 \leq X \leq 2) = 0.5438.$$

Now, we consider subsystem $B_1$. In Section 11.2, we learned how to find individual component reliabilities when the time to failure followed an exponential distribution. For subsystem $B_1$, the failure rate is 3 per year, so our individual component reliability is

$$R(0.5) = 1 - F(0.5) = 1 - (1 - e^{-(3)(0.5)}) = e^{-(3)(0.5)} = 0.2231.$$

Now, recall that subsystem $B_1$ is an active redundant system where two components of the three must work for the subsystem to work. If we let

$$Y = \text{the number of components that function for 6 months}$$

and recognize that $Y$ follows a binomial distribution with $n = 3$ and $p = 0.2231$, we can quickly compute the reliability of the subsystem $B_1$ as follows:

$$R_B(0.5) = P(Y \geq 2) = 1 - P(Y < 2) = 1 - P(Y \leq 1) = 1 - 0.8729 = 0.1271.$$

Finally, we can look at subsystem $B_2$. Again, we use the fact that the failure times follow an exponential distribution. The subsystem consists of two components; obviously they both need to work for the subsystem to work. The first component's reliability is

$$R(0.5) = 1 - F(0.5) = 1 - \left(1 - e^{-(2)(0.5)}\right) = e^{-(2)(0.5)} = 0.3679.$$

and for the other component the reliability is

$$R(0.5) = 1 - F(0.5) = 1 - \left(1 - e^{-(1)(0.5)}\right) = e^{-(1)(0.5)} = 0.6065.$$

Therefore, the reliability of the subsystem is

$$R_B(0.5) = (0.3679)(0.6065) = 0.2231.$$

Our overall system can now be drawn as shown in Figure 11.7 below.

From here we determine the reliability of subsystem $B$ by treating it as a system of two independent components in parallel where only one component must work. Therefore,

$$R_B(0.5) = R_B(0.5) + R_B(0.5) - R_B(0.5)R_B(0.5) = 0.1271 + 0.2231 - (0.1271)(0.2231) = 0.3218.$$

Finally, since subsystems $A$ and $B$ are in series, we can find the overall system reliability for 6 months by taking the product of the two subsystem reliabilities:

$$R_{System}(0.5) = R_A(0.5) * R_B(0.5) = (0.5438)(0.3218) = 0.1750$$

We have used a network reduction approach to determine the reliability for a large-scale system for a given time period. Starting with those subsystems, which consist of components independent of other subsystems, we reduced the size of our network by evaluating each subsystem reliability one at a time.

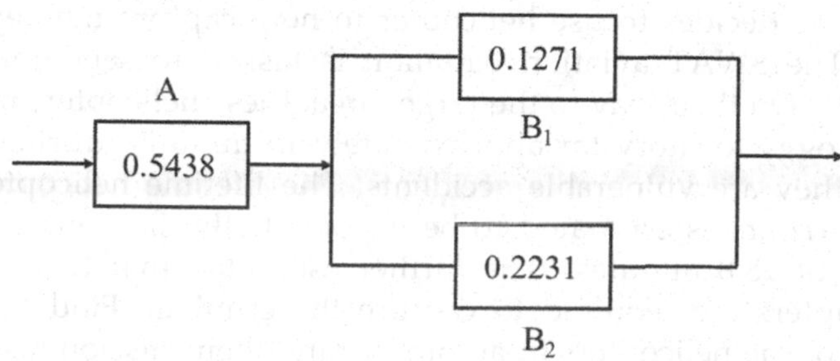


**FIGURE 11.7**
Simplified network example.

This approach works for any large-scale network consisting of basic subsystems of the type we have studied (series, active redundant, and standby redundant).

We have seen how methods from elementary probability can be used to model military reliability problems. The modeling approach presented here is useful in helping students simultaneously improve their understanding of both the military problems addressed and the mathematics behind these problems. The models presented also motivate students to appreciate the power of mathematics and its relevance to today's world.

## 11.7 Exercises

1. A continuous random variable, $Y$, representing the time to failure of a 0.50 mm tube, has a probability density function given by

$$f(y) = \begin{cases} 1/3e^{-y/3} & y \geq 0 \\ 0 & otherwise \end{cases}$$

   a. Find the reliability function for $Y$.
   b. Find the reliability for 1.2 time periods, $R$ (1.2).

2. The lifetime of a car engine (measured in time of operation) is exponentially distributed with a MTTF of 400 hours. You have received a mission that requires 12 hours of continuous operation. Your log book indicates that the car engine has been operating for 158 hours.
   Find the reliability of your engine for this mission.
   If your vehicle's engine had operated for 250 hours prior to the mission, find the reliability for the mission.

3. A criminal must be captured. You are on the police SWAT team when he decides to use helicopter to help capture the key criminal. The SWAT aviation battalion is tasked to send four helicopters. On their way to the target area, these helicopters must fly over foggy territory for approximately 15 minutes, during which time they are vulnerable accidents. The lifetime helicopter over this territory is estimated to be exponentially distributed with a mean of 18.8 minutes. It is further estimated that two or more helicopters are required to capture the criminal. Find the reliability of the helicopters in accomplishing their mission (assuming

the only reason a helicopter fails to reach the target is the foggy weather).

4. For the mission in Exercise 3, the police commissioner determines that, to justify risking the loss of helicopter, there must be at least an 80% chance of capturing the criminal. How many helicopters does the aviation battalion recommend to be sent? Justify your answer.
5. Mines are a dangerous obstacle. Most mines have three components – the firing device, the wire, and the mine itself (casing). If any of these components fail, the systems fails. These opponents of the mine start to "age" when they are unpacked from their sealed containers. All three components have MTTF that are exponentially distributed of 60 days, 300 days, and 35 days, respectively.
   a. Find the reliability of the mine after 90 days.
   b. What is the MTTF of the mine?
   c. What assumptions, if any, did you make?
6. You are a project manager for the new system being developed in Huntsville, Alabama. A critical subsystem has two components arranged in a parallel configuration. You have told the contractor that you require this subsystem to be at least 0.995 reliable. One of the subsystems came from an older system and has a known reliability of 0.95. What is the minimum reliability of the other component so that we meet our specifications?
7. You are in charge of stage lighting for an outdoor concert. There is some concern about the reliability of the lighting system for the stage. The lights are powered by a 1.5 kW generator that has a MTTF of 7.5 hours.
   a. Find the reliability of the generator for 10 hours if the generator's reliability is exponential.
   b. Find the reliability of the power system if two other identical 1.5 kW generators are available. First, consider as active redundant and then as stand-by redundant. Which would improve the reliability the most?
   c. How many generators would be necessary to ensure a 99% reliability?
8. Consider the system (Figure 11.8) with the reliability for each component as indicated. Assume all components are independent and the radars are active redundant.
   a. Find the system reliability for 6 months when $x = 0.96$.
   b. Find the system reliability for 6 months when $x = 0.939$.

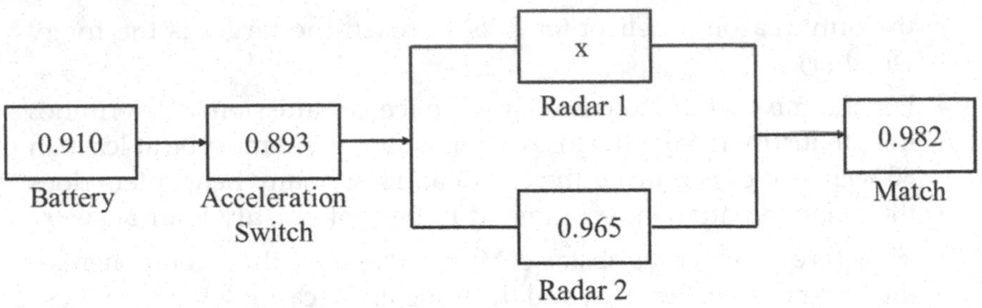


**FIGURE 11.8**
Radar system components and reliability.

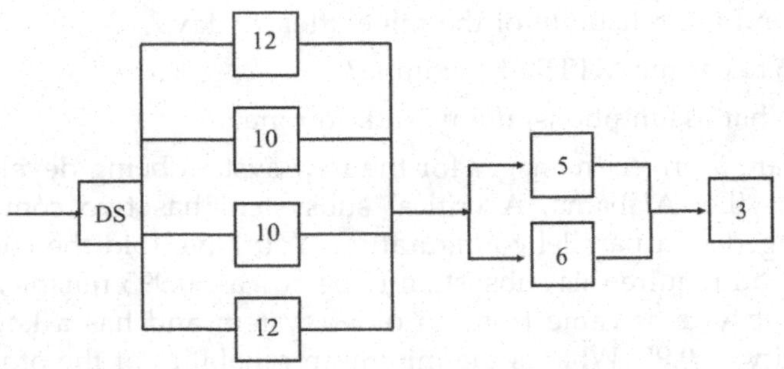


**FIGURE 11.9**
Major system components with mean time to failure.

9. A major system has the components as shown in Figure 11.9. All components have exponential times to failure with mean times to failure shown in hours. All components operate independently of each other. Find the reliability for this weapon system for 2 hours.

## Reference

Fox, W., and Horton, S. (1992). *Military Mathematical Modeling*, D. Arny (editor). USMA, West Point, NY.

# 12

# *Machine Learning and Unconstrained Optimal Process*

**OBJECTIVES**

1. Understand the processes for optimization with machine learning.
2. Understand "when" they are useful.
3. Understand advantages and disadvantages of methods.
4. Understand the use of training sets and test sets in linear regression.
5. Understand how to use simulated annealing and genetic algorithm for unconstrained optimization problems.

## 12.1 Introduction

Consider a situation where a process continually records data and we want to predict future yields based upon the given data. We might want to use machine learning to continually update the model in order to compute our predictions.

Machine learning is, in our opinion, another name for the use a computer to solve complex problems. In many ways, it is a method of testing and validating models in linear regression or other unconstrained optimization processes. We will not discuss constrained optimization process with machine learning here.

Many years ago in operation research programs, we were taught to break a large data set into parts, one for model building and the other for testing the model. In machine learning, these are now formalized into training sets and test sets. We suggest a method to do this. Create a random number between [0,1] and then use a partitioning of these numbers to select training and testing data points for these sets. For the most part, these methods of using training and testing sets are similar (if not identical) to the way we were taught.

DOI: 10.1201/9781003298762-12

**TABLE 12.1**

Gradient Data

| X | 1 | 3 | 5 |
|---|---|---|---|
| Y | 2.5 | 4.3 | 8 |

In practice, $x$ almost always represents multiple data points. So, for example, a housing price predictor might take not only square footage ($x_1$) but also number of bedrooms ($x_2$), number of bathrooms ($x_3$), size of yard ($x_4$), year built ($x_5$), zip code ($x_6$), and so forth. Determining which data inputs to use is an important part of machine language design. Again, we acknowledge that picking the variables is similar to stepwise regression, which includes and excludes variables based upon their ability to improve the model.

However, for the sake of our initial illustration of machine learning, it is easiest to assume a single input value is used. So, we can say that our simple predictor has this form: $h(x) = b_0 + b_1x$, where $b_0$ and $b_1$ are constants. Our goal is to find the best values of $b_0$ and $b_1$ to make our predictor value be as accurate as possible.

One method to use is a gradient search method. Gradient search has been previously covered, but we provide a slightly different algorithm here. We use a simple linear regression model (Radečić, 2020) of

$$h(x) = b_0 + b_1x$$

Assume for this illustration that we have the following data points (Table 12.1).

A scatter plot (Figure 12.1) would look like this, and we drew a subjective line through the points. This could have been any line, but we choose one that captured the trend and some data.

We estimate the equation of the line drawn in as $y = 1.5 + 2.5\ x$.

We certainly want our line to be better. The method we illustrate here uses a gradient search to minimize the sum of the squared error for linear regression. There are two approaches that might be used, so we briefly present each.

## 12.2 The Gradient Method

### 12.2.1 Gradient Method Algorithm 1

Choose an initial $b_0$ and $b_1$ values.

Compute the gradient of the function $S\ (y_i - (b_0 + b_1x_i))^2$.

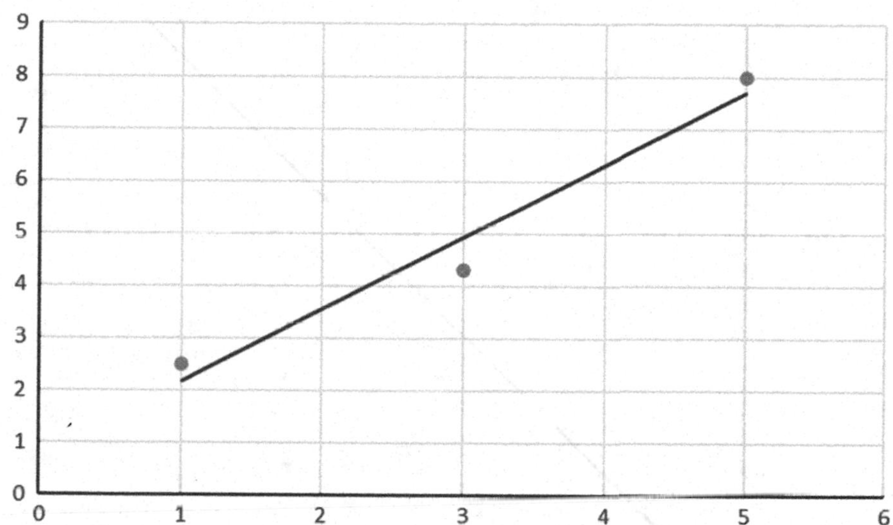

FIGURE 12.1
Excel scatter plot of gradient data.

Let the new point [new $b_0$, new $b_1$] = [$b_0$, $b_1$] – rate ∗ gradient

Iterate until the differences between $b_0$ and $b_1$ are very small.

Stop.

*Note:* The choice of the rate makes a big difference in the accuracy of the model.

**Example 12.1:** Gradient Search Method

We continue our analysis of the points and iterate to obtain better estimates for $b_0$ and $b_1$.

We need our sum of squared error function and its gradient.

Our function is

$$(2.5 + (-b_0 - b_1)^{\wedge 2} + (4.3 + (-b_0 - 3 * b_1)^{\wedge 2} + (8 + (-b_0 - 5 * b_1)^{\wedge 2} + (8 + (-b_0 - 4 * b_1)^{\wedge 2}$$

The gradient is $[-45.6 + 8b_0 + 26b_1, -174.8 + 26b_0 + 102b_1]$

After many experiments, we choose a rate of 0.005. We compute the sum of squared error and examine our loss curve (a parabola with a minimum at 1.2163); we find the minimum occurred at iteration 349 with $b_0 = 0.8080$ and $b_1 = 1.5075$ and a sum of squared error of 1.2163.

The model chosen by this method is $y = 0.8080 + 1.5076\ x$. We provide a visual of the data and model in Figure 12.2.

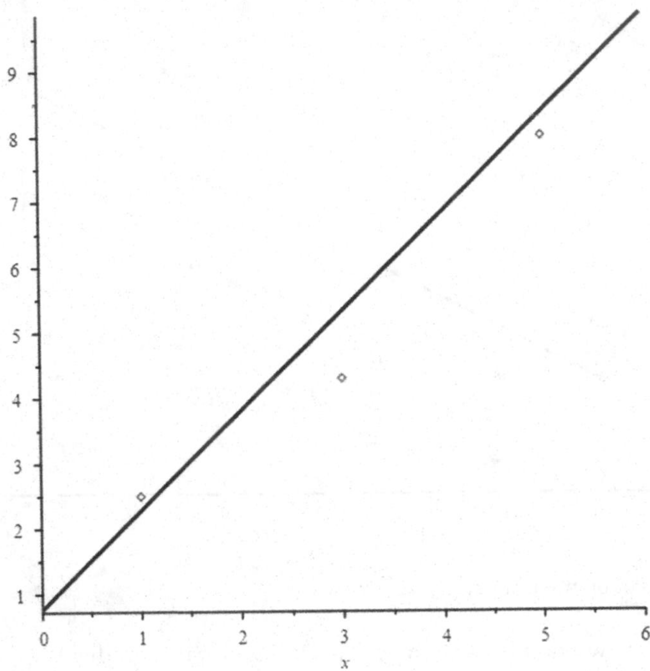


**FIGURE 12.2**
Plot of data and model for Example 12.1.

### 12.2.2 Gradient Method Algorithm 2

Most computers have software that will solve regression problems for simple linear regression. Using unconstrained minimization of the function with a better gradient search method, we find the "best" result is $y = 0.8083$ to $1.375\,x$ with a sum of squared error of 0.6017. First, we note that our sum of squared error is about half of what it was using method 1. Figure 12.3 provides a screen shot of the use of the Solver in Excel using RGR.

In regard to machine learning to obtain a regression model, we recommend using the functions available on the computer to obtain our "best" solution. If there does not exist a routine on the package such as exponential regression in Excel, then the previous methods work well.

We used a simple problem to illustrate the methods, but the reason machine learning exists is because, in the real world, the problems are much more complex. On a computer or a black board, we can draw you a picture of, at most, a 3D data set, but machine learning problems commonly deal with data with millions of dimensions, and very complex predictor functions. Machine learning solves problems that cannot be solved by implementing numerical means alone.

| b0 | b1 | x | y | |
|---|---|---|---|---|
| 0.808333 | 1.375 | 1 | 2.5 | 0.100278 |
| | | 3 | 4.3 | 0.401109 |
| | | 5 | 8 | 0.100279 |
| | | | | 0.601667 |

**FIGURE 12.3**
Screenshot from Excel for Example 12.1.

For example, in Excel we have only simple regression features and no features for exponential or sine regression. We need to understand the basics if we have to produce such models.

Optimizing the predictor *h(x)* is done using **training examples**. For each training example, we have an input value *x_train*, for which a corresponding output, *y*, is known in advance. For each example, we find the difference between the known, correct value *y*, and our predicted value *h (x_train)*. With enough training examples, these differences give us a useful way to measure the "differences or errors" of *h(x)*. We can then tweak *h(x)* by tweaking the values of $b_0$ and $b_1$ to make it "less wrong". This process is repeated over and over until the system has converged on the best values for $b_0$ and $b_1$. In this way, the predictor becomes trained, and is ready to do some real-world predicting.

### 12.2.3 Training and Test Sets in Machine Learning

We can tackle a larger problem using a two-step method. With that in mind, we can explore another simple example. Say we have the following training data, wherein company employees have rated their satisfaction on a scale of 1 to 100, shown in Figures 12.4–12.5:

We notice a *"reasonable"* (but not perfect) linear relationship between the data. That is, while we can see that there is a pattern to it (i.e. employee

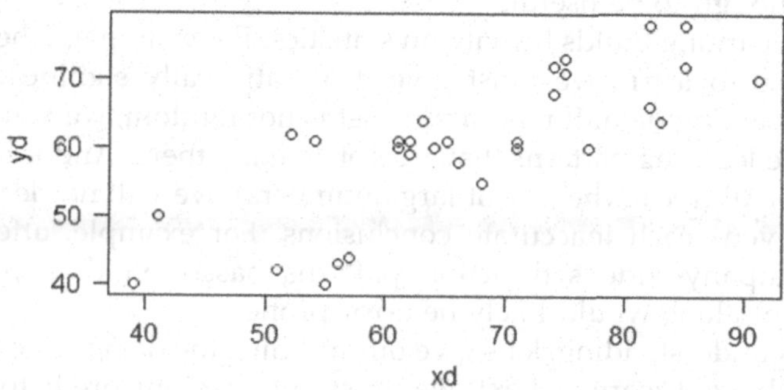


**FIGURE 12.4**
Scatterplot of satisfaction data.

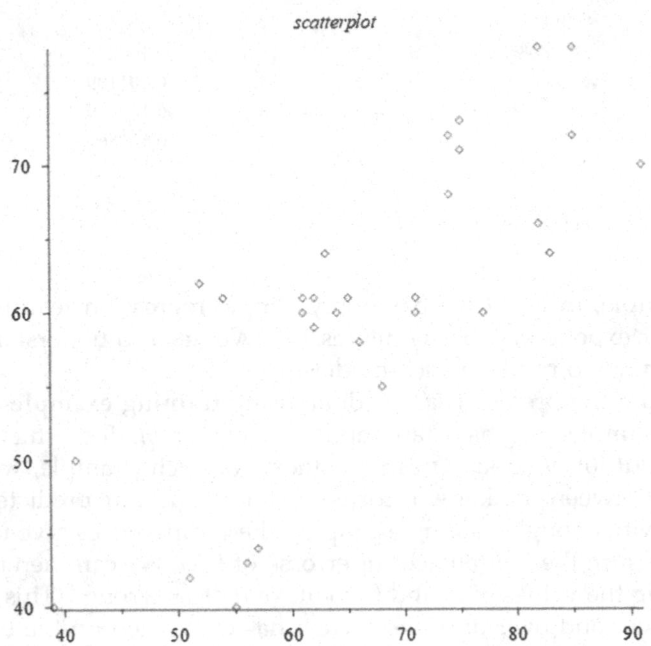


**FIGURE 12.5**
Scatterplot of satisfaction versus salary data.

satisfaction tends to go up as salary goes up), it does not all fit neatly on a straight line. This is usually true for all real data sets that the pattern will not be a perfect fit. This will always be the case with real-world data (and we absolutely want to train our machine using real-world data!). So then, how can we train a machine to perfectly predict an employee's level of satisfaction? The answer, of course, is that we can't. The goal of machine learning is never to make "perfect" guesses because machine learning deals in domains where there is no such thing. The goal is to make guesses that are good enough to be useful.

Machine learning builds heavily on statistics. For example, when we train our machine to learn, we must give it a statistically significant random sample as training data. If the training set is not random, we run the risk of the machine learning patterns that are not actually there. And if the training set is too small (recall the law of large numbers), we will not learn enough and may even reach inaccurate conclusions. For example, attempting to predict company-wide satisfaction patterns based on data from upper management alone would likely be error-prone.

With this understanding, let's give our machine the data we've been given above and have it learn it. First, we must initialize our predictor $h(x)$ with some reasonable values of $b_0$ and $b_1$. "Eyeball" methods might provide a reasonable guess, such as $h(x) = 30 + 0.5x$.

We could have picked a ridiculous initial equation, say $h(x) = 12 + 0.25 * x$. Now our predicting equation for the latter initial guess looks like this (Figure 12.6) when placed over our training set:

Using our better initial guess, we get a plot that looks like Figure 12.7 that captures the trend and the data more accurately than the other guess.

If we ask this predictor for the satisfaction of an employee making $60k, it will predict a rating of 60 for our better initial guess and about 27 for the poorer guess. The latter initial guess was a terrible guess and shows that this machine does not know very much.

So now, we can use *all* the salaries from our training set and take the differences between the resulting predicted satisfaction ratings and the actual satisfaction ratings of the corresponding employees. Using Method 1, with very high certainty, that values of 13.12 for $b_0$ and 0.61 for $b_1$ are going to give us a better predictor.

$$h(x) = 13.12 + 0.61x$$

And if we repeat this process, say 1,500 times, our predictor will end up looking like this (Figure 12.8):

$$h(x) = 15.54 + 0.75x$$

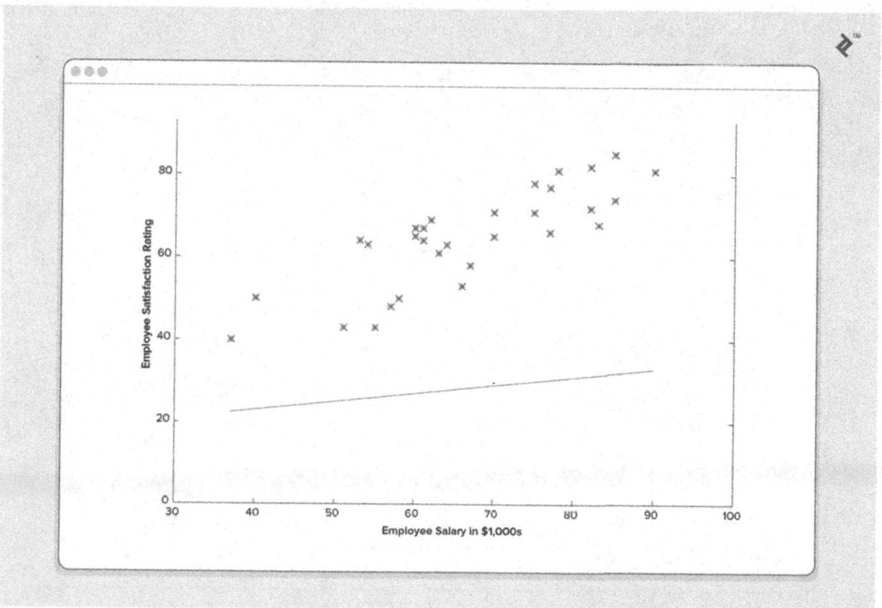


**FIGURE 12.6**
Data and initial poorer guess of an model.

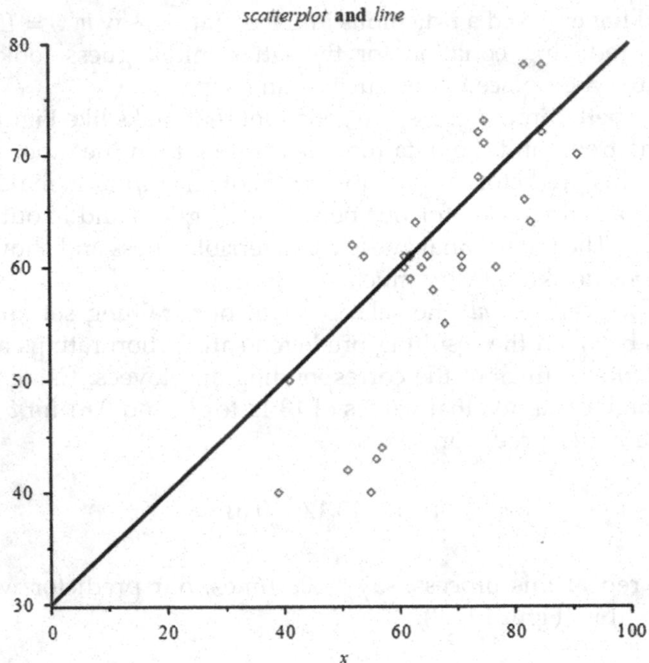


FIGURE 12.7
Data and better initial model.

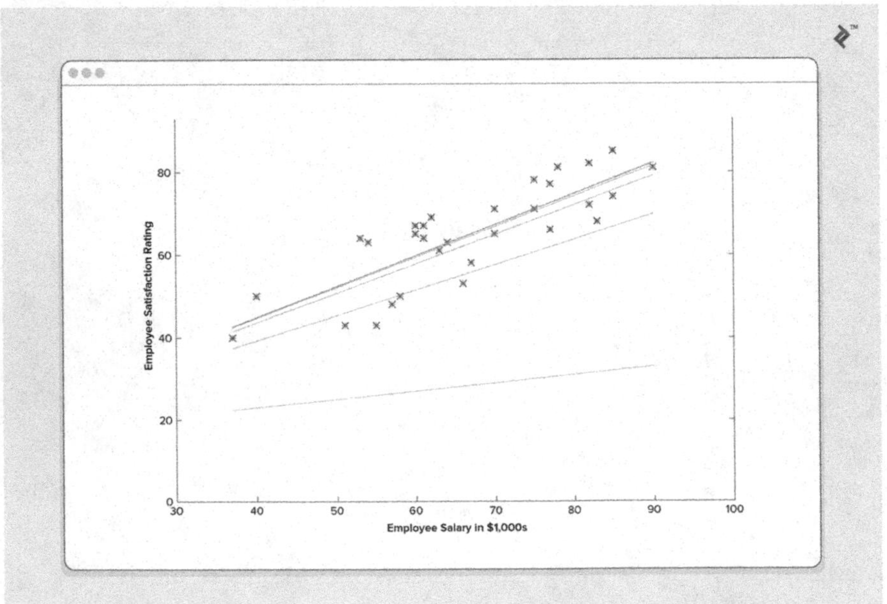


FIGURE 12.8
Better models emerge.

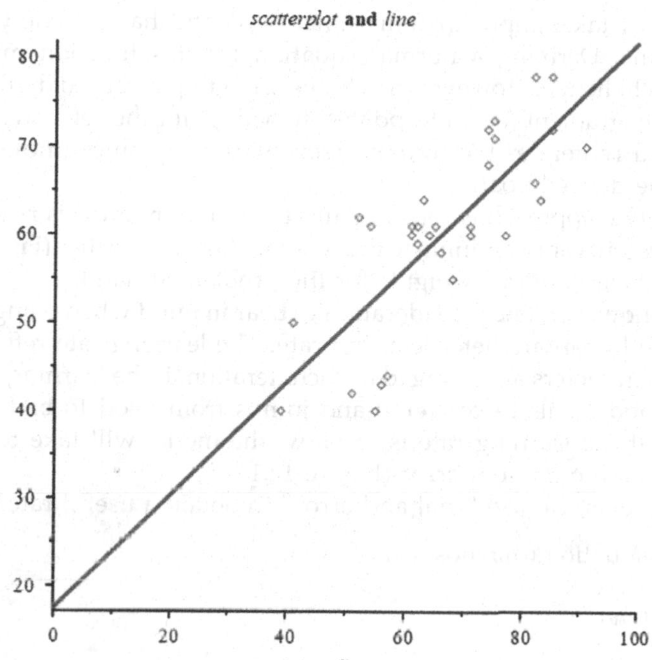


**FIGURE 12.9**
Data and least squares model.

At this point, if we repeat the process, we will find that $b_0$ and $b_1$ will not change by any appreciable amount anymore, and thus we see that the system has converged. If we have not made any mistakes, this means we've found the optimal predictor. Accordingly, if we now ask the machine again for the satisfaction rating of the employee who makes $60k, it will predict a rating of roughly 60.

Now, we are getting somewhere (Figure 12.9).

Using linear regression, our least squares model is $h(x) = 17.433 + 0.645(x)$, and our prediction would be $h(60) = 56.1246$.

## 12.3 Machine Learning Regression: A Note on Complexity

Example 12.1 is technically a simple problem of simple linear regression, which in reality can be solved by deriving a simple set of normal equation and skipping this "tuning" process altogether.

However, consider a predictor that looks like this

$$h(x_1, x_2, x_3, x_4) = \theta_0 + \theta_1 x_1 + \theta_2 x_3^2 + \theta_3 x_3 x_4 + \theta_4 x_1^3 x_2^2 + \theta_5 x_2 x_3^4 x_4^2:$$

This function takes input in four dimensions and has a variety of polynomial terms. Deriving a normal equation for this function might be a significant challenge. However, machines are set up nicely to do this. Based on the slope, gradient descent updates the values for the set of weights and the bias and re-iterates the training loop over new values (moving a step closer to the desired goal).

This iterative approach is repeated until a minimum error is reached, and gradient descent cannot minimize the cost function any further (Figure 12.10).

The results are optimal weights for the problem at hand.

There is, however, one consideration to bear in mind when using gradient descent: the hyperparameter learning rate. The learning rate refers to how much the parameters are changed at each iteration. If the learning rate is too high, the model fails to converge and jumps from good to bad cost optimizations. If the learning rate is too low, the model will take too long to converge to the minimum error (Figure 12.11).

In our models, we use "trial and error" to obtain a useful rate.

Model evaluation (diagnostics)

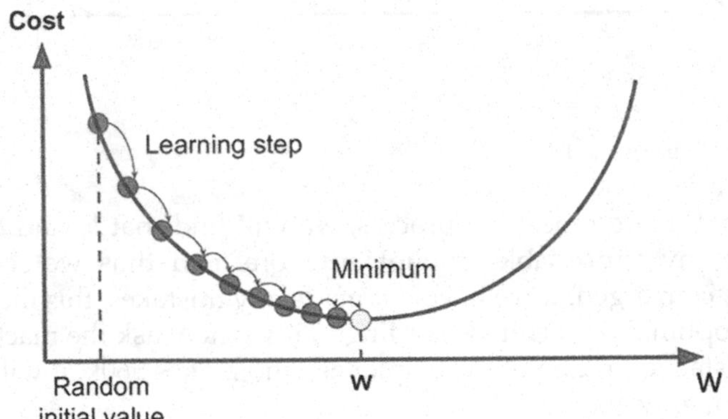


**FIGURE 12.10**
Loss function.

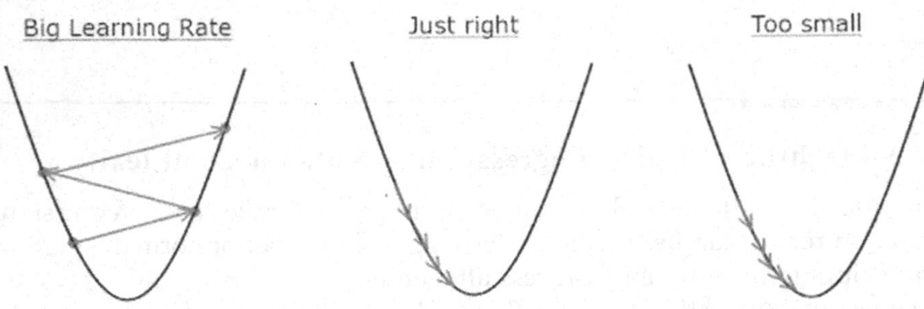


**FIGURE 12.11**
Effect of rate on loss function.

How do we evaluate the accuracy of our model? First of all, you need to make sure that you train the model on the training data set and build evaluation metrics on the test set to avoid overfitting. Afterward, you can check several evaluation metrics to determine how well your model performed. Among these are $R^2$, MSE, and RMSE.

There are various metrics to evaluate the goodness of fit:

Mean Squared Error (MSE). MSE is computed as RSS divided by the total number of data points, i.e. the total number of observations or examples in our given dataset. MSE tells us what the average RSS is per data point.

Root Mean Squared Error (RMSE). RMSE takes the MSE value and applies a square root over it. It is similar to MSE, but much more intuitive for error interpretation. It is equivalent to the absolute error between our linear regression line and any hypothetical observation point. Unlike MSE and RSS (which use squared values), RMSE can be directly used to interpret the "average error" that our prediction model makes.

R2 or R-squared or R2 score. R-squared is a measure of how much variance in the dependent variable that our linear function accounts for. This measure is more technical than the other two, so it's less intuitive for a non-statistician. As a rule of thumb, an R-squared value that is closer to 1 is better, because it accounts for more variance.

Once we have trained and evaluated our model, we improve it to make more accurate predictions.

## 12.4 Genetic Algorithm as Machine Learning in R

**Genetic Algorithms:** In the field of artificial intelligence, it is a search heuristic that mimics the process of natural selection. This heuristic is routinely used to generate useful solutions. Genetic algorithms belong to the larger class of evolutionary algorithms (EA), which generate solutions to optimization problems using techniques inspired by natural evolution.

This notion can be applied for a search problem. We consider a set of solutions for a problem and select the set of best ones out of them.

Five phases are considered in a genetic algorithm.

Initial population

Fitness function

Selection

Crossover

Mutation

## 12.5 Initial Population

The process begins with a set of individuals, which is called a **Population**. Each individual is a solution to the problem you want to solve.

We illustrate a couple of examples (Example 12.3 and 12.4) using source code in R by Luca Scrucca. The documentation and source code is obtained from the CRAN library.

(URL https://luca-scr.github.io/GA/BugReports https://github.com/luca-scr/GA/issues Repository CRAN ByteCompile true NeedsCompilation yes Author Luca Scrucca [aut, cre] () Maintainer Luca Scrucca)

We illustrate with two examples.

**Example 12.2:** Machine Learning 1

We will assume that $f(x) = (x^2 + x) * \cos(x)$, and we can easily plot this in R (Figure 12.12).

We see the maximum of this function is above 49 at a value of $x$ between 6 and 7, see Figure 12.12.

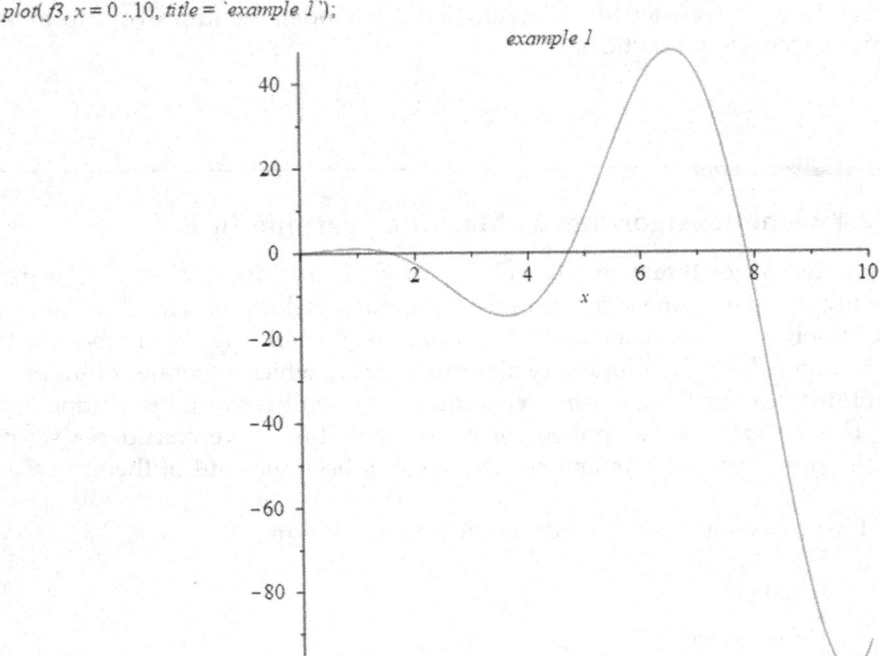


**FIGURE 12.12**
Plot of Example 12.2, $f(x) = (x^2+x) * cos(x)$.

We can now implement a Genetic Algorithm in R.

```
> ## — Genetic Algorithm ————————————
> ##
> ## GA settings:
> ## Type = real-valued
> ## Population size = 50
> ## Number of generations = 100
> ## Elitism = 2
> ## Crossover probability = 0.8
> ## Mutation probability = 0.1
> ## Search domain =
> ## th
> ## lower −10
> ## upper 10
> ##
> ## GA results:
> ## Iterations = 100
> ## Fitness function value = 47.70562
> ## Solution =
> ## th
> ## [1,] 6.560548
> plot (GA)
> install.packages ("GA")
> library (GA)
Loading required package: foreach
Loading required package: iterators
```

## Genetic Algorithm Output

```
Type 'citation("GA")' for citing this R package in publications.
Attaching package: 'GA'
The following object is masked from 'package:utils':
  de
```

```
Warning messages:
1: package 'GA' was built under R version 4.0.5
2: package 'foreach' was built under R version 4.0.5
3: package 'iterators' was built under R version 4.0.5
> f <-function(x) (x^2+x) * cos(x)
> lbound <- -10; ubound <- 10
> curve(f, from = lbound, to = ubound, n = 1000)
> GA <-ga(type = "real-valued", fitness = f, lower = c(th = lbound),
  upper = ubound)
GA | iter = 1  | Mean = 1.73261  | Best = 44.36230
GA | iter = 2  | Mean = 5.432003 | Best = 47.581546
GA | iter = 3  | Mean = 9.634557 | Best = 47.671815
GA | iter = 4  | Mean = 15.58913 | Best = 47.67182
GA | iter = 5  | Mean = 22.45189 | Best = 47.70322
GA | iter = 6  | Mean = 24.11626 | Best = 47.70526
GA | iter = 7  | Mean = 16.90515 | Best = 47.70526
GA | iter = 8  | Mean = 21.59528 | Best = 47.70526
GA | iter = 9  | Mean = 24.58188 | Best = 47.70526
GA | iter = 10 | Mean = 35.08532 | Best = 47.70526
GA | iter = 11 | Mean = 37.82867 | Best = 47.70549
GA | iter = 12 | Mean = 43.60740 | Best = 47.70549
GA | iter = 13 | Mean = 42.55550 | Best = 47.70549
GA | iter = 14 | Mean = 43.98262 | Best = 47.70549
GA | iter = 15 | Mean = 31.56686 | Best = 47.70549
GA | iter = 16 | Mean = 37.01424 | Best = 47.70549
GA | iter = 17 | Mean = 32.53054 | Best = 47.70549
GA | iter = 18 | Mean = 36.26340 | Best = 47.70549
GA | iter = 19 | Mean = 37.06366 | Best = 47.70549
GA | iter = 20 | Mean = 35.26582 | Best = 47.70549
GA | iter = 21 | Mean = 35.07788 | Best = 47.70549
GA | iter = 22 | Mean = 34.15623 | Best = 47.70549
```

GA | iter = 23 | Mean = 35.32545 | Best = 47.70549

GA | iter = 24 | Mean = 41.40310 | Best = 47.70549

GA | iter = 25 | Mean = 40.42410 | Best = 47.70549

GA | iter = 26 | Mean = 36.48381 | Best = 47.70549

GA | iter = 27 | Mean = 34.41219 | Best = 47.70549

GA | iter = 28 | Mean = 31.02465 | Best = 47.70549

GA | iter = 29 | Mean = 40.22317 | Best = 47.70549

GA | iter = 30 | Mean = 32.85867 | Best = 47.70549

GA | iter = 31 | Mean = 39.77480 | Best = 47.70549

GA | iter = 32 | Mean = 43.62557 | Best = 47.70549

GA | iter = 33 | Mean = 40.82960 | Best = 47.70549

GA | iter = 34 | Mean = 36.74009 | Best = 47.70549

GA | iter = 35 | Mean = 37.12876 | Best = 47.70549

GA | iter = 36 | Mean = 33.75108 | Best = 47.70549

GA | iter = 37 | Mean = 33.66556 | Best = 47.70561

GA | iter = 38 | Mean = 36.47489 | Best = 47.70561

GA | iter = 39 | Mean = 35.02810 | Best = 47.70561

GA | iter = 40 | Mean = 36.67345 | Best = 47.70561

GA | iter = 41 | Mean = 36.28483 | Best = 47.70561

GA | iter = 42 | Mean = 32.01288 | Best = 47.70561

GA | iter = 43 | Mean = 37.48237 | Best = 47.70561

GA | iter = 44 | Mean = 40.64395 | Best = 47.70561

GA | iter = 45 | Mean = 39.42224 | Best = 47.70561

GA | iter = 46 | Mean = 35.18709 | Best = 47.70561

GA | iter = 47 | Mean = 39.50457 | Best = 47.70561

GA | iter = 48 | Mean = 41.17369 | Best = 47.70561

GA | iter = 49 | Mean = 37.94046 | Best = 47.70561

GA | iter = 50 | Mean = 36.08178 | Best = 47.70561

GA | iter = 51 | Mean = 35.98765 | Best = 47.70561

GA | iter = 52 | Mean = 39.09733 | Best = 47.70561

GA | iter = 53 | Mean = 39.23104 | Best = 47.70561

GA | iter = 54 | Mean = 31.74235 | Best = 47.70561

```
GA | iter = 55 | Mean = 37.03128 | Best = 47.70561
GA | iter = 56 | Mean = 31.76054 | Best = 47.70561
GA | iter = 57 | Mean = 32.88019 | Best = 47.70561
GA | iter = 58 | Mean = 28.09433 | Best = 47.70561
GA | iter = 59 | Mean = 27.83565 | Best = 47.70561
GA | iter = 60 | Mean = 32.85265 | Best = 47.70561
GA | iter = 61 | Mean = 36.94046 | Best = 47.70561
GA | iter = 62 | Mean = 43.50529 | Best = 47.70561
GA | iter = 63 | Mean = 41.65105 | Best = 47.70561
GA | iter = 64 | Mean = 43.84563 | Best = 47.70561
GA | iter = 65 | Mean = 45.61793 | Best = 47.70561
GA | iter = 66 | Mean = 38.65638 | Best = 47.70561
GA | iter = 67 | Mean = 32.51981 | Best = 47.70561
GA | iter = 68 | Mean = 33.53754 | Best = 47.70561
GA | iter = 69 | Mean = 30.96679 | Best = 47.70561
GA | iter = 70 | Mean = 34.41130 | Best = 47.70561
GA | iter = 71 | Mean = 38.19444 | Best = 47.70561
GA | iter = 72 | Mean = 42.81060 | Best = 47.70561
GA | iter = 73 | Mean = 41.93909 | Best = 47.70561
GA | iter = 74 | Mean = 36.08957 | Best = 47.70561
GA | iter = 75 | Mean = 38.35492 | Best = 47.70561
GA | iter = 76 | Mean = 35.85461 | Best = 47.70561
GA | iter = 77 | Mean = 35.01293 | Best = 47.70561
GA | iter = 78 | Mean = 34.85520 | Best = 47.70561
GA | iter = 79 | Mean = 34.28683 | Best = 47.70561
GA | iter = 80 | Mean = 37.54808 | Best = 47.70561
GA | iter = 81 | Mean = 37.38300 | Best = 47.70562
GA | iter = 82 | Mean = 38.18646 | Best = 47.70562
GA | iter = 83 | Mean = 32.90659 | Best = 47.70562
GA | iter = 84 | Mean = 35.53161 | Best = 47.70562
GA | iter = 85 | Mean = 36.35022 | Best = 47.70562
GA | iter = 86 | Mean = 40.50274 | Best = 47.70562
```

```
GA | iter = 87 | Mean = 39.93693 | Best = 47.70562
GA | iter = 88 | Mean = 38.52067 | Best = 47.70562
GA | iter = 89 | Mean = 38.44509 | Best = 47.70562
GA | iter = 90 | Mean = 45.26107 | Best = 47.70562
GA | iter = 91 | Mean = 42.80682 | Best = 47.70562
GA | iter = 92 | Mean = 41.45643 | Best = 47.70562
GA | iter = 93 | Mean = 40.11646 | Best = 47.70562
GA | iter = 94 | Mean = 42.86939 | Best = 47.70562
GA | iter = 95 | Mean = 44.79701 | Best = 47.70562
GA | iter = 96 | Mean = 37.87120 | Best = 47.70562
GA | iter = 97 | Mean = 40.82283 | Best = 47.70562
GA | iter = 98 | Mean = 33.03996 | Best = 47.70562
GA | iter = 99 | Mean = 31.76110 | Best = 47.70562
GA | iter = 100 | Mean = 32.91774 | Best = 47.70562
> summary(GA)
-- Genetic Algorithm -------------------

GA settings:
Type                   = real-valued
Population size        = 50
Number of generations  = 100
Elitism                = 2
Crossover probability  = 0.8
Mutation probability   = 0.1
Search domain =
       th

lower –10
upper 10

GA results:
      Iterations = 100
      Fitness function value = 47.70562
      Solution =
```

```
th
[1,] 6.560527
> ## — Genetic Algorithm ——————————————
> ##
> ## GA settings:
> ## Type = real-valued
> ## Population size = 50
> ## Number of generations = 100
> ## Elitism = 2
> ## Crossover probability = 0.8
> ## Mutation probability = 0.1
> ## Search domain =
> ##        th
> ## lower −10
> ## upper 10
> ##
> ## GA results:
> ## Iterations = 100
> ## Fitness function value = 47.70562
> ## Solution =
> ## [1,] 6.560548
```

**Example 12.3:** Machine Learning 2

We will assume that $f(x) = e^{(x\wedge2)} * cos(x^2) * sqrt(x)$

We see the plot of the function in Figure 12.13 and the genetic algorithm in Figure 12.14.

```
> plot(GA)
> f <- function(x) exp(x^2) * cos(x^2) * sqrt(x)
> > lbound <–0; ubound <–1.25
Error: unexpected '>' in ">"
> > curve(f, from = lbound, to = ubound, n = 1000)
Error: unexpected '>' in ">"
```

```
> lbound<–0;ubound<–1.25
> curve(f, from = lbound, to = ubound, n = 1000)
> GA <- ga(type = "real-valued", fitness = f, lower = c(th = lbound), upper =
      ubound)
GA | iter = 1  | Mean = 0.9212282 | Best = 1.4817843
GA | iter = 2  | Mean = 1.155651 | Best = 1.488740
GA | iter = 3  | Mean = 1.266898 | Best = 1.489162
GA | iter = 4  | Mean = 1.437535 | Best = 1.489214
GA | iter = 5  | Mean = 1.419376 | Best = 1.489246
GA | iter = 6  | Mean = 1.420120 | Best = 1.489247
GA | iter = 7  | Mean = 1.372686 | Best = 1.489247
GA | iter = 8  | Mean = 1.365188 | Best = 1.489247
GA | iter = 9  | Mean = 1.447480 | Best = 1.489247
GA | iter = 10 | Mean = 1.425594 | Best = 1.489247
GA | iter = 11 | Mean = 1.439862 | Best = 1.489247
GA | iter = 12 | Mean = 1.408615 | Best = 1.489247
GA | iter = 13 | Mean = 1.390561 | Best = 1.489247
GA | iter = 14 | Mean = 1.394602 | Best = 1.489247
GA | iter = 15 | Mean = 1.434093 | Best = 1.489247
GA | iter = 16 | Mean = 1.394898 | Best = 1.489247
GA | iter = 17 | Mean = 1.443526 | Best = 1.489247
GA | iter = 18 | Mean = 1.436073 | Best = 1.489247
GA | iter = 19 | Mean = 1.433979 | Best = 1.489247
GA | iter = 20 | Mean = 1.402248 | Best = 1.489247
GA | iter = 21 | Mean = 1.342701 | Best = 1.489247
GA | iter = 22 | Mean = 1.368034 | Best = 1.489247
GA | iter = 23 | Mean = 1.422016 | Best = 1.489247
GA | iter = 24 | Mean = 1.424531 | Best = 1.489247
GA | iter = 25 | Mean = 1.367991 | Best = 1.489247
GA | iter = 26 | Mean = 1.439773 | Best = 1.489247
GA | iter = 27 | Mean = 1.418078 | Best = 1.489247
GA | iter = 28 | Mean = 1.353344 | Best = 1.489247
```

GA | iter = 29 | Mean = 1.391622 | Best = 1.489247

GA | iter = 30 | Mean = 1.376262 | Best = 1.489247

GA | iter = 31 | Mean = 1.402512 | Best = 1.489247

GA | iter = 32 | Mean = 1.424066 | Best = 1.489247

GA | iter = 33 | Mean = 1.452097 | Best = 1.489247

GA | iter = 34 | Mean = 1.418564 | Best = 1.489247

GA | iter = 35 | Mean = 1.469314 | Best = 1.489247

GA | iter = 36 | Mean = 1.430905 | Best = 1.489247

GA | iter = 37 | Mean = 1.424344 | Best = 1.489247

GA | iter = 38 | Mean = 1.447633 | Best = 1.489247

GA | iter = 39 | Mean = 1.439482 | Best = 1.489247

GA | iter = 40 | Mean = 1.376854 | Best = 1.489247

GA | iter = 41 | Mean = 1.381366 | Best = 1.489247

GA | iter = 42 | Mean = 1.390957 | Best = 1.489247

GA | iter = 43 | Mean = 1.363667 | Best = 1.489247

GA | iter = 44 | Mean = 1.445264 | Best = 1.489247

GA | iter = 45 | Mean = 1.359032 | Best = 1.489247

GA | iter = 46 | Mean = 1.430421 | Best = 1.489247

GA | iter = 47 | Mean = 1.415601 | Best = 1.489247

GA | iter = 48 | Mean = 1.410380 | Best = 1.489247

GA | iter = 49 | Mean = 1.431569 | Best = 1.489247

GA | iter = 50 | Mean = 1.423205 | Best = 1.489247

GA | iter = 51 | Mean = 1.429360 | Best = 1.489247

GA | iter = 52 | Mean = 1.395978 | Best = 1.489247

GA | iter = 53 | Mean = 1.382839 | Best = 1.489247

GA | iter = 54 | Mean = 1.442466 | Best = 1.489247

GA | iter = 55 | Mean = 1.385382 | Best = 1.489247

GA | iter = 56 | Mean = 1.341015 | Best = 1.489247

GA | iter = 57 | Mean = 1.380760 | Best = 1.489247

GA | iter = 58 | Mean = 1.367469 | Best = 1.489247

GA | iter = 59 | Mean = 1.459464 | Best = 1.489247

GA | iter = 60 | Mean = 1.452287 | Best = 1.489247

```
GA | iter = 61 | Mean = 1.474943 | Best = 1.489247
GA | iter = 62 | Mean = 1.459396 | Best = 1.489247
GA | iter = 63 | Mean = 1.446360 | Best = 1.489247
GA | iter = 64 | Mean = 1.361241 | Best = 1.489247
GA | iter = 65 | Mean = 1.446171 | Best = 1.489247
GA | iter = 66 | Mean = 1.429072 | Best = 1.489247
GA | iter = 67 | Mean = 1.382108 | Best = 1.489247
GA | iter = 68 | Mean = 1.357921 | Best = 1.489247
GA | iter = 69 | Mean = 1.420132 | Best = 1.489247
GA | iter = 70 | Mean = 1.433171 | Best = 1.489247
GA | iter = 71 | Mean = 1.454016 | Best = 1.489247
GA | iter = 72 | Mean = 1.387553 | Best = 1.489247
GA | iter = 73 | Mean = 1.407271 | Best = 1.489247
GA | iter = 74 | Mean = 1.436927 | Best = 1.489247
GA | iter = 75 | Mean = 1.442956 | Best = 1.489247
GA | iter = 76 | Mean = 1.426882 | Best = 1.489247
GA | iter = 77 | Mean = 1.401205 | Best = 1.489247
GA | iter = 78 | Mean = 1.388928 | Best = 1.489247
GA | iter = 79 | Mean = 1.403899 | Best = 1.489247
GA | iter = 80 | Mean = 1.366423 | Best = 1.489247
GA | iter = 81 | Mean = 1.364520 | Best = 1.489247
GA | iter = 82 | Mean = 1.272635 | Best = 1.489247
GA | iter = 83 | Mean = 1.420868 | Best = 1.489247
GA | iter = 84 | Mean = 1.362094 | Best = 1.489247
GA | iter = 85 | Mean = 1.287870 | Best = 1.489247
GA | iter = 86 | Mean = 1.376006 | Best = 1.489247
GA | iter = 87 | Mean = 1.421613 | Best = 1.489247
GA | iter = 88 | Mean = 1.431365 | Best = 1.489247
GA | iter = 89 | Mean = 1.372076 | Best = 1.489247
GA | iter = 90 | Mean = 1.371904 | Best = 1.489247
GA | iter = 91 | Mean = 1.389290 | Best = 1.489247
GA | iter = 92 | Mean = 1.407730 | Best = 1.489247
```

```
GA | iter = 93 | Mean = 1.433815 | Best = 1.489247
GA | iter = 94 | Mean = 1.416462 | Best = 1.489247
GA | iter = 95 | Mean = 1.417751 | Best = 1.489247
GA | iter = 96 | Mean = 1.441168 | Best = 1.489247
GA | iter = 97 | Mean = 1.444353 | Best = 1.489247
GA | iter = 98 | Mean = 1.434745 | Best = 1.489247
GA | iter = 99 | Mean = 1.413124 | Best = 1.489247
GA | iter = 100 | Mean = 1.414792 | Best = 1.489247
> summary(GA)
-- Genetic Algorithm -------------------

GA settings:
Type = real-valued
Population size = 50
Number of generations = 100
Elitism = 2
Crossover probability = 0.8
Mutation probability = 0.1
Search domain =
    th
lower 0.00
upper 1.25

GA results:
Iterations = 100
Fitness function value = 1.489247
Solution =
    th
[1,] 0.9518592
> plot(GA)
```

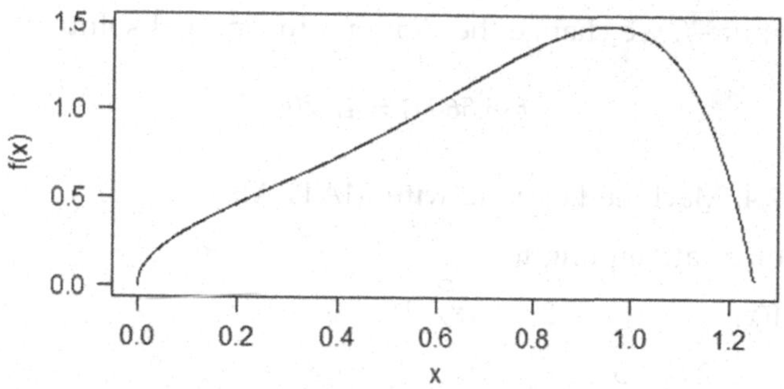


**FIGURE 12.13**
Plot of $f(x) = e^{(x^\wedge 2)} * cos(x^2) * sqrt(x)$.

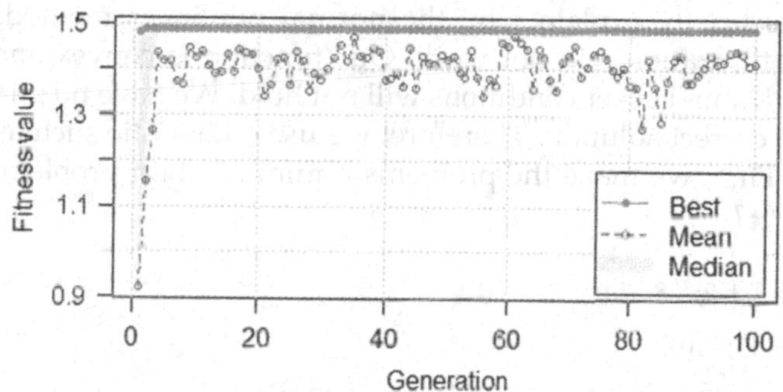


**FIGURE 12.14**
Plot from the genetic algorithm bounding at the optimal value.

Thus far, we prefer the R code as it provides the iterative results.

In MATLAB, there is an internal genetic algorithm within the Optimization Toolbox.

The commands are straight forward;

```
ga(function, nvars,[],[],[],[], LB, UB)
```

Students requiring constrained genetic algorithm should refer to MATLAB help commands. The default in MATLAB is to minimize the function. We repeat example 1 but since MATLAB requires us to minimize we input $-f$.

```
>> [x fval] = ga(f, nars, [], [], [], []. LB, UB

x = 6.5606

fval = −47.7056
```

Since we input $-f$, we change the sign on our optimal solution.

$$f(6.5606) = 47.7056$$

**Example 12.4:** Machine Learning with MATLAB

We will look at attempting to

Maximize $f(x_1, x_2) = (x_1 - 2)^2 + (x_2 - 2)^2$

Subject to $x_1 + 2x_2 \leq 3$

$8x_1 + 5x_2 \geq 10$

$x_1, x_2 \geq 0$

First, let us briefly explain why the normal nonlinear methods for constrained optimization will not work. Our function is convex and not concave so the Kuhn-Tucker conditions will not hold. We have no guarantee for finding the correct solution. Therefore, we use a heuristic such as a genetic algorithm. First, we make the problems a minimization problem by multiplying $f$ by $-1$.

```
>> a = [ 1 2; -8 -5];
>> b = [3;-10];
>> LB = [0,0];
>> UB = [3,3];
>> ObjectiveFunction=@simple_fitness;
>> nvars = 2;
>> [x,fval] = ga(ObjectiveFunction, nvars, a,b,[],[],LB,UB)
```

Optimization terminated: average change in the fitness value less than options.FunctionTolerance.

```
x =
  3.0000 0.0000

fval =
  -5.0000

>>
```

Do not forget to multiple $f$ by $-1$ to return to our original problem.

We obtain the correct solution, [3, 0] and a functional evaluation of 5.

## 12.6 Simulated Annealing

**Simulated annealing (SA)** is a probabilistic technique for approximating the global optimum of a given constrained optimization problem that was described by Scott Kirkpatrick, C. Daniel Gelatt and Mario P. Vecchi in 1983. Specifically, it is a metaheuristic to find good approximate global optimization solution in large search spaces. It is often used when the search space is discrete (e.g. all tours that visit a given set of cities). For problems where finding the precise global optimum is less important than finding an acceptable local optimum in a fixed amount of time, simulated annealing may be preferable to other alternatives. Simulated annealing interprets slow cooling as a slow decrease in the probability of accepting worse solutions as it explores the solution space. Accepting worse solutions is a fundamental property of metaheuristics because it allows for a more extensive search for the optimal solution (Fox, 2012).

Simulated annealing code from R is used in our examples.

**Example 12.5:** Simulated Annealing in R

One variable example, $f(x) = x^2 - x + 2$.

We can obtain a good estimate from this plot, Figure 12.15 as $f(x) = 1.75$ when $x$ is about 0.5.

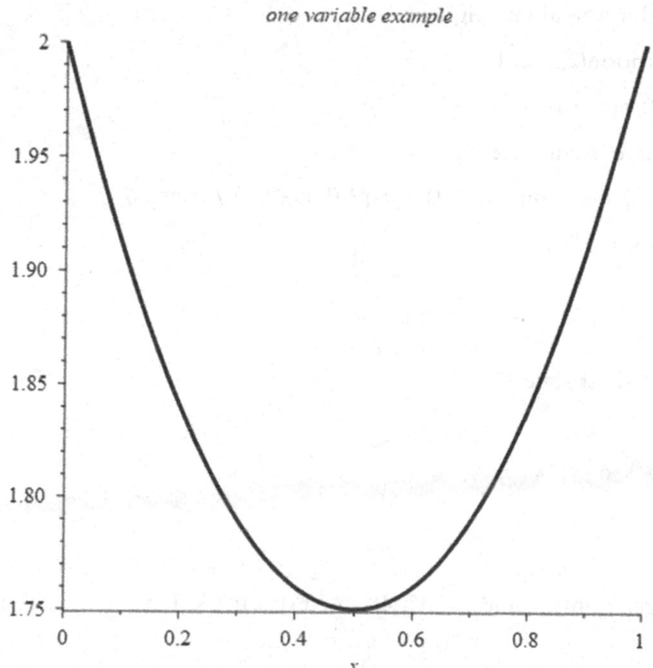


**FIGURE 12.15**
Plot of $f(x) = x^2 - x + 2$.

Now, let's use simulated annealing in R.

```
> simulated_annealing <- function(func, s0, niter = 10, step = 0.1) {
+
+ # Initialize
+ ## s stands for state
+ ## f stands for function value
+ ## b stands for best
+ ## c stands for current
+ ## n stands for neighbor
+ s_b <- s_c <- s_n <- s0
+ f_b <- f_c <- f_n <- func(s_n)
+ message("It\tBest\tCurrent\tNeigh\tTemp")
+ message(sprintf("%i\t%.4f\t%.4f\t%.4f\t%.4f", 0L, f_b, f_c, f_n, 1))
+
+ for (k in 1:niter) {
+    Temp <- (1 - step)^k
+    # consider a random neighbor
+    s_n <- rnorm(2, s_c, 1)
+    f_n <- func(s_n)
+    # update current state
+    if (f_n < f_c || runif(1, 0, 1) < exp(-(f_n - f_c) / Temp)) {
+      s_c <- s_n
+      f_c <- f_n
+    }
+    # update best state
+    if (f_n < f_b) {
+      s_b <- s_n
+      f_b <- f_n
+    }
+    message(sprintf("%i\t%.4f\t%.4f\t%.4f\t%.4f", k, f_b, f_c, f_n, Temp))
+ }
```

```
+ return(list(iterations = niter, best_value = f_b, best_state = s_b))
+ }
> test2 <- function(xx)
   {x1 <- xx[1]
   fact1 <- x1^2-x1+2
   y <- (fact1)
   return(y)
   }
>
> sol <- simulated_annealing(test2, s0 = c(0))
test2 <- function(xx)
+ {x1 <- xx[1]
+ fact1 <-x1^2-x1+2
+ y <- (fact1)
+ return(y)
+ }
>
> sol <- simulated_annealing(test2, s0 = c(0))
```

| It | Best | Current | Neigh | Temp |
|---|---|---|---|---|
| 0 | 2.0000 | 2.0000 | 2.0000 | 1.0000 |
| 1 | 1.7754 | 1.7754 | 1.7754 | 0.9000 |
| 2 | 1.7754 | 1.7754 | 1.9551 | 0.8100 |
| 3 | 1.7754 | 1.7754 | 3.0687 | 0.7290 |
| 4 | 1.7754 | 1.7754 | 4.6183 | 0.6561 |
| 5 | 1.7537 | 1.7537 | 1.7537 | 0.5905 |
| 6 | 1.7537 | 1.9344 | 1.9344 | 0.5314 |
| 7 | 1.7537 | 1.9344 | 4.5308 | 0.4783 |
| 8 | 1.7502 | 1.7502 | 1.7502 | 0.4305 |
| 9 | 1.7502 | 2.0531 | 2.0531 | 0.3874 |
| 10 | 1.7502 | 2.0531 | 2.8822 | 0.3487 |

```
>
```

In R, we obtain the estimated value of the function as 1.7502.

Now, we ran out for 50 iterations, we obtain a functional value of 1.750127 when $x = 0.5112$ a fair approximation for 0.5.

sol

$iterations

[1] 50

$best_value

[1] 1.750127

$best_state

[1] 0.5112899 −1.2663339

**Example 12.6:** Simulated Annealing 2

Using simulated annealing to find Minimize $f(x_1, x_2) = (x_1 - 2)^2 + (x_2 - 2)^2$

After 50 iterations, we have a functional value of 0.1094762 when $x_1 = 1.732$ and $x_2 = 2.192132$. We obviously need more iteration to converge closer to the actual solution.

$iterations

[1] 50

$best_value

[1] 0.1084762

$best_state

[1] 1.732490 2.192132

We increased the iterations to 500, and obtained the solution:

sol

$iterations

[1] 500

$best_value

[1] 0.00689246

$best_state

[1] 1.991642 2.082599

This is a much-improved solution.

See (https://codereview.stackexchange.com/questions/84688/simulated-annealing-in-r/84718)

**Simulated Annealing in MATLAB**

```
x = simulannealbnd(fun,x0)
x = simulannealbnd(fun,x0,lb,ub)
x = simulannealbnd(fun,x0,lb,ub,options)
x = simulannealbnd(problem)
[x,fval] = simulannealbnd(___)
[x,fval,exitflag,output] = simulannealbnd(___)
```

x = simulannealbnd(fun,x0) finds a local minimum, *x*, to the function handle fun that computes the values of the objective function. *x0* is an initial point for the simulated annealing algorithm, a real vector.

**Example 12.7:** Simulated Annealing with MATLAB 1

We want to Minimize $f(x) = x^2 - x + 2$
Figure 12.16 provide a MATLAB screenshot for Example 12.7
In MATLAB, we found that at $x = 0.5$, the minimum value of $f(x*)$ is 1.75.

**Example 12.8:** Simulated annealing with MATLAB 2

We want to Minimize $f(x_1, x_2) = (x_1-2)^2 + (x_2-2)^2$
Figure 12.17 provide a MATLAB screenshot for Example 12.8

## 12.7 Exercises

Given the data points in Table 12.2 to sue in Exercises 1–6:

1. Obtain a scatter plot of the data and comment on the trend.
2. Build the function for the least squares model for $y = b_0 + b_1x$.
3. Take the derivatives of your equation in (2) and set equal to zero. Solve for the parameters, $b_0$ and $b_1$.
4. Find the gradient of the function in (2) and start with an estimate for $b_0$ and $b_1$ such as [1, 1]. Choose a rate and iterate to find the minimum value of the equation.
5. Solve the unconstrained function in (2) with simulated annealing.

```
>> fun=@objfunx1;
>> x0=[0];
>> [x,fval]=simulannealbnd(fun,x0)
Optimization terminated: change in best function value less than options.FunctionTolerance.

x =

    0.5000


fval =

    1.7500

>>
```

**FIGURE 12.16**
MATLAB screenshot of solution to Example 12.7.

```
f=(x(1)-2)^2+(x(2)-2)^2;

>> fun=@objfunx;
>> x0=[0,0];
>> [x,fval]=simulannealbnd(fun,x0,LB,UB)
Optimization terminated: change in best function value less than options.FunctionTolerance.

x =

    1.9987    1.9938


fval =

   4.0436e-05
```

**FIGURE 12.17**
MATLAB screenshot of solution to Example 12.8.

**TABLE 12.2**

Exercise 1–6 Data

| x | 2 | 6 | 10 | 14 |
|---|---|---|---|---|
| y | 3.9 | 7.2 | 12.5 | 19.2 |

6. Solve the unconstrained function in (2) with genetic algorithm.
7. Use a genetic algorithm, to Maximize $(x_1^2 + 2 * x_2^2)$ subject to $(x_1 + x_2 \leq 3, 2\ x_1 + x_2 \leq 4)$.
8. Use simulated annealing to Minimize $f(x) = x^2 - 25x + 100$.

## Reference

Radečić, Dario (2020, Sept 25). https://towardsdatascience.com/machine-learning-with-r-linear-regression-558fa2edaaf0

# Index

Printed in the United States
by Baker & Taylor Publisher Services